# Microorganisms as Model Systems for Studying Evolution

*MONOGRAPHS IN EVOLUTIONARY BIOLOGY*
Series Editors:

MAX K. HECHT  BRUCE WALLACE  GHILLEAN T. PRANCE

Queens College of the  Virginia Polytechnic Institute  New York Botanical Garden
City University of New York  and State University  Bronx, New York
Flushing, New York  Blacksburg, Virginia

MACROMOLECULAR SEQUENCES IN SYSTEMATIC AND
EVOLUTIONARY BIOLOGY
Edited by Morris Goodman

EVOLUTIONARY GENETICS OF FISHES
Edited by Bruce J. Turner

MICROORGANISMS AS MODEL SYSTEMS FOR STUDYING
EVOLUTION
Edited by Robert P. Mortlock

# Microorganisms as Model Systems for Studying Evolution

Edited by
**Robert P. Mortlock**
Cornell University
Ithaca, New York

**PLENUM PRESS • NEW YORK AND LONDON**

**Library of Congress Cataloging in Publication Data**

Main entry under title:

Microorganisms as model systems for studying evolution.

(Monographs in evolutionary biology)
Includes bibliographical references and index.
1. Microorganisms—Evolution. 2. Microbial genetics. 3. Biological models.
I. Mortlock, Robert P., 1931-    . II. Series.
QR13.M53    1984                    576'.138                    84-17938
ISBN 0-306-41788-X

©1984 Plenum Press, New York
A Division of Plenum Publishing Corporation
233 Spring Street, New York, N.Y. 10013

All rights reserved

No part of this book may be reproduced, stored in a retrieval system, or transmitted in any form or by any means, electronic, mechanical, photocopying, microfilming, recording, or otherwise, without written permission from the Publisher

Printed in the United States of America

# Contributors

**Patricia H. Clarke** Department of Biochemistry, University College London, London WC1E 6BT, England

**Barry G. Hall** Microbiology Section, Biological Sciences Group, University of Connecticut, Storrs, Connecticut 06268

**B. S. Hartley** Department of Biochemistry, Imperial College of Science and Technology, London SW7 2AZ, England

**Jost Kemper**† Institute of Molecular Biology, University of Texas at Dallas, Richardson, Texas 75080

**E. C. C. Lin** Departments of Microbiology and Molecular Genetics, Harvard Medical School, Boston, Massachusetts 02115

**Robert P. Mortlock** Department of Microbiology, New York State College of Agriculture and Life Sciences, Cornell University, Ithaca, New York 14853

**Monica Riley** Department of Biochemistry, State University of New York at Stony Brook, Stony Brook, New York 11794

**Christopher Wills** Department of Biology, University of California at San Diego, La Jolla, California 92093

**T. T. Wu** Departments of Biochemistry and Molecular Biology, Northwestern University, Evanston, Illinois 60201

# *Preface*

The microorganisms present on the earth today possess a vast range of metabolic activities and are often able to demonstrate their surprising versatility by gaining both new enzyme activities and new metabolic pathways through mutations. It is generally assumed that the earliest microorganisms were very limited in their metabolic abilities, but as time passed they gradually expanded their range of enzymatic activities and increased both their biosynthetic and catabolic capacity. It is also believed that these primitive microorganisms increased the amount of genetic material they possessed by duplicating their existing genes and possibly by acquiring genetic material from other organisms.

A small group of scientists has been exploring the means by which existing microorganisms are capable of mutating to expand their biochemical abilities. In recent years, more attention has been focused on this type of research, sometimes called "evolution in a test tube." The recent advances in biotechnology and modern techniques of genetic transfer have generated new interest in the methods by which a microorganism's metabolic activities can be improved or deliberately changed in some specific manner.

The work reported in this book describes a type of "genetic engineering" whereby microorganisms are challenged to develop new metabolic activities, not by the acquisition of genetic material from external sources, but by altering their own existing genetic information. This approach to studying evolution has been surprisingly fruitful and has clearly illustrated, for example, the importance of regulatory mutations in the establishment of new metabolic pathways. This area of research holds much more promise than previously realized, and it is hoped and anticipated that as we learn more about the process by which mutations create new metabolic abilities we will be better able to control and direct that process.

Many of the early studies involved the use of novel sugars of five-carbon structure as potential growth substrates and were inspired by the observation in several laboratories that the genetic mechanism establishing the growth of *Klebsiella* strains on the pentitol xylitol, was a single mutation in the regulation of the inducible pathway for another pentitol, ribitol. The first three chapters in this book describe investigations into how bacteria, especially those in the genera *Klebsiella* and *Escherichia,* are able to gain the ability to establish and then improve a pathway for the catabolism of xylitol.

Chapter 1 contains a general description of the routes of pentitol degradation by coliform bacteria and shows how such bacteria can acquire, through mutation, the ability to utilize some of the less common pentitols as substrates for both carbon and energy. Once the genetic and biochemical changes leading to the establishment of growth on xylitol were clearly understood, further experiments were designed to improve the new xylitol pathway, and these experiments are described by Hartley in Chapter 2. Two different approaches for increasing the activity of a newly established enzymatic reaction are discussed. For an enzyme "borrowed" from an existing pathway for use in catalyzing a similar reaction for a novel substrate in a new pathway, the activity of the enzyme for the new substrate may be poor and rate limiting, as it is for the new xylitol pathway. In such a case it can be demonstrated that two types of genetic change, either altering the enzyme to make it more efficient for the new substrate or making larger amounts of the unaltered enzyme by mechanisms such as gene amplification, will both improve the growth rate on the novel substrate. Chapter 3, also by Hartley, continues the discussion of the mutations permitting growth on xylitol with an elegant study of the sequence of the regulatory and structural genes involved in the establishment of the new xylitol pathway in the xylitol-positive mutants.

Chapter 4 examines some of the five-carbon aldopentose sugars and the mutations which establish new catabolic pathways for the degradation of such carbohydrates as D-arabinose, D-lyxose, L-xylose, and L-lyxose. For most of these new pathways the critical enzyme activities needed for the metabolism of the novel substrate are shown to be borrowed, via regulatory mutations, from pathways established by evolution for the degradation of more common substrates. For one of the uncommon aldopentoses, however, D-lyxose, the new enzyme activity mobilized to initiate the D-lyxose catabolic pathway may indeed come from what was previously an unexpressed and unknown "silent gene."

# PREFACE

Chapter 5, by Lin and Wu, also shows how enzymes can be "borrowed" from an existing pathway to establish a new catabolic pathway. It describes how the establishment of a new pathway for the degradation of propanediol results from a regulatory mutation that converts an enzyme normally employed as an anaerobic reductase into a dehydrogenase that can function under aerobic conditions. Chapter 5 further illustrates how mutations leading to the establishment of a new metabolic pathway can result in the elimination of another pathway.

Chapter 6 describes a different approach to studying the means by which microorganisms can establish new metabolic activity. In these experiments, an existing ability for the utilization of a substrate (lactose) is removed by genetic deletion and the mutant cells are challenged to find and modify other genes to replace the function of those deleted. In his studies reported in this chapter, Hall shows the step-by-step development of an entire new operon to replace the *lac* operon, an operon that responds to new regulatory elements. The location of the new operon suggests the recruitment of "silent genes" on the cell chromosome to replace those of the normal *lac* operon.

In Chapter 7, Clarke describes experiments designed to expand the range of amide substrates for the aliphatic amidase enzyme of *Pseudomonas aeruginosa*, experiments that led to the establishment of new catabolic pathways. She examines the role of both regulatory and structural gene mutations in establishing pathways for new amide substrates. In Chapter 8, Wills documents how selective pressure can be placed upon an enzyme, in this case an alcohol dehydrogenase of yeast, to select for deliberate alterations in the enzyme's kinetic properties. Chapter 9, by Kemper,* describes the most interesting situation, where the loss of a subunit of one enzyme actually results in the replacement of a missing subunit for a different enzyme. In such a manner, mutants auxotrophic for the leucine biosynthetic pathway regain the ability to synthesize leucine as the result of a deletion in a different location of the chromosome that makes available a product of another gene, the *new leu* gene, to take the place of the missing subunit.

Chapter 10 is slightly different from the others, for there Riley describes her fascinating studies on the current structure of the *E. coli*

---

*As this volume was in the final stages of preparation I learned, to my great sorrow, of the death of Dr. Jost Kemper. He will be deeply missed by all of his colleagues.

chromosome, studies that not only suggest how silent genes may originate but also illustrate the role past duplications have played in the evolutionary development of the modern cell's DNA.

It is not an exaggeration to state that the work presented in this volume, most of it the result of the investigations of the authors themselves, represents the pioneering studies on the means by which microorganisms can mutate to increase their metabolic capabilities.

*Ithaca, New York*                                                                Robert P. Mortlock

# Contents

Chapter 1
**The Utilization of Pentitols in Studies of the Evolution of Enzyme Pathways**
Robert P. Mortlock

| | |
|---|---:|
| 1. Introduction | 1 |
| 2. The Pentitols | 2 |
| 3. The Utilization of Pentitols by *Klebsiella* Species | 3 |
|    3.1. Growth on Xylitol and L-Arabitol | 3 |
|    3.2. The Nature of the Mutations Establishing Growth on the New Pentitol Substrates | 7 |
|    3.3. The Origin of the Xylitol Dehydrogenase Activity | 9 |
|    3.4. The Origin of the D-Xylulokinase of the New Xylitol Pathway | 12 |
| 4. The Origin of the L-Arabitol Dehydrogenase Activity | 13 |
| 5. Mutations Improving the Growth Rate on Xylitol | 13 |
|    5.1. Alterations in the Dehydrogenase Activity | 13 |
|    5.2. Mutants with Increased Amounts of Ribitol Dehydrogenase | 14 |
|    5.3. Utilization of the D-Arabitol Transport System to Facilitate the Transport of Xylitol | 15 |
| 6. The Growth of *Escherichia Coli* Strains on Xylitol | 17 |
|    6.1. The Utilization of Ribitol Dehydrogenase to Establish Growth on Xylitol | 17 |
|    6.2. The Construction of a Different Route for Xylitol Catabolism | 18 |
| 7. The Utilization of Xylitol by a Mutant in the Genus *Erwinia* | 18 |
| 8. Summary | 19 |
|    References | 19 |

## Chapter 2
**Experimental Evolution of Ribitol Dehydrogenase**
B. S. Hartley

| | |
|---|---|
| 1. Introduction | 23 |
| 1.1. Evolutionary Lessons from Protein Structures | 23 |
| 1.2. Microbial Enzyme Evolution | 27 |
| 2. Pentitol Metabolism in *Klebsiella aerogenes* | 27 |
| 3. Chemostat Culture of *Klebsiella aerogenes* on Xylitol | 29 |
| 4. Evolution of Ribitol Dehydrogenase in the Chemostat | 33 |
| 4.1. Enzyme Superproduction | 34 |
| 4.2. Mechanisms of Enzyme Superproduction | 37 |
| 4.3. Improved Xylitol Dehydrogenases | 38 |
| 5. Fluctuating Selective Pressure | 45 |
| 6. Transfer of the *Klebsiella aerogenes* Ribitol Dehydrogenase Gene into *Escherichia coli* K12 | 46 |
| 6.1. Evolution of Ribitol Dehydrogenase in *Escherichia coli* | 49 |
| 7. Evolutionary Lessons from the Chemostat Studies | 51 |
| References | 52 |

## Chapter 3
**The Structure and Control of the Pentitol Operons**
B. S. Hartley

| | |
|---|---|
| 1. Introduction | 55 |
| 1.1. Construction of $\lambda p$ *rbt* | 55 |
| 1.2. Construction of $\lambda p$ *rbt dal* | 56 |
| 2. The Structure of $\lambda p$ *rbt* and $\lambda p$ *rbt dal* | 58 |
| 2.1. Genetic Analysis | 58 |
| 2.2. Physical Analyses | 59 |
| 2.3. How Did $\lambda p$ *rbt* and $\lambda p$ *rbt dal* Arise? | 59 |
| 3. Bipolar Transcription of the Pentitol Operons | 61 |
| 4. The Pentitol Operon Enzymes | 63 |
| 4.1. D-Arabitol Dehydrogenase | 63 |
| 4.2. D-Ribulokinase | 67 |
| 4.3. D-Xylulokinase | 70 |
| 5. Substrate Specificity of the Pentitol Operon Enzymes | 70 |
| 6. *rbt* Messenger RNA | 72 |
| 6.1. A Switch in *rbt* mRNA Translation in Mid Log Phase | 72 |

|   |   |
|---|---|
| 6.2. Purification and Properties of *rbt* mRNA | 78 |
| 6.3. Is *rbt* mRNA Superstable? | 79 |
| 7. DNA Sequencing of the Pentitol Operons | 80 |
| 7.1. The Ribitol Dehydrogenase Gene | 82 |
| 7.2. The D-Arabitol Dehydrogenase Gene | 85 |
| 7.3. The *rbt* Repressor Protein | 85 |
| 7.4. The *dal* Repressor Protein | 87 |
| 7.5. The *dal* Promoter | 87 |
| 7.6. The *rbt* Promoter | 91 |
| 7.7. The *rbt* Repressor Promoter | 91 |
| 7.8. The *dal* Repressor Promoter | 93 |
| 8. Translation of the Two Kinases | 93 |
| 9. Invert Repeat Sequences Enclose the Two Operons | 96 |
| 10. Structure of an Experimentally Evolved Gene Duplication | 97 |
| 11. Evolutionary Lessons from the Pentitol Operons | 99 |
| References | 104 |

## Chapter 4
### The Development of Catabolic Pathways for the Uncommon Aldopentoses
Robert P. Mortlock

|   |   |
|---|---|
| 1. The Structure of the Aldopentoses and Their Occurrence in Nature | 109 |
| 1.1. The Structure of the Aldopentoses | 109 |
| 1.2. D-Ribose and L-Ribose | 109 |
| 1.3. D-Xylose and L-Xylose | 110 |
| 1.4. D-Arabinose and L-Arabinose | 110 |
| 1.5. D-Lyxose and L-Lyxose | 110 |
| 2. The Pathways of Degradation of Aldopentoses by Coliform Bacteria | 111 |
| 2.1. Pathways for the Degradation of Those Sugars Commonly Found in Nature | 111 |
| 2.2. Pathways for the Degradation of Those Sugars Not Commonly Found in Nature | 112 |
| 2.3. Enzyme Activities Establishing Growth on the New Aldopentose Substrates | 115 |
| 3. The Biochemical and Genetic Bases for the Establishment of New Enzymatic Pathways for the Degradation of Aldopentoses | 116 |

3.1. The Utilization of D-Lyxose .................................... 116
3.2. The Utilization of D-Arabinose .............................. 120
3.3. The L-Lycose and L-Xylose Pathways in *Klebsiella pneumoniae* ................................................................. 129
4. Summary ............................................................................ 130
References .............................................................................. 132


Chapter 5

**Functional Divergence of the L-Fucose System in Mutants of *Escherichia coli***

E. C. C. Lin and T. T. Wu

1. Introduction ........................................................................ 135
2. Reversibility of NAD-Linked Reactions ........................ 136
3. A Mutant That Uses an NAD-Linked Dehydrogenase to Grow on L-1,2-Propanediol .................................... 136
   3.1. Characterization of the Novel Biochemical Pathway in the Mutant ............................................................. 137
   3.2. Identifying the Original Role of a Recruited Enzyme ... 138
   3.3. Connection of the Propanediol Oxidoreductase with the Fucose System .................................................... 139
4. Biochemistry of the Fucose System ................................ 140
5. Enzymic Changes in the Fucose System in Mutants and Revertants ........................................................................ 140
   5.1. Propanediol-Positive Mutants Exploit Both Branches of the Fucose System ................................................ 140
   5.2. A Primary Stage Mutant ........................................... 143
   5.3. A Secondary Mutant ................................................. 143
   5.4. Pseudorevertants That Regained the Ability to Grow on Fucose ................................................................. 145
   5.5. A Mutant with Superior Scavenger Power for Propanediol ............................................................... 145
   5.6. Changes in the Property of the Oxidoreductase ....... 147
6. Genetic Organization and Regulation of the Fucose System ... 148
   6.1. A Regulon Comprised of Closely Linked Operons .... 148
   6.2. Positive Control ......................................................... 149
   6.3. The Inducer ............................................................... 149
   6.4. Lactaldehyde Dehydrogenase under Separate Control ... 150
   6.5. Post Transcriptional Control of the Oxidoreductase Activity ..................................................................... 150

7. Sequential Mutations Changing Propanediol and Fucose
   Utilization .................................................................. 151
8. Relationship of the Fucose and the Rhamnose Systems ..... 152
9. Conversion of the Fucose System for D-Arabinose
   Utilization .................................................................. 153
10. Propanediol-Positive Mutants as Evolutionary Vanguards... 154
    10.1. Mutants That Grow on D-Arabitol ........................ 154
    10.2. Mutants That Grow on Xylitol ............................. 155
    10.3. Mutants That Grow on Ethylene Glycol ................ 157
11. Retrospective and Prospective Views ........................... 157
    References................................................................. 161

## Chapter 6
### The Evolved β-Galactosidase System of *Escherichia coli*
Barry G. Hall

1. Introduction .................................................................. 165
2. Development of the Evolved β-Galactosidase System as a
   Tool for Studying Evolution ........................................... 167
3. Evolution of Multiple Functions for Evolved β-
   Galactosidase Enzyme: An Evolutionary Pathway............. 171
4. Kinetic Analysis of Evolved β-Galactosidase Enzymes....... 174
5. Evolution by Intragenic Recombination.......................... 176
6. Allolactose Synthesis: Another New Function for Class IV
   Enzyme........................................................................ 177
7. The Role of Regulatory Mutations in the Evolution of
   Lactose Utilization........................................................ 179
8. Directed Evolution of a Repressor ................................. 180
9. The Fully Evolved EBG Operon .................................... 181
10. A Model for Evolution in Diploid Organisms.................. 182
11. Future Perspectives...................................................... 184
    References................................................................... 184

## Chapter 7
### Amidases of *Pseudomonas aeruginosa*
Patricia H. Clarke

1. Introduction .................................................................. 187
   1.1. Biochemical Activities of *Pseudomonas* Species......... 187
   1.2. Choice of Enzyme System ........................................ 188

1.3. Growth of *Pseudomonas aeruginosa* on Acetamide ..... 189
1.4. The Wild-Type Amidase of *Pseudomonas aeruginosa* PAC1 ......................................................... 190
2. Amidase Regulatory Mutants.................................... 191
   2.1. Isolation of Mutants from Succinate/Formamide Medium.................................................... 194
   2.2. Isolation of Mutants from Succinate/Lactamide Medium 198
   2.3. Isolation of Mutants with Altered Inducibility............ 199
3. Amidase-Negative Mutants....................................... 199
   3.1. Isolation of Acetamide-Negative Mutants ................ 199
   3.2. Mutations in the *amiE* Gene.............................. 200
   3.3. Mutations in the *amiR* Gene.............................. 200
   3.4. Promoter Mutations ....................................... 202
4. Mutants with Altered Enzymes ................................. 205
   4.1. Butyramide-Utilizing Mutants: B Group.................. 205
   4.2. Valeramide-Utilizing Mutants: V Group ................. 207
   4.3. Phenylacetamide-Utilizing Mutants: Ph Group........... 207
   4.4. Acetanilide-Utilizing Mutants: AI Group ................. 210
5. Properties of Wild-Type and Mutant Amidases................ 214
   5.1. Enzyme Structure ......................................... 214
   5.2. Catalytic Activities ........................................ 217
6. Amidase Genes and Enzymes.................................... 222
   6.1. Gene Mapping ............................................. 222
   6.2. Mutation.................................................... 223
   6.3. Role of Recombination .................................... 224
   6.4. Alignment of *amiE* Gene and Protein .................... 224
   6.5. Amidase Gene Capture.................................... 226
   6.6. How Many More Amidases? ............................. 226
References................................................................ 228

## Chapter 8
**Structural Evolution of Yeast Alcohol Dehydrogenase in the Laboratory**
Christopher Wills

1. Introduction ....................................................... 233
2. The Biochemistry and Regulation of Yeast Alcohol Dehydrogenase .................................................. 236

|   |   |
|---|---|
| 3. The Mechanism of Allyl Alcohol Resistance | 243 |
| 4. Amino Acid Substitutions in the Mutant ADHs | 245 |
| 5. Altered Kinetics of the Mutants | 248 |
| 6. Evolutionary Implications | 251 |
| References | 252 |

## Chapter 9
### Gene Recruitment for a Subunit of Isopropylmalate Isomerase
Jost Kemper

|   |   |
|---|---|
| 1. The Leucine Operon in *Salmonella typhimurium* Wild-Type Strains | 255 |
| 2. The Wild-Type Isopropylmalate Isomerase | 255 |
| 3. Strains Carrying *leuD* Mutations Revert to Leucine Prototrophy | 257 |
| 4. Model for Leucine Biosynthesis in *leuD–supQ* Mutant Strains | 257 |
| 5. Leucine Biosynthesis in *leuD–supQ* Mutant Strains | 258 |
| 6. Genetic Characterization of the *leuC–newD* Isopropylmalate Isomerase | 259 |
|    6.1. A Free *leuC* Polypeptide Is Needed | 259 |
|    6.2. *supQ* Mutations Result in Availability of the *newD* Gene Product | 260 |
|    6.3. Nature of *supQ* Mutations | 263 |
|    6.4. Direction of Transcription in the *supQ newD* Region | 267 |
|    6.5. What Is the Original Function of *supQ newD*? | 268 |
|    6.6. Are *supQ newD* Genes Existing or Functional in *Escherichia coli*? | 272 |
| 7. Biochemical Characterization of the *leuC–newD* Isopropylmalate Isomerase | 274 |
|    7.1. *In Vitro* Specific Activity of the Hybrid *leuC–newD* Isopropylmalate Isomerase | 274 |
|    7.2. Mutant Isopropylmalate Isomerase Activity Is Limited by the Endogenous Concentration of $\alpha$-Isopropylmalate | 275 |
|    7.3. Growth Limitations in Strains with an Unbridled Leucine Biosynthesis Pathway | 277 |
| 8. Theoretical Steps in the Evolution of a Complex Enzyme | 278 |
| 9. Characterization of the *newD* (and *supQ*) Gene(s) | 280 |
| References | 282 |

## Chapter 10
### Arrangement and Rearrangement of Bacterial Genomes
Monica Riley

| | |
|---|---|
| 1. Introduction | 285 |
| 2. Chromosomal Rearrangements: Mechanisms of Change | 285 |
|    2.1. Duplications | 285 |
|    2.2. Transpositions | 296 |
|    2.3. Inversions | 300 |
|    2.4. Additions and Deletions | 301 |
| 3. Conservation of Global Gene Order: Mechanisms of Stability | 306 |
|    3.1. Integrity of Taxonomic Groupings | 306 |
|    3.2. Genetic Maps | 307 |
|    3.3. Possible Stabilizing Factors | 308 |
| 4. Conclusion | 310 |
|    References | 310 |
| **Glossary** | 317 |
| **Index** | 321 |

CHAPTER 1

# The Utilization of Pentitols in Studies of the Evolution of Enzyme Pathways

ROBERT P. MORTLOCK

## 1. Introduction

Microbiologists generally agree that the earliest microorganisms to evolve were simple in their metabolic capabilities and had to be supplied with many preformed, complex molecules to satisfy their nutritional requirements. As time passed, certain organisms developed new metabolic pathways to synthesize such required compounds from smaller, less complex molecules and thus became more versatile in their nutritional requirements, utilizing a wider range of food sources. Changes in the metabolism of higher organisms must have produced new types of organic compounds that entered the environment upon the death of the organisms. Microorganisms, in turn, evolved metabolic pathways for the degradation of these new molecules and used them to obtain carbon and energy to satisfy their growth requirements. The extra DNA provided by the duplication of previously existing genes may have been modified through mutation to eventually become the structural and regulatory genes for new degradative pathways.

Although studies comparing the amino acid sequences in proteins or nucleotide sequences in nucleic acids can provide some information about the evolutionary relationship between enzyme pathways, they do not provide details on the exact mutational events that led to the differentia-

---

ROBERT P. MORTLOCK • Department of Microbiology, New York State College of Agriculture and Life Sciences, Cornell University, Ithaca, New York 14853.

tion and evolution of the new pathways. In recent years, a small number of laboratories have been utilizing the rapid growth rates and high cell densities of microorganisms to obtain mutants that have gained the ability to utilize a novel compound as a growth substrate or that have otherwise increased their metabolic ability. Once obtained, such mutants can be studied in order to elucidate the nature of the genetic and biochemical events that permitted the cells to acquire the new ability. Perhaps through chance, some of the earliest of these experiments utilized sugars of five-carbon structure as the potential growth substrates.

## 2. The Pentitols

Those polyalcoholic sugars that contain a five-carbon backbone are known as pentitols and there are four different structures, as shown in Fig. 1. These four sugars, which differ in the location of their hydroxyl groups, are named ribitol, xylitol, D-arabitol, and L-arabitol.

Ribitol (sometimes called adonitol), occurs naturally in some plants, is a constituent of riboflavin, and is found in the capsular polysaccharides or associated within the cell walls of some bacteria (Schaffer, 1972). D-Arabitol (D-lyxitol) has been found in many lichens, certain mushrooms, some plants, and some plant seeds. L-Arabitol does not occur naturally, except in the urine of patients suffering from pentosuria. Xylitol occurs naturally in mushrooms (Schaffer, 1972), and can be found in small amounts

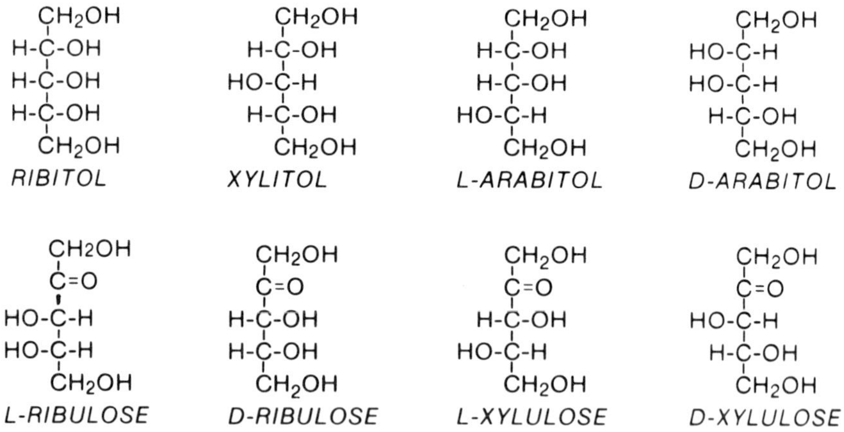

```
   CH2OH            CH2OH            CH2OH            CH2OH
   |                |                |                |
 H-C-OH           H-C-OH           H-C-OH          HO-C-H
   |                |                |                |
 H-C-OH          HO-C-H            H-C-OH          HO-C-H
   |                |                |                |
 H-C-OH           H-C-OH          HO-C-H            H-C-OH
   |                |                |                |
   CH2OH            CH2OH            CH2OH            CH2OH
  RIBITOL          XYLITOL        L-ARABITOL       D-ARABITOL


   CH2OH            CH2OH            CH2OH            CH2OH
   |                |                |                |
   C=O              C=O              C=O              C=O
   |                |                |                |
 HO-C-H           H-C-OH            H-C-OH         HO-C-H
   |                |                |                |
 HO-C-H           H-C-OH          HO-C-H            H-C-OH
   |                |                |                |
   CH2OH            CH2OH            CH2OH            CH2OH
 L-RIBULOSE      D-RIBULOSE       L-XYLULOSE       D-XYLULOSE
```

**Figure 1.** The structures of the pentitols and pentuloses. The oxidation of a pentitol at the C-2 position results in the formation of a 2-ketopentose or pentulose.

in material from plants and in a wide variety of fruits (Washuttl *et al.*, 1973). It is currently being produced in certain Scandinavian countries for use as a sweetening agent (Makinen and Scheinin, 1982). Pentitols are sometimes synthesized during the metabolism of carbohydrates by certain fungi (Gong *et al.*, 1981).

Ribitol and D-arabitol are relatively common in nature and many bacteria have evolved enzyme pathways for their degradation. As an example, among the family of bacteria known as the Enterobacteriaceae, many members of the genus *Klebsiella* have been found to possess enzyme pathways for metabolizing both ribitol and D-arabitol and can use either of these pentitols as a sole source of both carbon and energy (Wood *et al.*, 1961). The other two pentitols, xylitol and L-arabitol, are less common, and the ability to degrade these pentitols is not found as often among bacteria (Mortlock and Wood, 1964a; Mortlock, 1976).

## 3. The Utilization of Pentitols by *Klebsiella* Species

### 3.1. Growth on Xylitol and L-Arabitol

A number of years ago, scientists working at the Prairie Regional Laboratories in Saskatoon, Saskatchewan, Canada, examined an organism (then termed *Aerobacter aerogenes* strain PRL-R3, now renamed *Klebsiella pneumoniae* var. *oxytoca* strain PRL-R3) that could not only grow on ribitol or D-arabitol, but also appeared to be capable of growth on both of the less common pentitols, xylitol and L-arabitol (Mortlock, 1976). Subsequent research established the pathways of catabolism of all four of these pentitols by this strain of *Klebsiella* (Wood *et al.*, 1961; Fossitt *et al.*, 1964). These pathways are illustrated in Fig. 2.

In each case the pentitol was probably transported through the cell membrane and into the cell cytoplasm by a protein transport system or permease, although this still has not been carefully studied for all of the pentitols (Mays and Mortlock, 1983). After the pentitol was within the cell an enzyme called a dehydrogenase catalyzed an oxidation at the C-2 position to form the corresponding 2-ketopentose (or pentulose) with the reduction of NAD. There are also four possible structures for these pentuloses, termed D-ribulose, L-ribulose, D-xylulose, and L-xylulose, as shown in Fig. 1.

The second reaction in each pathway involved a kinase to catalyze the transfer of phosphate from ATP to the C-5 of the pentulose, resulting in the formation of a pentulose 5-phosphate. In such a manner, the ribitol catabolic pathway was shown to consist of a ribitol dehydrogenase and

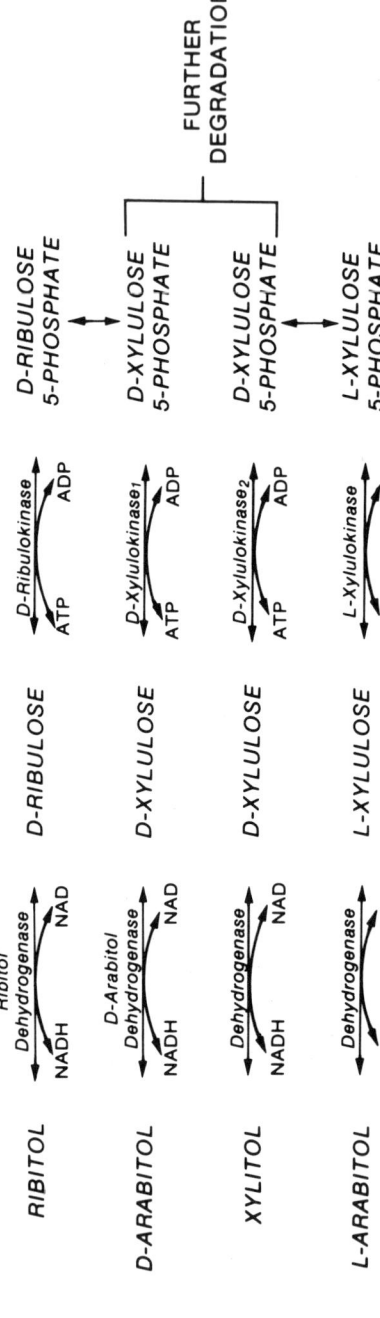

**Figure 2.** The pathways of degradation of the four pentitols by *Klebsiella pneumoniae* PRL-R3. For each pathway, an enzyme dehydrogenase catalyzes the oxidation of the pentitol to form a 2-ketopentose or pentulose. A kinase then catalyzes the phosphorylation of the pentulose in the C-5 position. D-Ribulose 5-phosphate and L-xylulose 5-phosphate are converted to D-xylulose 5-phosphate by epimerization reactions. D-Xylulose 5-phosphate is further degraded by other cellular enzymes. The pathways for ribitol and D-arabitol are inducible in the parent *Klebsiella* strain, but the pathways for xylitol and L-arabitol degradation must be established by mutation.

a D-ribulokinase, with the conversion of ribitol to D-ribulose 5-phosphate, while the D-arabitol pathway consisted of a D-arabitol dehydrogenase and a D-xylulokinase, with the conversion of D-arabitol to D-xylulose 5-phosphate, as shown in Fig. 1 (Wood et al., 1961).

Studies on the regulation of the pathways for the degradation of ribitol and D-arabitol demonstrated that the enzymes of both pathways were inducible. The bacteria synthesized very low amounts of the enzymes in the absence of the pentitol substrate, but the activity of the enzymes of a pathway was greatly increased in response to the presence of the substrate of that pathway in the growth medium. As shown in Table I, when cells were grown in the absence of ribitol on a substrate containing amino acids such as peptone, the cells synthesized only very low levels of the enzymes of the ribitol pathway; less than 0.02 units per mg protein (U/mg protein) in cell-free extracts. However, when ribitol was added to the medium the levels of ribitol dehydrogenase in the cells were greatly increased after incubation for only a few hours, with the enzyme activity ranging from 1.1 to 3.7 U/mg protein, as determined in separate experiments. Both ribitol dehydrogenase and D-ribulokinase were induced when ribitol was added to the growth medium, while the addition of D-arabitol resulted in the induction of D-arabitol dehydrogenase and D-xylulokinase (Table I).

Thus, the cells possessed the necessary genetic information to permit them to synthesize the enzymes needed to catalyze the degradation of the pentitol, but they did so only when they received a chemical signal that the pentitol was present as a potential growth substrate. The induction of these enzymes was complete and actual growth was established by the bacteria, using either ribitol or D-arabitol as the energy source, within a

Table I
Enzyme Activities for the Catabolism of Ribitol and D-Arabitol

| Growth substrate | Specific activity[a] (U/mg protein) | | | |
| --- | --- | --- | --- | --- |
| | Ribitol dehydrogenase | D-Ribulokinase | D-Arabitol dehydrogenase | D-Xylulokinase |
| Casein hydrolysate | <0.02 | <0.02 | <0.02 | <0.02 |
| D-Glucose | <0.02 | <0.02 | <0.02 | <0.02 |
| Ribitol | 1.08–3.71 | 0.08–0.51 | <0.02–0.05 | <0.02 |
| D-Arabitol | 0.05–0.07 | <0.02 | 1.13–4.27 | 0.23–0.27 |

[a] The value <0.02 indicates that activity was not detectable, but that, if present, it was less than 0.02 U enzyme activity/mg protein in the cell-free extracts. [Data are from Mortlock and Wood (1964a), Bisson et al. (1968), and Wilson and Mortlock (1972).]

few hours after the pentitol had been added to the growth medium (Table II) (Mortlock and Wood, 1964a; Bisson et al., 1968).

In contrast to these data, when growth experiments were done using the less common pentitols xylitol and L-arabitol, much longer periods of incubation were required before any growth could be observed (Table II). In similar experiments that measured enzyme activity rather than growth, D-arabitol dehydrogenase activity could be detected only 1 hr after cells had been exposed to D-arabitol, and ribitol dehydrogenase activity could be detected only 2 hr after cells were incubated in the presence of ribitol. When cells were incubated with L-arabitol, however, 23 hr was required before dehydrogenase activity for L-arabitol could be detected and 45 hr of incubation with xylitol was required before any dehydrogenase activity for xylitol could be detected in the cells (Mortlock and Wood, 1964b).

Careful investigation showed that the wild-type or parent strain of *K. pneumoniae* PRL-R3 was really not capable of growth on xylitol or L-arabitol. However, when a sufficient number of cells (about $10^7$) were inoculated into a medium containing either of these two pentitols as the potential growth substrate, selection occurred for cells that had gained the ability to degrade the pentitol through spontaneous mutation. The time period required before growth of the culture was observed was actually the time required for the selection and growth of these mutants. Once a pure culture of the mutant cells had been established, growth upon xylitol could be observed within a few hours after inoculation, rather than within days as for the parent strain (Mortlock and Wood, 1964a).

Examination of the xylitol-positive mutants grown on xylitol showed that they possessed a new pathway that would slowly degrade xylitol, a pathway illustrated in Fig. 2. The same strategy was observed for the metabolism of xylitol and L-arabitol as had been observed for ribitol and D-arabitol. The new xylitol pathway consisted of a transport system to bring xylitol into the cells, a dehydrogenase to oxidize it, and a kinase to

Table II
The Growth of *Klebsiella pneumoniae* PRL-R3 on Pentitols[a]

| Growth substrate | Time to reach complete growth (days) | Lag period before growth was observed |
|---|---|---|
| D-Glucose | 0.5 | 1–4 hr |
| Ribitol | 1.0 | 1–4 hr |
| D-Arabitol | 0.5 | 1–4 hr |
| Xylitol | 4.0 | 3 days |
| L-Arabitol | 2.0 | 15 hr |

[a]Each culture, consisting of minimal-salts supplemented with 0.05% of the indicated carbohydrate, was inoculated with 9% by volume of D-glucose-grown cells. Incubation was aerobic at 26°C. [Data are from Mortlock and Wood (1964a).]

form the ketopentose 5-phosphate. This new enzymatic pathway could not be detected in the parent xylitol-negative strain, even when the cells of that strain were incubated in the presence of xylitol for 24 hr.

Similar results were found for the L-arabitol-positive mutants, in that mutant cells grown on L-arabitol were found to possess a pathway for L-arabitol degradation, also shown in Fig. 2. This pathway was not found in wild-type cells that had been incubated in the presence of L-arabitol for many hours, but was found in L-arabitol-positive mutant cells that had been grown on L-arabitol.

### 3.2. The Nature of the Mutations Establishing Growth on the New Pentitol Substrates

The question of interest was how a mutation, perhaps a single mutation occurring at relatively high frequency, could establish a complete degradative pathway for a new substrate. It was quite apparent how mutations could result in the loss of the activity of an enzyme so that a pathway might cease to function, but it was not clear how mutations could equip the cells with new enzymatic activities and new metabolic pathways. In the case of the xylitol-positive mutants, a mutation had permitted the cells to synthesize two enzymes, a dehydrogenase and a kinase, enabling the cells to convert xylitol to D-xylulose 5-phosphate. In an attempt to answer these questions, studies focused on the nature of the biochemical and genetic bases of the origin of these new pathways.

Horowitz (1945, 1965) had postulated that the evolution of biochemical pathways in primitive bacteria occurred by a process of retrograde evolution. According to this concept, when the organisms had almost depleted a required metabolic substrate from their environment, selection for mutants would occur, including mutants that possessed multiple copies of the gene coding for the enzyme utilizing that substrate. Such gene duplication would increase the amount of the enzyme that was synthesized and available in each cell and permit more efficient scavenging of low levels of the metabolite. One of the duplicate genes was then postulated to undergo mutation so that it would code for a modified enzyme that was capable of converting a new and hitherto unused compound to the required metabolite. With the eventual exhaustion of this new substrate from the environment, the process could be repeated, and the sequence of gene duplication and modification might eventually result in the development of a completely new metabolic pathway. This postulated sequence is illustrated in Fig. 3. This procedure, however, would require a number of different mutational events, and both the xylitol-positive and the L-arabitol-positive mutants appeared to occur at relatively high frequencies, too high to be the result of multiple mutations.

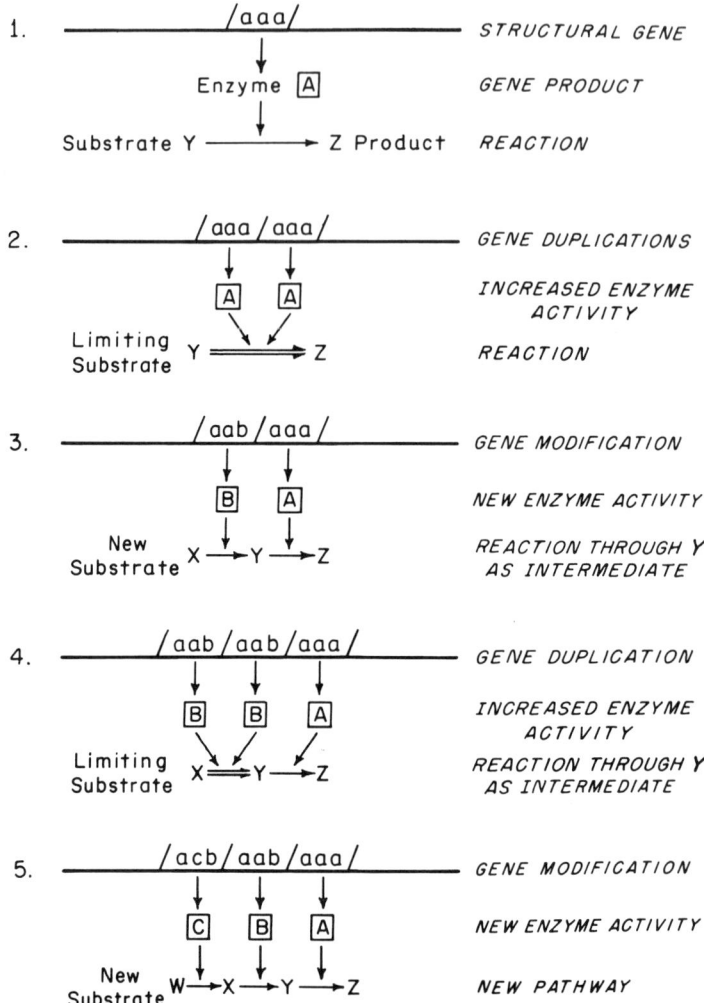

**Figure 3.** Possible steps leading to the evolution of an operon *in situ*. (1) A structural gene (*aaa*) codes for an enzyme (enzyme A) that catalyzes the conversion of a substrate (Y) to a metabolizable product (Z). (2) Substrate Y becomes limiting, leading to selection for increased synthesis of enzyme A and selection for duplication of gene *aaa*. (3) A mutation in one of the duplicated genes results in gene *aab*, which produces a modified enzyme (enzyme B) capable of catalyzing the conversion of new substrate (X) to Y. Enzyme A still catalyzes the conversion of Y to Z. (4) Substrate X becomes limiting, leading to selection for duplication in gene *aab* and increased enzyme activity for the conversion of X to Y. (5) A modification in gene *aab* leads to the formation of gene *acb* that produces a modified enzyme capable of catalyzing the conversion of a new substrate (W) to X. Genes *acb*, *aab*, and *aaa* constitute an operon producing the enzymes of the pathway converting substrate (W) to product (Z). [Taken from Mortlock (1982).]

## 3.3. The Origin of the Xylitol Dehydrogenase Activity

In order to convert xylitol to D-xylulose or L-arabitol to L-xylulose, the mutants had to acquire enzymatic activity for that particular pentitol, activity to catalyze an oxidation at the C-2 position (Fig. 2). The xylitol-positive mutant cells clearly possessed low levels of such a xylitol dehydrogenase activity, an activity that could not be found in the parent xylitol-negative strain even when the cells had been incubated in the presence of xylitol. It was obvious that some mutational event must have occurred that gave the mutant cells this enzymatic activity.

In addition to the low levels of xylitol dehydrogenase and D-arabitol dehydrogenase found in the xylitol-positive mutants, they were also shown to possess extremely high amounts of ribitol dehydrogenase (Mortlock and Wood, 1964a), as shown in Table III. Not only did the xylitol-grown cultures possess high levels of ribitol dehydrogenase activity, but the xylitol-positive mutants isolated from those cultures continued to synthesize high levels of ribitol dehydrogenase activity, between 18 and 24 U/mg cell protein, when cultured in the absence of either ribitol or xylitol.

Originally it was believed that xylitol metabolism might act for the accidental or gratuitous induction of the ribitol pathway enzymes, but, while purification techniques were able to separate the D-arabitol dehydrogenase enzyme from the other dehydrogenase activities, the ribitol and xylitol dehydrogenase activities could not be separated. Eventually, biochemical, immunological, and genetic experiments showed that these lat-

Table III
Constitutive Ribitol Dehydrogenase Activity in Xylitol-Positive and D-Arabitol-Positive Mutants of *Klebsiella aerogenes* PRL-R3[a]

| Strain | Growth substrate | Ribitol dehydrogenase (U/mg protein) |
|---|---|---|
| Parent | D-Glucose | <0.02 |
|  | Ribitol | 1.10–3.71 |
|  | D-Arabitol | 0.05–0.06 |
|  | Peptone | <0.02 |
| Xylitol-grown culture | Xylitol | 3.01–19.3 |
| L-Arabitol-grown culture | L-Arabitol | 0.08–0.38 |
| Pure culture of xylitol-positive mutant | Peptone | 18.07–24.1 |
| Pure culture of D-arabitol-positive mutant | Peptone | 15.66–20.48 |

[a]Each culture consisted of minimal-salts supplemented with 0.5% of the indicated carbohydrate or peptone. [Data are from Mortlock and Wood (1964a) and Mortlock *et al.* (1965b).]

ter two activities were the result of a common enzyme (Lerner et al., 1964; Mortlock et al., 1965a,b).

Because of similarities in the structure of the two sugars, the enzyme that had apparently evolved for the oxidation of ribitol, ribitol dehydrogenase, had a natural but weak ability to catalyze the oxidation of xylitol. If the bacterial cells possessed high enough levels of this dehydrogenase, its weak, gratuitous activity for the oxidation of xylitol permitted the slow oxidation of xylitol to D-xylulose. D-Xylulose could be further degraded by other existing cellular enzymes, and growth of the bacteria with xylitol as the sole carbon source was possible.

If the information presented above was true and the organism was already capable of synthesizing an enzyme that could catalyze the oxidation of xylitol to D-xylulose, then why should any mutation be required to permit growth on xylitol? A mutation was still required because the regulation of the enzymes of the ribitol operon, including ribitol dehydrogenase (RDH), had evolved such that these enzymes would be synthesized only when ribitol was present as a potential growth substrate. The cells recognized no regulatory signal to synthesize RDH when only xylitol was available. In order to program the cells with sufficient RDH to permit growth on xylitol, a mutation in the regulation of the ribitol operon was required, a mutation that allowed the synthesis of high levels of the enzymes of the ribitol operon in the absence of any inducer (Lerner et al., 1964; Mortlock et al., 1965a,b).

The genetic determinants of the ribitol pathway are located close to one another on the bacterial chromosome and constitute an operon; the genes coding for the structure of RDH and D-ribulokinase are transcribed by a single strand of messenger RNA. In the genetic map illustrated in Fig. 4, the structural gene for D-ribulokinase is represented by *rbtK*, while *rbtD* is the gene possessing the information for the structure of ribitol dehydrogenase. The *rbtC* and *rbtB* genes are involved in the regulation of the ribitol operon, with *rbtB* representing the regulator gene for the operon and *rbtC* probably representing the operator/promoter region of the operon, a region for the binding of the repressor protein and the initiation of the synthesis of messenger RNA. The ribitol operon appears to be under negative control, in that the function of the regulator gene *rbtB* is to synthesize a repressor protein that normally prevents the expression of the structural genes and the synthesis of ribitol dehydrogenase and D-ribulokinase. The function of the inducer of the pathway formed when ribitol is present as a potential growth substrate is to overcome this repression and permit the synthesis of the enzymes. Once the ribitol has been exhausted from the growth medium, the inducer concentration rapidly drops and the repressor again prevents synthesis of the enzymes.

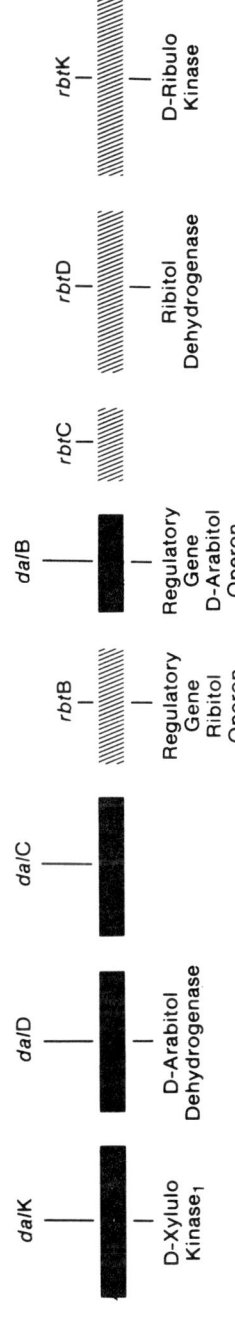

**Figure 4.** The genes representing the ribitol and D-arabitol operons. The genes of the D-arabitol operon are the structural gene for D-xylulokinase (*dalK*), the structural gene for D-arabitol dehydrogenase (*dalD*), a regulatory site probably representing an operator–promoter region (*dalC*), and a probable regulatory gene (*dalB*). The genes of the ribitol operon are the structural gene for D-ribulokinase (*rbtK*), the structural gene for ribitol dehydrogenase (*rbtD*), a regulatory site probably representing an operator–promoter region (*rbtC*), and a regulatory gene (*rbtB*). Most of the mutants permitting constitutive synthesis of the enzymes of the ribitol operon and growth on xylitol are located in *rbtB*. The mutations resulting in an altered ribitol dehydrogenase with improved activity for the oxidation of xylitol are located in *rbtD*.

However, if the repressor gene is damaged by mutation so that it cannot synthesize an active repressor, inducer is not required and the enzymes of the operon are synthesized continuously. Such mutants, which synthesize high amounts of the enzyme in the absence of substrate and inducer, are termed constitutive mutants (Charnetsky and Mortlock, 1974a,b; Neuberger and Hartley, 1979).

The organism did not require any new enzymatic activity to establish a pathway for growth on xylitol. Because of the similarity in structure between ribitol and xylitol, the dehydrogenase that evolved for the ribitol pathway was capable of supplying the needed oxidizing activity to convert xylitol to D-xylulose. To grow on xylitol, the cells required a regulatory mutation permitting the dehydrogenase of the ribitol pathway to be synthesized in the absence of ribitol and used for the degradation of xylitol.

Most of the mutations permitting growth on xylitol were found to occur in the *rbtB* gene of the ribitol operon, while a few were in the other regulatory gene, the *rbtC* gene. All of these mutations resulted in the constitutive synthesis of the enzymes of the operon. Since the ribitol operon was under negative control, the normal function of the *rbtB* regulator gene was to prevent the synthesis of the enzymes of the operon. The mutations that permitted growth on xylitol and were located in the *rbtB* gene prevented the gene from functioning properly so that it was unable to inhibit the synthesis of the enzymes of the operon. The result was that high levels of RDH were continuously synthesized and growth on xylitol was possible. Growth on the new substrate, xylitol, was actually made possible through a mutation that caused a loss of the function of a previously existing gene, the regulator gene of the ribitol pathway. Since numerous types of mutations can result in the loss of the function of a gene, it suddenly became clear why mutations for growth on xylitol occurred at such a high frequency.

### 3.4. The Origin of the D-Xylulokinase of the New Xylitol Pathway

The new xylitol catabolic pathway consisted of ribitol dehydrogenase to catalyze the oxidation of xylitol to D-xylulose, and a D-xylulokinase to phosphorylate D-xylulose at the C-5 position, as illustrated in Fig. 2. Cells grown on xylitol were found to be induced for low levels of the enzymes of the D-arabitol pathway, and the kinase used for the new xylitol pathway was reported to be the D-xylulokinase of the D-arabitol pathway. However, the *Klebsiella* cells were later shown to be capable of the synthesis of another D-xylulokinase, one associated with a D-xylose catabolic pathway. Apparently either of these kinases could function during the catabolism of xylitol, since mutants obtained from a xylitol-positive strain that had lost either of these two kinases were still capable of growth on xylitol. If both kinases were lost from the strain by mutation, however, the cells also lost the ability to phosphorylate D-xylulose and could not grow on xylitol (Wilson and Mortlock, 1972).

## 4. The Origin of the L-Arabitol Dehydrogenase Activity

With the elucidation of the concept that the new xylitol catabolic pathway consisted of enzymes borrowed from pathways already established by evolution, the L-arabitol-positive mutants were next examined to see if a similar situation existed for them. In about 90% of the L-arabitol-positive mutants examined, the mutation permitting growth was also a mutation resulting in the constitutive synthesis of RDH. It was this enzyme that oxidized L-arabitol to L-xylulose. The formation of L-xylulose within the cells caused the induction of an L-xylulokinase activity to catalyze the formation of L-xylulose 5-phosphate. The latter was converted to L-ribulose 5-phosphate and then D-xylulose 5-phosphate by other, presumably inducible, enzymes (LeBlanc and Mortlock, 1973).

Thus, both of the new pathways, the one for xylitol degradation as well as the one for L-arabitol degradation, used the existing RDH enzyme to oxidize the pentitol substrate and establish growth. About 10% of the L-arabitol-positive mutants examined were not constitutive for the synthesis of RDH, and the nature of the mutation permitting growth on L-arabitol for these mutants has not been elucidated.

## 5. Mutations Improving the Growth Rate on Xylitol

### 5.1. Alterations in the Dehydrogenase Activity

The growth of these mutants on xylitol was slow, apparently limited by the poor activity of the enzyme for xylitol. It could be predicted that continuous cultivation with xylitol might select for additional mutations that increased the rate of oxidation of xylitol and thus the growth rate with xylitol as a substrate. Two obvious types of mutation would increase the intercellular xylitol-oxidizing activity. The first would improve the catalytic activity of the enzyme for xylitol as a substrate, while the second would simply increase the quantity of the enzyme synthesized by each cell. Both types of mutation have been observed, as described below.

The parent strain for these initial experiments by Lin and co-workers was a xylitol-positive, ribitol dehydrogenase-constitutive mutant with a doubling time on xylitol of 4 hr at 37°C. The cells were treated with a mutagenic agent and cultured on xylitol until a strain was isolated that grew on xylitol at about twice the rate as the first strain, with a new doubling time of about 2 hr with 0.2% xylitol as the substrate. Upon examination, the mutation in this second strain was found to have resulted in the production of an altered dehydrogenase with improved activity for the oxidation of xylitol. The binding ability of the altered enzyme had improved by at least twofold, while the maximum velocity of the enzyme for xylitol was two and one-half times that of the dehydrogenase of the parent strain (Wu *et al.*, 1968; Lin *et al.*, 1976).

Mutants producing an altered dehydrogenase with improved activity for oxidation of xylitol have also been directly isolated from chemostats where the growth on the organism was limited by cultivation on low levels of xylitol (Burleigh *et al.*, 1974). Such experiments will be described in Chapter 2.

### 5.2. Mutants with Increased Amounts of Ribitol Dehydrogenase

Another way in which the bacteria might improve their growth rate on xylitol despite poor activity of the enzyme for the oxidation of xylitol would be for them to synthesize higher quantities of the inefficient enzyme. Mutants of this type, hyperproducing ribitol dehydrogenase, have been isolated from chemostat experiments where the growth of the cells was limited by low concentrations of xylitol (Hartley *et al.*, 1976; Inderlied and Mortlock, 1977). One relatively stable type of mutant that was isolated repeatedly from such chemostat experiments was found to possess about fourfold elevated levels of dehydrogenase. This was believed to be a mutation of the "promoter-up" type, where increased enzyme levels resulted from an increased rate of synthesis of messenger RNA. This is illustrated in Table IV by the strain possessing a dehydrogenase activity of 19.7 U/mg protein. A second type of mutant isolated was found to possess much higher levels of ribitol dehydrogenase, but these mutants were "unstable" in that the very high enzyme levels were lost when the cells were cultured under other growth conditions. These latter mutants are represented in Table IV by the strain possessing a dehydrogenase activity of 90 U/mg protein, over 22 times the activity of the original xylitol-positive mutant, and were believed to contain duplicate copies of the structural gene for ribitol dehydrogenase, the *rbtD* gene. Similar mutants have also been isolated from chemostat experiments with a mixture of xylitol and ribitol as substrate (Thompson and Krawiec, 1983). More extensive chemostat experiments have been done by Brian Hartley and co-workers and are described in Chapter 2.

Table IV
Constitutive and Hyperconstitutive Levels of Ribitol Dehydrogenase in Xylitol-Positive Mutants of *Klebsiella aerogenes* W70[a]

| Isolate | Growth substrate | Ribitol dehydrogenase (U/mg protein) | Phenotype |
|---|---|---|---|
| Parent strain | Ribitol | 1.5 | Inducible |
| | Casein hydrolysate | <0.02 | Inducible |
| Xylitol-positive | Casein hydrolysate | 4.0 | Constitutive |
| Chemostat isolate | Casein hydrolysate | 19.7 | Hyperconstitutive |
| Chemostat isolate | Casein hydrolysate | 90.4 | Hyperconstitutive |

[a] Data are from Inderlied and Mortlock (1977).

### 5.3. Utilization of the D-Arabitol Transport System to Facilitate the Transport of Xylitol

Wu et al. (1968) reported that xylitol-positive mutants of *Klebsiella aerogenes* strain 1033, utilizing ribitol dehydrogenase for the oxidation of xylitol to D-xylulose, were dependent upon slow, simple diffusion for the transport of xylitol through the cell membrane and into the cells. These workers obtained a modification in the structural gene for ribitol dehydrogenase permitting more efficient oxidation of xylitol. This modified strain was then cultured with very low levels of xylitol, a condition where the rate of diffusion of xylitol into the cells limited the growth rate. Under these selective conditions, they successfully obtained an additional mutation in their strain, improving the transport of xylitol into the cells. Upon examination, this new mutant was shown to possess an energy-utilizing, active transport system to transport xylitol through the cell membrane, a transport system that permitted more rapid growth of the cells with low concentrations of xylitol in the growth medium. This transport system was found to be the same normally employed by the parent strain during growth on the more common pentitol, D-arabitol (Lin et al., 1976).

D-Arabitol is metabolized by *Klebsiella* strains by an inducible enzyme pathway consisting of a transport system to bring D-arabitol into the cells, a D-arabitol dehydrogenase to oxidize D-arabitol to D-xylulose, and a D-xylulokinase to catalyze the phosphorylation to form D-xylulose 5-phosphate, as shown in Fig. 2. The enzymes of this entire D-arabitol catabolic pathway were found to be synthesized constitutively in these new xylitol mutants. The D-arabitol transport system was found to be capable of transporting xylitol in addition to D-arabitol, and a mutation in the regulation of the transport system, permitting its constitutive synthesis, allowed it to be synthesized and used during growth of the mutant cells on xylitol. This regulatory mutation in the D-arabitol catabolic pathway affected all elements of the pathway, and the dehydrogenase and kinase were also synthesized constitutively.

This 1033 strain first underwent a mutation in the regulator gene of the ribitol pathway, permitting the constitutive synthesis of ribitol dehydrogenase and establishing growth on xylitol. Next it underwent modification of the ribitol dehydrogenase structural gene to permit the dehydrogenase coded by that gene to have improved activity for the oxidation of xylitol as a substrate. Finally, it underwent a mutation in the regulation of the D-arabitol pathway that permitted the constitutive synthesis of the elements of that pathway, with the utilization of the D-arabitol transport permease to increase the rate of transport of xylitol into the cells.

With the PRL-R3 strain of *K. pneumoniae*, growth on xylitol resulted in the induction of low levels of D-arabitol dehydrogenase and D-xylulokinase (Mortlock and Wood, 1964a). It is not known if this induction resulted from the reduction of some of the D-xylulose produced from xylitol to D-arabitol, the apparent natural inducer of the D-arabitol pathway enzymes, or if xylitol itself could function as a weak inducer of the D-arabitol pathway enzymes. In any event, cells growing on xylitol possessed low levels of the D-arabitol catabolic pathway enzymes. If the transport system for D-arabitol was coordinately controlled along with the rest of the enzymes of the D-arabitol catabolic pathway in this organism, mutants oxidizing xylitol should have been weakly induced for the D-arabitol transport system and should have been able to utilize it to aid the transport of xylitol into the cells without requiring any mutation in the regulation in the D-arabitol pathway.

Mutants of *K. pneumoniae* PRL-R3 that were constitutive for the ribitol enzymes were not found to be constitutive for a ribitol transport system. However, ribitol constitutive mutants of *K. aerogenes* W70 were also constitutive for a ribitol transport system, showing coordinate regulation between the ribitol transport system and the ribitol operon in that strain. Recent data indicate that some form of common regulation may exist for the transport systems for ribitol and D-arabitol in the PRL-R3 strain (Mays and Mortlock, 1983).

The structural genes for the dehydrogenase and the kinase of the D-arabitol pathway were found in an operon that was located adjacent to the ribitol operon, as shown in Fig. 4. These inverted operons with apparent overlapping regulatory regions have been postulated to have evolved from one another by a process of gene duplication and inversion (Charnetzky and Mortlock, 1974b). They have also been found in some strains of *Escherichia coli* C, where it has been postulated they were acquired by transfer from yet another organism (Link and Reiner, 1982).

In the experiments for the improved xylitol-utilizing strains of *K.*

*aerogenes* 1033 published by Lin and his colleagues, the mutation first establishing growth on xylitol was in the *rbtB* regulatory gene of the ribitol operon and permitted the constitutive synthesis of the enzymes of the ribitol operon, including ribitol dehydrogenase, which catalyzed the oxidation of xylitol. The second mutation established was in the *rbtD* structural gene for ribitol dehydrogenase so that the dehydrogenase gene product had improved activity for the oxidation of xylitol. The third mutation was in the regulation of the D-arabitol operon, perhaps in the *dalB* regulatory region. This permitted the constitutive synthesis of D-arabitol dehydrogenase, D-xylulokinase, and a D-arabitol transport protein whose structural gene location was unknown, but was part of the same regulon (controlled by the same regulator gene) as the other enzymes of the D-arabitol pathway. Since this transport system could be utilized to bring xylitol into the cells, the high levels of the transport system in the constitutive mutant permitted more efficient growth when xylitol concentrations in the growth medium were very low (Lin *et al.*, 1976). The two genes from the ribitol and D-arabitol operons not involved in establishing this improved xylitol pathway were the structural genes for D-arabitol dehydrogenase and D-ribulokinase (*dalD* and *rbtK*). These genes did not appear to serve any function during growth on xylitol. It can be postulated that continued culture of cells on xylitol without exposure to ribitol of D-arabitol as substrate might result in the modification of these genes through genetic drift and the eventual loss of those enzymatic activities.

It should also be noted that a genetic alteration replacing the D-arabitol dehydrogenase structural gene (*dalD*) with the ribitol dehydrogenase structural gene (*rbtD*) would result in a modified D-arabitol operon containing structural genes for a dehydrogenase to oxidize xylitol to D-xylulose and a kinase to convert D-xylulose to D-xylulose 5-phosphate. Furthermore, a single mutation permitting the constitutive synthesis of this modified operon not only would program the cells with the dehydrogenase and kinase needed to establish a new xylitol pathway, but would also permit the constitutive synthesis of a transport system to aid in the transport of xylitol into the cells through the cell membrane.

## 6. The Growth of *Escherichia coli* Strains on Xylitol

### 6.1. The Utilization of Ribitol Dehydrogenase to Establish Growth on Xylitol

Although naturally occurring strains of *E. coli* K12 or *E. coli* B have never been shown to possess either the ribitol or D-arabitol operons nor utilize any pentitols as growth substrates, some strains of *E. coli* C have been found to use both ribitol and D-arabitol (Reiner, 1975). The arrangement of the operons was the same as observed for *Klebsiella* and mutants could be obtained that would grow on xylitol by employing a constitutive ribitol dehydrogenase to oxidize xylitol to D-xylulose. Both of the operons could be transferred by a bacteriophage-mediated transduction into *E. coli* K12. Although the direct selection of K12 mutants for growth on xylitol was always unsuccessful, the strains transduced to gain the ribitol operon were then able to mutate for growth on xylitol and would employ a constitutive ribitol dehydrogenase to catalyze the oxidation of the new pentitol substrate (Scangos and Reiner, 1978).

### 6.2. The Construction of a Different Route for Xylitol Catabolism

Mutations establishing different enzymes for the catabolism of xylitol have been deliberately programmed into *E. coli* K12. These new mutations permitted the degradation of xylitol by an enzyme borrowed from the natural pathway for utilization of D-xylose and a fermentation pathway that normally produced 1,2-propanediol as a product (Wu, 1976a). A pathway for the degradation of D-arabitol by *E. coli* K12 has also been constructed by utilizing a transport system normally used for the transport of 1,2-propanediol. A dehydrogenase then catalyzed the oxidation of D-arabitol to D-xylulose, which could be further degraded by the existing enzymes of the D-xylose catabolic pathway (Wu, 1976b). A complete discussion of the mutations effecting propanediol dehydrogenase and the establishment of growth on new substrates will be found in Chapter 5.

### 7. The Utilization of Xylitol by a Mutant in the Genus *Erwinia*

Recently a strain of *Erwinia uredovora* was studied that also possessed the ability to mutate to gain the ability to utilize xylitol as a growth substrate. Xylitol-positive mutants occurred with a frequency of about $9 \times 10^{-8}$. The mutation establishing growth on xylitol resulted in the constitutive synthesis of a novel, NAD-linked xylitol-4-dehydrogenase that oxidized xylitol at the C-4 position to produce L-xylulose rather than D-xylulose. The $K_m$ of this dehydrogenase for xylitol was 48 mM. The enzyme had no activity for the oxidation of ribitol, the organism was unable to utilize ribitol as a growth substrate, and ribitol-positive mutants could not be obtained (Doten and Mortlock, 1983).

## 8. Summary

The evolution of the catabolic pathways for the two more common pentitols, ribitol and D-arabitol, has apparently been closely related, at least among the bacteria in the coliform group. The genetic determinants of the enzymes for the degradation of these pentitols are found in adjacent but inverted operons on the bacterial chromosome in many species of *Klebsiella* and some strains of *E. coli*. The other two pentitols, xylitol and L-arabitol, do not appear to be as common, and bacteria that have evolved the ability to utilize these sugars as growth substrates are not as frequently found in nature. However, mutants can often be isolated that have gained the ability to grow on xylitol or L-arabitol and, in all cases studied, the initial mutation permitting growth on either xylitol or L-arabitol has been shown to be a regulatory mutation that permitted an enzyme that had apparently evolved for a different purpose to catalyze the initial reaction required for the degradation of the new substrate. Once growth on the new carbohydrate had been established, selection could then be expected to occur for additional mutations that might improve the growth rate. Eventually such a process might lead to the formation of specific operons regulated for efficient catabolism of the new substrate.

These studies are only preliminary investigations into the mechanisms by which bacteria acquire new metabolic pathways and are able to broaden their range of utilizable growth substrates. They, and many of the investigations reported in other chapters in this book, clearly show the importance of mutations in the regulation of existing metabolic pathways in order to accommodate the formation of new pathways.

ACKNOWLEDGMENT. The author's research on this topic has been supported by Public Health Services Grant AI 15328 from the National Institute of Allergy and Infectious Diseases and United States Department of Agriculture Hatch Project 144-436.

## References

Bisson, T. M., Oliver, E. J., and Mortlock, R. P., 1968, Regulation of pentitol metabolism by *Aerobacter aerogenes*. II. Induction of the ribitol pathway, *J. Bacteriol.* **95**:932–936.

Burleigh, B. D. Jr., Rigby, P. W. J., and Hartley, B. S., 1974, A comparison of wild-type and mutant ribitol dehydrogenases from *Klebsiella aerogenes*, *Biochem. J.* **143**:341–352.

Charnetsky, W. T., and Mortlock, R. P., 1974a, Ribitol catabolic pathway in *Klebsiella aerogenes*, *J. Bacteriol.* **119**:162–169.

Charnetzky, W. T., and Mortlock, R. P., 1974b, Close genetic linkage of the determinants of the ribitol and D-arabitol catabolic pathways in *Klebsiella aerogenes*, *J. Bacteriol.* **119**:176-182.

Doten, R. C., and Mortlock, R. P., 1983, NAD-linked xylitol-4-dehydrogenase in *Erwinia*, *Abstr. Annu. Meet. Am. Soc. Microbiol.* K 244, p. 217.

Fossitt, D. D., Mortlock, R. P., Anderson, R. L., and Wood, W. A., 1964, Pathways of L-arabitol and xylitol metabolism in *Aerobacter aerogenes*, *J. Biol. Chem.* **239**:2110-2115.

Gong, Cheng-Shung, Chen, L. F., and Tsao, G. T., 1981, Quantitative production of xylitol from D-xylose by a high-xylitol producing yeast mutant *Candida tropicalis* HXP2, *Biotechnol. Lett.* **3**:125-130.

Hartley, B. S., Altosaar, I., Dothie, J. M., and Neuberger, M. S., 1976, Experimental evolution of a xylitol dehydrogenase, in: *Proceedings of the Third John Innes Symposium* (R. Markham and R. W. Horne, eds.), North-Holland, Amsterdam, pp. 191-200.

Horowitz, N. H., 1945, On the evolution of biochemical synthesis, *Proc. Natl. Acad. Sci. USA* **31**:153-157.

Horowitz, N. H., 1965, The evolution of biochemical synthesis—Retrospect and prospect, in: *Evolving Genes and Proteins* (V. Bryson and H. J. Vogel, eds.), Academic Press, New York, pp. 15-23.

Inderlied, C. B., and Mortlock, R. P., 1977, Growth of *Klebsiella aerogenes* on xylitol: Implications for bacterial enzyme evolution, *J. Mol. Evol.* **9**:181-190.

LeBlanc, D. J., and Mortlock, R. P., 1973, Regulation of the L-arabinose catabolic pathway in *Aerobacter aerogenes*, *Arch. Biochem. Biophys.* **156**:390-396.

Lerner, S. A., Wu, T. T., and Lin, E. C. C., 1964, Evolution of a catabolic pathway in bacteria, *Science* **146**:1313-1315.

Lin, E. C. C., Hacking, A. J., and Aguilar, J., 1976, Experimental models of acquisitive evolution, *BioScience* **26**:548-555.

Link, C. D., and Reiner, A. M., 1982, Inverted repeats surround the ribitol-arabitol genes of *E. coli* C, *Nature* **298**:94-96.

Makinen, K. K., and Scheinin, A., 1982, Xylitol and dental caries, *Annu. Rev. Nutr.* **2**:133-150.

Mays, J. P., and Mortlock, R. P., 1983, Inducer exclusion and ribitol transport in *Klebsiella*, *Abstr. Annu. Meet. Am. Soc. Microbiol.* K 113, p. 195.

Mortlock, R. P., 1976, Catabolism of unnatural carbohydrates by micro-organisms, *Adv. Microb. Physiol.* **13**:2-53.

Mortlock, R. P., 1982, Regulatory mutations and the development of new metabolic pathways by bacteria, in: *Evolutionary Biology*, Vol. 14 (M. K. Hecht, B. Wallace, and G. T. Prance, eds.), Plenum Press, New York, pp. 205-268.

Mortlock, R. P., and Wood, W. A., 1964a, Metabolism of pentoses and pentitols by *Aerobacter aerogenes*. I. Demonstration of pentose isomerase, pentulokinase, and pentitol dehydrogenase enzyme families, *J. Bacteriol.* **88**:835-844.

Mortlock, R. P., and Wood, W. A., 1964b, Metabolism of pentoses and pentitols by *Aerobacter aerogenes*. II. Mechanism of acquisition of kinase, isomerase, and dehydrogenase activity, *J. Bacteriol.* **88**:845-849.

Mortlock, R. P., Fossitt, D. D., Petering, D. H., and Wood, W. A., 1965a, Metabolism of pentoses and pentitols by *Aerobacter aerogenes* III. Physical and immunological properties of pentitol dehydrogenases and pentulokinases, *J. Bacteriol.* **89**:129-135.

Mortlock, R. P., Fossitt, D. D., and Wood, W. A., 1965b, A basis for utilization of unnatural pentoses and pentitols by *Aerobacter aerogenes*, *Proc. Natl. Acad. Sci. USA* **54**:572-579.

Neuberger, M. S., and Hartley, B. S., 1979, Investigations into the *Klebsiella aerogenes* pentitol operons using specialized transducing phages p*rbt* and p*rbt dal*, *J. Mol. Biol.* **132**:435-470.

Reiner, A. M., 1975, Genes for ribitol and D-arabitol catabolism in *Escherichia coli*: Their loci on C strains and absence in K-12 and B strains, *J. Bacteriol.* **123**:530–536.

Scangos, G. A., and Reiner, A. M., 1978, Ribitol and D-arabitol catabolism in *Escherichia coli*, *J. Bacteriol.* **134**:492–500.

Schaffer, R., 1972, Occurrence, properties, and preparation of naturally occurring monosaccharides (including 6-deoxy sugars), in: *The Carbohydrates*, Vol. 1A (W. Pigman and D. Horton, eds.), Academic Press, New York, pp. 69–111.

Thompson, L. W., and Krawiec, S., 1983, Acquisitive evolution of ribitol dehydrogenase in *Klebsiella pneumoniae*, *J. Bacteriol.* **154**:1027–1031.

Washuttl, J., Riederer, P., and Bank, E., 1973, A qualitative and quantitative study of sugar-alcohols in several foods, *J. Food Sci.* **38**:1262–1263.

Wilson, B. L., and Mortlock, R. P., 1972, Regulation of D-xylose and D-arabitol catabolism by *Aerobacter aerogenes*, *J. Bacteriol.* **113**:1404–1411.

Wood, W. A., McDonough, M. J., and Jacobs, L. B., 1961, Ribitol and D-arabitol utilization by *Aerobacter aerogenes*, *J. Biol. Chem.* **236**:2190–2195.

Wu, T. T., 1976a, Growth of a mutant of *Escherichia coli* K-12 on xylitol by recruiting enzymes for D-xylose and L-1,2-propanediol metabolism, *Biochim. Biophys. Acta* **428**:656–663.

Wu, T. T., 1976b, Growth on D-arabitol of a mutant strain of *Escherichia coli* K-12 using a novel dehydrogenase and enzymes related to L-1,2 propanediol and D-xylose metabolism, *J. Gen. Microbiol.* **94**:246–256.

Wu, T. T., Lin, E. C. C., and Tanaka, S., 1968, Mutants of *Aerobacter aerogenes* capable of utilizing xylitol as a novel carbon, *J. Bacteriol.* **96**:447–456.

CHAPTER 2

# Experimental Evolution of Ribitol Dehydrogenase

## B. S. HARTLEY

## 1. Introduction

### 1.1. Evolutionary Lessons from Protein Structures

This project began in 1968, when the impact of amino acid sequencing and protein crystallography had revealed a flood of data with great impact on evolutionary theory. These facts allowed the following important conclusions:

1. The conformation of the same protein from different species is carefully conserved throughout evolution.
2. Internal amino acid residues that clearly contribute to that conformation are generally, but not invariably, conserved.
3. Surface amino acids are more variable, except for those that clearly contribute to the catalytic activity. This encouraged the concept of "neutral mutations."
4. Evolutionary trees can be constructed from a matrix of species differences, either overall or by translating back to DNA sequences via the known genetic code or by assuming "invariant" and "variable" regions. These evolutionary trees can be made to bear a satisfying resemblance to the known fossil record.
5. The rate of sequence variation appears to correlate with time rather than with the assumed number of generations between species. This led to the "neutral drift" theory of Kimura (1969).

*B. S. HARTLEY* • Department of Biochemistry, Imperial College of Science and Technology, London SW7 2AZ, England.

6. Protein families with similar functions, such as myoglobin and hemoglobin, or the pancreatic serine proteases, also have almost superimposable conformations. Divergence from a common ancestor via gene duplication was clearly implied.
7. Sequence variations between these similar proteins in the same individual follow the same pattern as species differences for a single protein. Hence, evolutionary trees implying distance from a common ancestor could be constructed.
8. Specificity differences between chymotrypsin, trypsin, and elastase appeared to require only one or two amino acid changes, respectively, with no significant conformational change in the specificity sites (Hartley and Shotton, 1971).
9. Convergent evolution to a common enzyme mechanism from two completely different protein ancestors was obvious from the structures of chymotrypsin and subtilisin (Kraut *et al.*, 1971).

However, the concepts engendered by these data initially relied on the implicit assumption of random point mutation as the predominant source of sequence variation. These mutations were assumed to be fixed in a population by continuous Darwinian selection for the "fitter" protein or against the less fit protein, with the additional concept of "neutral drift" to fill in the gray area. Further lessons from protein structure cast doubt on these simplistic assumptions:

10. Deletions or insertions of several amino acids are occasionally observed in different species, and very commonly between members of a protein family. Structural studies showed that these deletions or insertions invariably occurred at $NH_2$- or COOH-terminal tails, or at external loops, with little or no effect on the rest of the conformation.
11. Clusters of internal hydrophobic residues often vary between two proteins in such a way that it would be difficult to achieve step-by-step replacement without disturbing the conformation of the chain and thereby produce an inactive product (Hartley, 1974).
12. Three classes of serine protease structure can be recognized: a "trypsin class" found in animals; a "*Myxobacter* class" of related but distinct structure found in some microorganisms (McLachlan and Shotton, 1971); and the "subtilisin class" typical of bacilli, which have a totally unrelated sequence and conformation (Wright *et al.*, 1969).
13. *Streptomyces griseus* contains genes for all three of these classes of serine protease: a trypsinlike enzyme with 43% homology to bovine trypsin, and elastaselike enzyme with 30% homology to

the *Myxobacter* enzyme, and a "subtilisin" with a Thr-Ser-Met-Ala active site sequence that is characteristic of that class (Awad et al., 1972).

Stepwise point mutations followed by continuous selection of the protein product could hardly account for these latter structures. Deletions might be achieved by looping out small segments of DNA during replication and this would lead to a viable protein product only if the deleted regions were structurally unimportant, e.g., $NH_2$- or COOH-terminal extensions or redundant loops in $\beta$ structures. A pair of compensating frameshift mutations would radically change the intervening protein sequence and this might be tolerated in regions that contribute little to protein structure or activity.

However, most deletions or frameshifts within a gene, as with most point mutations, would lead to a protein product that could not fold correctly. Would such genes be eliminated rapidly from a population or might they persist for a sufficient time to allow compensating mutations to accumulate? Such "silent genes" might be the most potent source of evolutionary change, since the rate of accumulation of sequence variation would no longer be subject to screening by selection for the protein product. A population would therefore rapidly become heterogeneous in the progeny of the original "silent gene," since sequence changes would occur at a "mutagenesis rate" rather than the rate imposed by selection for an essential protein activity. A compensating mutation that restored the capacity of the protein to fold would bring with it the sequence variations accumulated during the silent phase.

It is obvious that a diploid organism could more readily carry a "silent allele" than a haploid organism. If we consider that most genes have spent some appreciable part of their evolutionary history as "silent genes," it is not surprising that species differences in a particular protein are proportional to chronological time rather than generation time, but that the rate of fixation varies widely from one protein to another. The concept of "neutral drift" therefore becomes superfluous, or, more fairly, becomes embedded in this assumption, since all mutations accumulated in the silent phase are "neutral."

Moreover, the silent gene concept is of particular relevance to the evolution of protein families. Here we have several genes within one individual that have undoubtedly diverged from a single gene. Gene duplication must have been the first step, and it is obvious that the duplication was conserved sufficiently long for a new enzyme activity to evolve. Figure 1 shows the probable history of this evolutionary divergence. It illustrates that selection operates to conserve amino acid sequence and

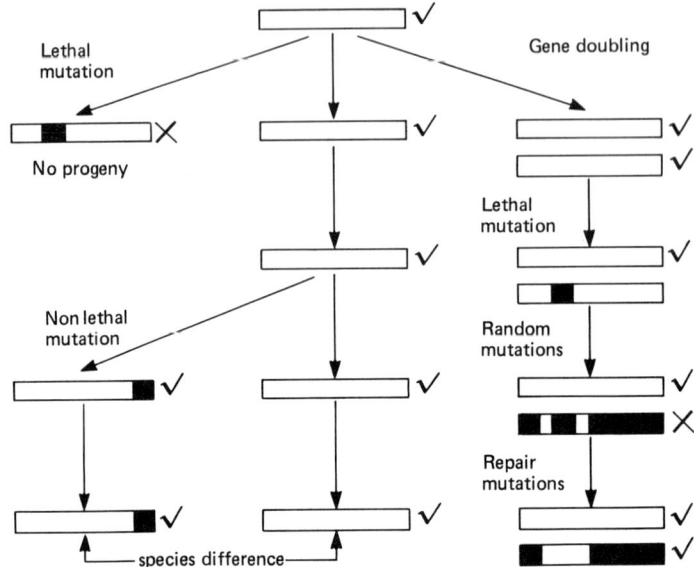

**Figure 1.** Gene doubling in evolution. A check indicates an active enzyme, a cross an inactive enzyme. Shaded areas show mutations in the genes.

implies that sequence variations in a single essential gene in a haploid organism are rare. However, following fixation of a gene duplication, the organism has a second copy of the gene that is no longer subject to the conservative discipline of natural selection. Mutation to a protein product that cannot fold is likely to be the second step, followed by rapid accumulation of mutations in the "silent gene" pool within the population. This will be balanced by the selective pressure to eliminate the silent gene from the population. A reversion of the original mutation might restore the capacity of the protein to fold, and in this case the random mutations accumulated during the silent phase will be revealed as if in a sudden multiple mutation event.

In seeking to explain the evolution of novel enzyme specificities, it seemed to me that this pathway was the most probable (Hartley, 1966). In most cases the repair mutations would produce proteins of, at best, activity equivalent to the viable protein ancestor: these might be seen as isozymes. In some cases, however, the multiple mutations elicited in the silent phase would generate a novel enzyme specificity, and subsequent positive selection pressure would quickly establish this gene in the population.

## 1.2. Microbial Enzyme Evolution

Such speculations about the history of protein evolution must be of the nature of a Kipling "just-so" story, since there can be no fossil record and we would need a time machine to unearth the necessary facts. We can, however, begin to study evolutionary mechanisms at work today and to use this evidence to temper our views about the past. The fast generation time of microorganisms and their enormous range of metabolic potentials, combined with the ease with which mutants can be obtained and fixed in large populations, make microorganisms an obvious target for such studies of experimental evolution.

However, one might make the criticism that the experimental models do not adequately mimic the selection processes at work in the real world today, let alone those that prevailed in the past. The tools of the modern microbiologist include techniques for the isolation of cultures derived from a single organism and for the selection of mutants of these by plating techniques in which the experimenter, rather than the blind force of natural competition, is the agent of selection. For this reason our group was attracted to the concept of using a continuous culture in a chemostat as the agency of selection for "fitter" microorganisms. Here the external parameters are precisely controlled in a way somewhat remote from a true ecological niche, and one generally maintains a sterile environment for a pure culture that limits competition from outside. Even so, there is internal competition between mutants and the parental population, and the final takeover is a truly Darwinian event.

Could one use the chemostat to study molecular events as they took place in the evolution of a new microbial enzyme allowing growth on a novel substrate? For this purpose one requires a system in which the acquisition of the new enzyme activity would increase the growth rate of the organism that contained it. In practice this means that the growth rate of the parental organism must be controlled by the activity of a poor enzyme. Otherwise, the organism would not grow at all on the new substrate. It is not easy to find natural examples of such a system, since natural selection rapidly tends to optimize new enzyme activity once it has arisen.

## 2. Pentitol Metabolism in *Klebsiella aerogenes*

A possible system for experimentally studying enzyme evolution in the chemostat came to our attention from the work of Mortlock and his colleagues on pentitol metabolism in *K. aerogenes* (Mortlock *et al.*, 1965).

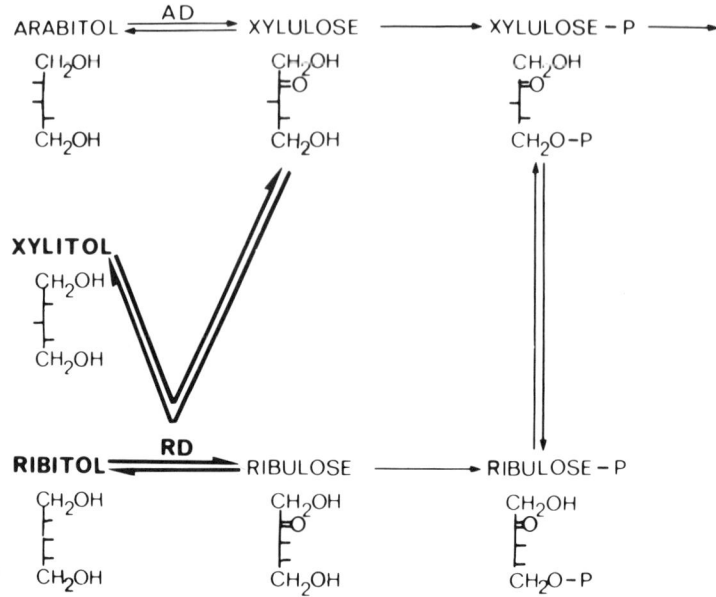

**Figure 2.** Pentitol metabolism in *Klebsiella aerogenes*. (AD) D-Arabitol dehydrogenase; (RD) ribitol dehydrogenase.

They showed that this organism metabolizes ribitol and D-arabitol—two pentitols of relatively high natural abundance—by the pathways shown in Fig. 2. Ribitol is imported by a pentitol permease and converted to D-ribulose by an NAD-linked ribitol dehydrogenase and then to D-ribulose 5-phosphate by D-ribulokinase + ATP. D-Arabitol similarly requires pentitol permease, a D-arabitol dehydrogenase (also NAD-linked), and a D-xylulokinase. The induction of these enzymes is relevant to our purpose. Both ribitol dehydrogenase and D-ribulokinase are induced by D-ribulose (Charnetzky and Mortlock, 1974a). D-Arabitol induces both D-arabitol dehydrogenase and D-xylulokinase (Charnetzky and Mortlock, 1974b). The permease appears to be induced by D-arabitol (Wu *et al.*, 1968). The sugar phosphates enter mainstream glycolytic metabolism by pathways common to most microorganisms.

Wild-type *K. aerogenes* will not grow on the two less common pentitols, xylitol and L-arabitol, which do not induce these pathways. However, mutants constitutive for synthesis of ribitol dehydrogenase ($rbt^c$) *will* grow on xylitol, since this is a poor substrate for that enzyme, which converts it to D-xylulose (Lerner *et al.*, 1964). The D-xylulose then induces the D-arabitol operon by reduction to D-arabitol and is then converted to

D-xylulose 5-phosphate by D-xylulokinase. However, xylitol is such a poor substrate for ribitol dehydrogenase [initial velocity $V$ is 31 U/mg for ribitol and 0.084 U/mg for xylitol at 5 mM pentitol, 1 mM NAD, 28°C (Burleigh *et al.*, 1974)] that the activity of this enzyme becomes growth-limiting in continuous culture on xylitol. Hence, mutants with increased xylitol dehydrogenase activity should grow faster and take over the culture.

## 3. Chemostat Culture of *Klebsiella aerogenes* on Xylitol

The way was therefore open to apply a precisely defined selective pressure to a single gene in a single pure culture of a microorganism and to analyze the products of Darwinian takeover events at the molecular level (Rigby *et al.*, 1974). The selective pressure is precisely defined because microbial growth in a chemostat is governed by the Monod equation:

$$\mu = \frac{\mu_m S}{K_s + S} = D$$

where the growth rate $\mu$ depends on the concentration of growth-limiting substrate $S$ plus two constants: $\mu_m$, the maximum growth rate on saturating substrate, and $K_s$, an experimental parameter defined as the substrate concentration giving half-maximum growth rate. In steady state, $\mu = D$, the dilution rate for the chemostat, which is defined as the flow rate divided by the volume. Hence, the limiting substrate concentration, which determines the growth rate of the cells, is under experimental control by adjustment of the dilution rate:

$$S = \frac{\mu K_s}{\mu_m - \mu} = \frac{D K_s}{\mu_m - D}$$

For a good growth substrate at reasonable dilution rates, $K_s$ is low and $\mu$ is high, so $S$ is very small. In other words, the chemostat culture is an efficient ecological niche that maximizes utilization for food input. However, xylitol has such a high $K_m$ for ribitol dehydrogenase that this governs the Monod constant $K_s$ for the chemostat culture, and the effluent substrate concentration $S$ is very high. This represents inefficient utilization of the food source, so the ratio of influent to effluent substrate concentration is a measure of the "biological fitness" of the organism. A mutation that increases the activity of ribitol dehydrogenase at the intracellular xylitol level, itself controlled by the experimental dilution rate, will decrease $K_s$ and/or increase $\mu_m$. Hence the mutant will grow faster and take over the culture at a new and lower steady-state substrate concentration, thereby signalling an increased "biological fitness."

An opportunity to demonstrate this takeover in a constructed experiment was available since severe nitrosoguanidine mutagenesis of *K. aerogenes* strain X1 (*arg gua rbt$^c$*) had yielded a mutant strain X2 (*arg gua rbt$^c$ rbtB*) that grew better on xylitol plates and appeared to have an altered ribitol dehydrogenase (Wu *et al.*, 1968). This enzyme (RDH-B) was subsequently purified and proved to have 5.83 times the specific activity of wild-type enzyme (RDH-A) toward 5 mM xylitol at 1 mM NAD (Burleigh *et al.*, 1974).

About 20 cells of a derivative of strain X2 (strain B: *gua rbt$^c$ rbtB*) were added to a Porton chemostat (Herbert *et al.*, 1965) containing $10^{12}$ cells of a derivative of strain X1 (strain A: *arg rbt$^c$ rbtA*) in steady-state growth at $D = 0.2$ hr$^{-1}$ on 0.2% xylitol/M9 medium supplemented with traces of guanine and arginine. Figure 3 shows that there was a rapid takeover by strain B within 2–4 days after the addition, with an increase in biomass and a decrease in effluent xylitol. It is interesting to note, however, that in this and most other experiments there was no significant change in viable cell count: the cells merely became bigger.

Since our main concern was the number of takeover events rather than a precise description of the microbial physiology at each stage, we constructed simple small chemostats (Fig. 4) (Rigby, 1971) analogous to the larger vessels described by Herbert *et al.* (1965). These consisted of 500-ml culture vessels (Quickfit and Quartz FV 500) with a stainless steel lid containing five entry ports, sealed by an O-ring and three G-clamps. Medium and air are introduced through the magnetic impeller from a 20-liter Nalgene vessel via a Watson–Marlow peristaltic pump. The overflow controls the working volume at 250 ml and a water jacket controls temperature, normally at 37°C. Alternatively, four such chemostats can be mounted in a stirred water bath. Two ports are used for inoculation or sampling and the third carries a plane quartz window (from a Beckman Model E analytical ultracentrifuge) to allow *in situ* irradiation by a small UV lamp from an LKB Uvicord.

At the commencement of each run the chemostat was assembled in three units—the medium reservoir, the culture vessel with attached inlets and outlets and medium delivery system, and the overflow reservoir. These were sterilized separately by autoclaving and assembled aseptically. Two hundred milliters of medium (M9/0.2% xylitol, pH 6.7) was then pumped into the culture vessel, which was inoculated with 10 ml of a culture of the strain under study grown overnight on the same medium. Stirrer speed was set to 300 rpm, and air flow to 600 ml/min, and the

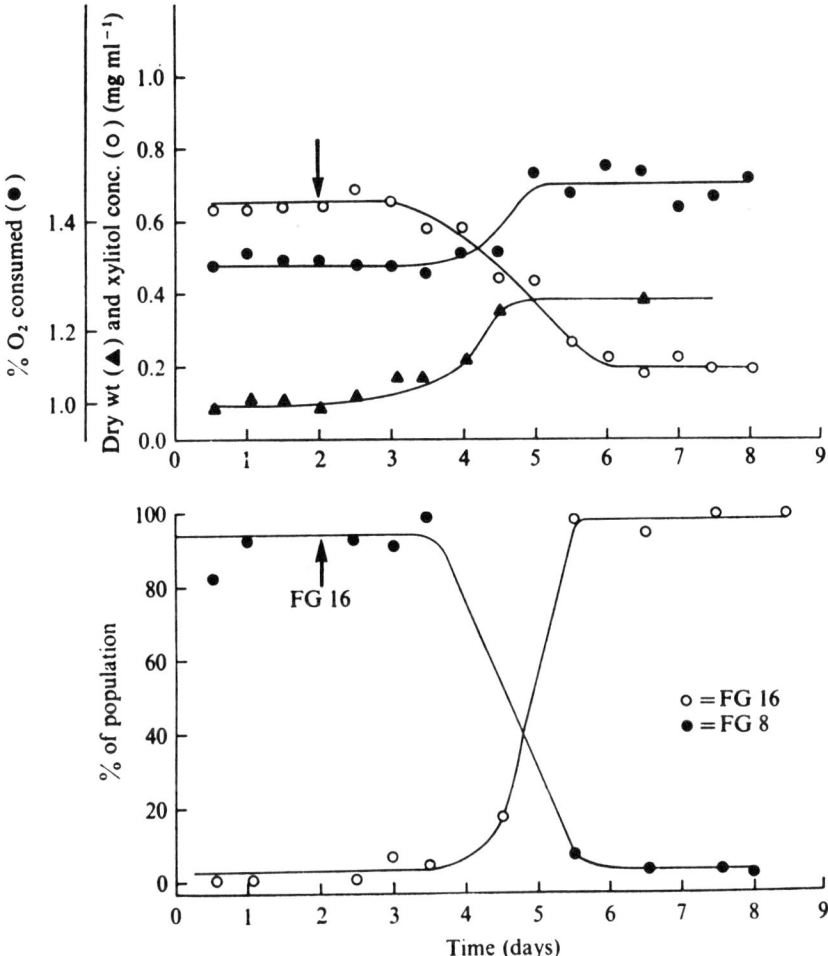

Figure 3. Takeover of *Klebsiella aerogenes* in a chemostat by an evolvant. At the point indicated by the arrow, approximately 20 cells of strain 1B (FG16, a guanine auxotroph of a mutant with improved xylitol dehydrogenase) were added to a steady-state culture containing approximately $10^{12}$ cells of strain 2A (FG8, an arginine auxotroph of the ancestral strain) grown on xylitol. The composition of the population was determined by plating on appropriate media (Hartley et al., 1972).

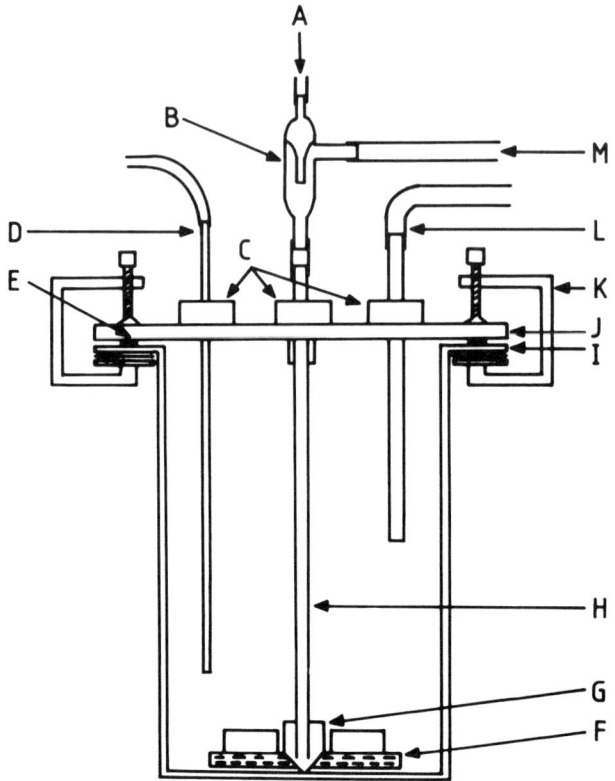

**Figure 4.** Diagram of chemostat vessel. (A) Medium inlet; (B) medium drip-feed; (C) ports; (D) sampling tube; (E) Neoprene rubber gasket; (F) Teflon impeller; (G) impeller support collar; (H) inlet tube; (I) ground glass flange; (J) stainless steel lid; (K) G clamp; (L) overflow tube; (M) air inlet.

culture was grown batchwise for at least 12 hr. The medium pump was then turned on to give a dilution rate of 0.2 $hr^{-1}$ and the culture left for at least 1 day before sampling was begun.

Chemostats were routinely monitored each day for dilution rate, and 12-ml samples were removed for analysis of pH, optical density, viable cell count, bacterial dry weight, and residual substrate level. We found that bacterial dry weight and residual substrate level are the most reliable indicators of takeovers. Optical density values are greatly influenced by the degree of flocculation in the culture, and viable cell counting is quite insensitive to the appearance of evolved strains. Following takeovers, the dilution rate is increased to restore the original xylitol level. The principal problem encountered during our continuous culture experiments resulted from the "stickiness" of this organism. Several runs had to be terminated because of uncontrollable wall growth and we occasionally observed takeovers by mutants that formed macrocolonies that would not wash out.

## 4. Evolution of Ribitol Dehydrogenase in the Chemostat

The history of a number of such experiments carried out in the M.R.C. Laboratory of Molecular Biology in Cambridge from 1970 to 1974 is summarized in Fig. 5. Each new strain represents an organism isolated on xylitol plates from a sample taken from a chemostat following a putative takeover that was signalled by a significant drop in effluent xylitol concentration. Note that not all the fitter mutants that arise in the chemostat will be selected, since there is a statistical chance of washout before takeover can be established (Powell, 1965) and the time for takeover will reflect the relative increase in growth rate of the mutant (Pechurkin, 1969). In some cases (see below) we observed multiple "evolvants" within the same pot and it is not clear whether these arose from parallel or consecutive events. Nevertheless, the genealogy shown in Fig. 5 does by and large represent a true history of parallel or consecutive evolutionary events.

Spontaneous takeovers from wild type occurred in 50–200 generations. To avoid long delays, low-level UV irradiation for periods of 1–2 hr was frequently used to increase mutation rates. Following a putative

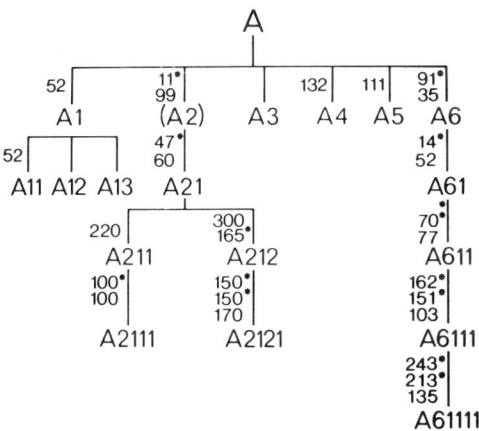

Figure 5. Genealogy of spontaneous and UV-induced evolvants of *Klebsiella aerogenes*. The strain numbers indicate their evolutionary history, e.g., A1, A2, etc., are derived directly from A; A11, A12, etc., from A1; and so on. The numbers between strains are the generations that elapsed before isolation of the new strain. Asterisk indicates a step of UV mutagenesis.

takeover the maximum specific growth rate of the "evolvant" was determined by inoculating a separate chemostat, setting up steady state, increasing $D$ to 0.75 hr$^{-1}$, and following the rate of decrease in biomass $x$. According to Tempest (1970),

$$\frac{d}{dt}(\ln x) = \mu_m - D$$

Following each putative takeover, the single colony isolates were taken from the chemostat sample and cell-free extracts made from cultures of each grown to stationary phase on M9/0.2% xylitol. In most cases, a Lineweaver–Burk plot was made of the xylitol dehydrogenase activity in this extract, but in some cases the ratio of activity at standard xylitol and ribitol concentrations was used to detect enzymes with changed specificity for xylitol.

### 4.1. Enzyme Superproduction

Table I shows that in all the strains shown in Fig. 5 there has been no change in the kinetic properties of the ribitol dehydrogenase in the extract. In some cases the enzyme was purified to homogeneity and was kinetically indistinguishable from wild type. Even when UV mutagenesis was used, none of the strains showed an altered enzyme. This was a surprise, since strain B, obtained by Wu *et al.* (1968) after nitrosoguanidine mutagenesis, did indeed have an improved xylitol dehydrogenase.

In contrast, each of the takeover events appeared to be due to an increased intracellular level of the original inefficient enzyme. In overnight batch cultures strain A makes ribitol dehydrogenase as about 2% of its total soluble protein, while there are stepwise increases to about 10–15% in strains A1, A3, A4, and A6, followed by further steps to 35% in strain A6111 and 48% in strain A112. Moreover, the maximum specific growth rate correlates closely with the level of enzyme in extracts of overnight batch cultures.

Figure 6 demonstrates the microbial physiology of the ancestral strain and two superproducing strains. Both of the latter show a chemostat response consistent with that predicted from the Monod equation (Herbert *et al.*, 1956), in that effectively all of the substrate is utilized at low dilution rates and washout is approached as $D$ approximates $\mu_m$. In these cases, plots of $1/D$ versus $1/S$ yielded $K_s$ values of 1.7 and 2.8 mM and $\mu_m$ values of 0.76 and 0.90 for A11 and A211, respectively. The latter compare reasonably with the values determined by the washout technique ($\mu_m$ = 0.82 and 0.84, respectively). However, the chemostat behavior of strain A is anomalous, in that only 70% of the substrate is utilized even at the

## Table I
### Properties of Strains That Superproduce RDH-A

| Strain | Maximum specific growth rate ($\mu_m$) (hr$^{-1}$) | Enzyme activity in cell-free extracts[a] | |
|---|---|---|---|
| | | RDH activity (U/mg protein) | Apparent $K_m$ for xylitol (M) |
| A | 0.53 | 2.0 | 1.2 |
| A1 | 0.64 | 9.0 | 1.4 |
| A11 | 0.82 | 32.0 | 1.3 |
| A12 | — | 27.8 | 1.3 |
| A112 | 1.20 | 48.5 | — |
| A2 | — | 2.4 | 1.4 |
| A21 | 0.64 | 11.0 | 1.6 |
| A211 | 0.84 | 27.0 | 1.2 |
| A212 | 0.81 | — | 1.3 |
| A2121 | — | — | 1.3 |
| A3 | — | 13.0 | 1.3 |
| A4 | 0.60 | 15.8 | 1.4 |
| A5 | 0.70 | — | 1.2 |
| A6 | — | 12.8 | 1.5 |
| A61 | — | 18.2 | 1.5 |
| A611 | — | 18.7 | 1.0 |
| A6111 | — | 35.0 | 1.5 |
| A61111 | 1.61 | 32.0 | 1.5 |

[a]Ribitol dehydrogenase activities were measured at pH 7.0 and 28°C in 50 mM ribitol, 0.83 mM NAD, 100 mM phosphate; 1 unit = 1 $\mu$mole, NADH/min. The apparent $K_m$ for xylitol is determined at similar [NAD] from Lineweaver–Burk plots.

lowest dilution rate. It is instructive to try to relate the xylitol uptake shown in Fig. 6 to the maximum yield of ribitol dehydrogenase found in cell extracts. By assuming that soluble cell protein is 50% of bacterial dry weight and that intracellular [NAD] is 1 mM, we can calculate the maximum flux of xylitol through this enzyme from the kinetic data of Burleigh et al. (1974). Table II shows such calculations for various assumed intracellular xylitol concentrations. It is clear that a passive permease cannot account for the observed xylitol uptakes. For strain A, active transport is needed to pump from 4 mM to about 1 M xylitol: for strain A11 an internal xylitol concentration of 100 mM would suffice. The anomalous growth kinetics of strain A is probably related to the severe energy demand required to achieve such high intracellular xylitol concentrations. Note, however, that there was no significant change in growth yields (mg bacterial dry weight/mg xylitol used): these were 0.40 for strain A, 0.45 for strain A11, and 0.44 for strain B.

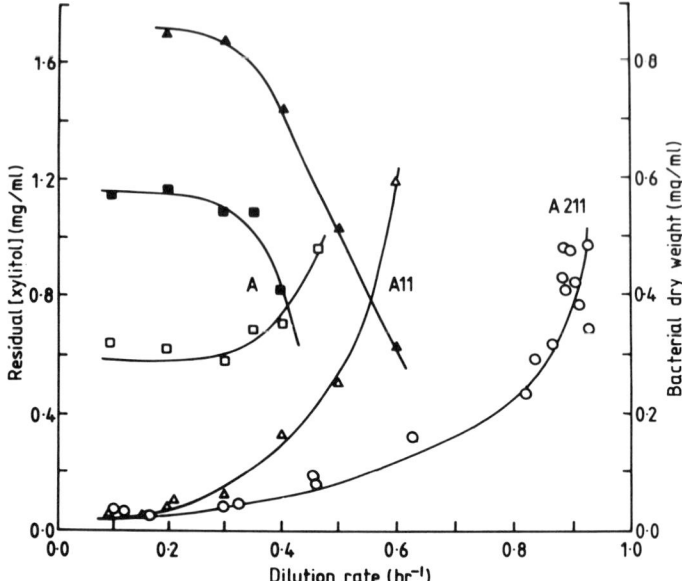

**Figure 6.** Steady-state continuous culture parameters for *Klebsiella aerogenes* growing on xylitol (2 mg/ml) as limiting substrate. Residual xylitol concentrations are shown as open symbols: (□), strain A; (△), strain A11; (○), strain A211. Bacterial dry weights are in solid symbols: (■), strain A; (▲), strain A11.

Table II
Comparison of Intracellular Xylitol Dehydrogenase Activity and Xylitol Uptake[a]

| | Strain A | | | Strain A11 | | |
|---|---|---|---|---|---|---|
| Dilution rate (hr$^{-1}$) | 0.2 | 0.3 | 0.4 | 0.2 | 0.4 | 0.6 |
| $X_s$ (mM) | 4.0 | 4.1 | 4.9 | 0.5 | 2.0 | 7.6 |
| [Cells] (mg/ml) | 0.56 | 0.56 | 0.41 | 0.85 | 0.71 | 0.31 |
| RDH/cell (%) | 0.9 | 0.9 | 0.9 | 14.5 | 14.5 | 14.5 |
| RDH ($\mu$g/ml) | 5.2 | 5.0 | 3.7 | 123 | 103 | 45 |
| $V_x(X_i = X_s)$ ($\mu$M/min) | 0.3 | 0.3 | 0.3 | 0.9 | 3.4 | 5.7 |
| $V_x(X_i = 100$ mM$)$ | 7.6 | 7.4 | 5.4 | 181 | 151 | 66 |
| $V_x(X_i = 1$ M$)$ | 34.8 | 34.5 | 25.3 | 838 | 702 | 307 |
| Xylitol uptake ($\mu$M/min) | 30.7 | 45.4 | 54.8 | 152 | 75 | 55 |

[a] Steady-state data are taken from Fig. 6 and enzyme data from Table I and Burleigh *et al.* (1974). [$X_s$], steady-state xylitol concentration; the xylitol uptake is $D(X_i - X_s)$, where $X_i$ is the influent xylitol (13.2 mM) and $D$ is the dilution rate.

## 4.2. Mechanisms of Enzyme Superproduction

What are the mechanisms by which the organism has evolved this massive superproduction of an inefficient enzyme? There are several possibilities:

1. *Improved transcription.* "Up-promoter mutations" have been described that increase expression of the *lac* repressor from less than 0.1% to 5% of the total soluble protein (Muller-Hill et al., 1975). These would at first sight seem a plausible response to the selective pressure. However, the *rbt* promoter and translation signals are already efficient, since the constitutive strain A makes RDH as 2% of the soluble protein. For comparison, the *lac* promoter makes β-galactosidase as up to 5% of the soluble protein (Zabin and Fowler, 1980), but this should be adjusted for the relative subunit weights: 27,000 for ribitol dehydrogenase and 120,000 for β-galactosidase. Thus, on a molar basis the wild-type *rbt* promoter is almost twice as efficient as the *lac* promoter.
2. *Improved translation.* Mutations in sequences controlling initiation of protein synthesis might increase the number of protein molecules synthesized from a single mRNA transcript. A mutation to increase mRNA stability would have the same effect. However, such explanations for the events observed in the chemostats suffer from the above argument that both transcription and translation are already efficient.
3. *Increased gene dosage.* This was the preferred explanation for enzyme superproduction, since Horiuchi et al. (1963) had shown that selective pressure on *E. coli* for growth on limiting lactose in a chemostat produced strains that superproduced wild-type β-galactosidase. Such strains were postulated to contain multiple tandem copies of the constitutive *lac* operon, since they gave rise to segregant clones at high frequency under nonselective conditions that had the same enzyme levels as the parental constitutive strain. Similar segregation frequencies had been observed with contiguous gene duplications of glycyl-tRNA (Hill et al., 1969), tyrosyl-tRNA Russell et al., 1970), and glycyl-tRNA synthetase (Folk and Berg, 1971). Therefore the frequency of segregation to lower RDH production was measured for some of the strains shown in Fig. 5. Single colonies were grown on rich medium (B broth) and spread on xylitol plates. The proportion of small colonies was counted and several of these were examined for RDH levels in their cell-free extracts after growth on 3% casamino acids. Strain

A1 yielded no segregants, but strain A11 yielded 0.14% of segregants with parental RDH levels. Strain A211 yielded 0.68% of segregants with lower enzyme levels and these were of three classes: 0.31% like A11, 0.6% like A, and 0.21% with intermediate levels of RDH. These segregation frequencies are higher than for the *lac* or tRNA gene duplications noted above, but lower than those for an *E. coli* strain in which 5% of the genome was duplicated (Folk and Berg, 1971). Since the frequency of segregation of a duplication would be expected to be inversely proportional to the size of the duplication, this suggests that strain A11 contains a duplication involving a number of genes surrounding that for ribitol dehydrogenase, while A211 probably contains three or more copies of this same duplication that have arisen by unequal crossing over during recombination—an event of probably higher frequency than the genesis of the original duplication. Strain A1 did not segregate small colonies on xylitol plates and may not be a gene duplication. This conclusion was reinforced by determining the frequency of mutants lacking ribitol dehydrogenase (RDH$^-$) in the above strains. If the frequency of RDH$^-$ in a single-copy strain is $p$ in a standard mutagenesis procedure, then that in a gene-duplicated strain should be $p^2$. Observed values were $2.6 \times 10^{-4}$ for strain A, $6.0 \times 10^{-4}$ for strain A1, and $<3 \times 10^{-8}$ for strain A11. Hence, A11 is probably gene-amplified, while A1 is not.

### 4.3. Improved Xylitol Dehydrogenases

It was surprising that all of the events depicted in Fig. 5 appeared to arise from gene amplification rather than changes in enzyme specificity, since Wu *et al.* (1968) had isolated strain B on plates after severe nitrosoguanidine mutagenesis of strain A. As noted above, this strain makes an enzyme RDH-B with improved specificity for xylitol and reduced thermal stability (Table III), which must result from amino acid replacements. The number of generations screened in the original chemostat studies was such as to suggest that no single amino acid replacement could produce a "fitter," improved xylitol dehydrogenase (Rigby *et al.*, 1974).

Some further chemostat selection experiments were therefore carried out on strain A and strain B after nitrosoguanidine mutagenesis, the rationale being that strains containing multiple amino acid replacements might then be selected. About $3 \times 10^9$ cells grown to exponential phase in B broth were washed and incubated at 37°C for 1 hr with *N*-methyl-

### Table III
Properties of Wild-Type and Mutant Ribitol Dehydrogenases in Cell-Free Extracts of Various Strains[a]

| Strain | 50 mM ribitol SA | 50 mM ribitol $K_{NAD}$ | $K_m$ (mM) at 1 mM NAD Ribitol | $K_m$ (mM) at 1 mM NAD Xylitol | $K_m$ (mM) at 1 mM NAD L-Arabitol | $t_{1/2}$ at 55°C (min) | Mobility $R_m$ |
|---|---|---|---|---|---|---|---|
| A | 2 | 0.16 | 5 | 1020 | 250 | 21 | 0.40 |
| B | 2 | 0.16 | 8 | 520 | 180 | 4 | 0.40 |
| D1 | 20 | 0.16 | 9 | 660 | 280 | 36 | 0.40 |
| E | 14 | 0.14 | 7 | 830 | 290 | 17 | 0.40 |
| F | 10 | 0.11 | 4 | 420 | 140 | 30 | 0.40 |
| G | 10 | 0.30 | 11 | 660 | 220 | 18 | 0.44 |
| H1 | 6 | 0.10 | 11 | 550 | 170 | 18 | 0.40 |
| J1 | 4 | 0.10 | 20 | 360 | 260 | 17 | 0.46 |
| K | 5 | 0.10 | 7 | 600 | 270 | 19 | 0.40 |
| L | 0.2 | 0.10 | 26 | 400 | 220 | 10 | 0.40 |
| EA | 4 | 0.20 | 8 | 1000 | 220 | 20 | 0.40 |
| EB | 8 | 0.10 | 12 | 660 | 260 | 10 | 0.46 |
| EC | 4 | 0.15 | 6 | 330 | 180 | 19 | 0.46 |
| ED | 0.4 | 0.12 | 29 | 460 | 400 | 8 | 0.41 |
| $E_cA$ | 16 | 0.15 | 6 | 830 | 260 | 7 | 0.56 |
| $E_cB$ | 10 | 0.13 | 16 | 580 | 330 | 4 | 0.56 |

[a] Strain A is the RDH-constitutive parent *K. aerogenes* and EA is an *E. coli* K12 strain containing the *K. aerogenes rbt* operon (Rigby *et al.*, 1976). Strains B–L and EB–ED1 are evolved from these parents in chemostats on 0.2% xylitol: D1, F1, etc., superproduce the enzyme in question. The specific activity (SA) and apparent $K_m$ for NAD (mM) on 50 mM ribitol and the apparent $K_m$ for ribitol, xylitol, and L-arabitol at 1 mM NAD are measured according to Burleigh *et al.* (1974). The half-life of RDH activity at 55°C ($t_{1/2}$, min) and electrophoretic mobility of native enzyme relative to the dye front ($R_m$) are as in Hartley *et al.* (1976).

*N*-nitro-*N*-nitrosoguanidine (NG), 0.2 or 1.0 mg/ml. The washed, mutagenized cells were grown overnight in 20 ml of B broth, washed again with M9 medium, and used to inoculate a chemostat grown on 0.2% xylitol/M9 under standard conditions.

Figure 7 summarizes the results of these experiments. At low levels of NG mutagenesis, takeovers of either strain A or strain B clearly represented increased enzyme production rather than improved xylitol specificity. This was also true when selection was carried out at 27°C, on the hypothesis that more unstable enzymes might then be tolerated. However, when powerful NG mutagenesis was employed, strains with altered enzymes arose, signalled by the increased ratio of activity to xylitol over ribitol in cell-free extracts. One such mutagenesis yielded three different strains with improved xylitol activity in the original chemostat takeover

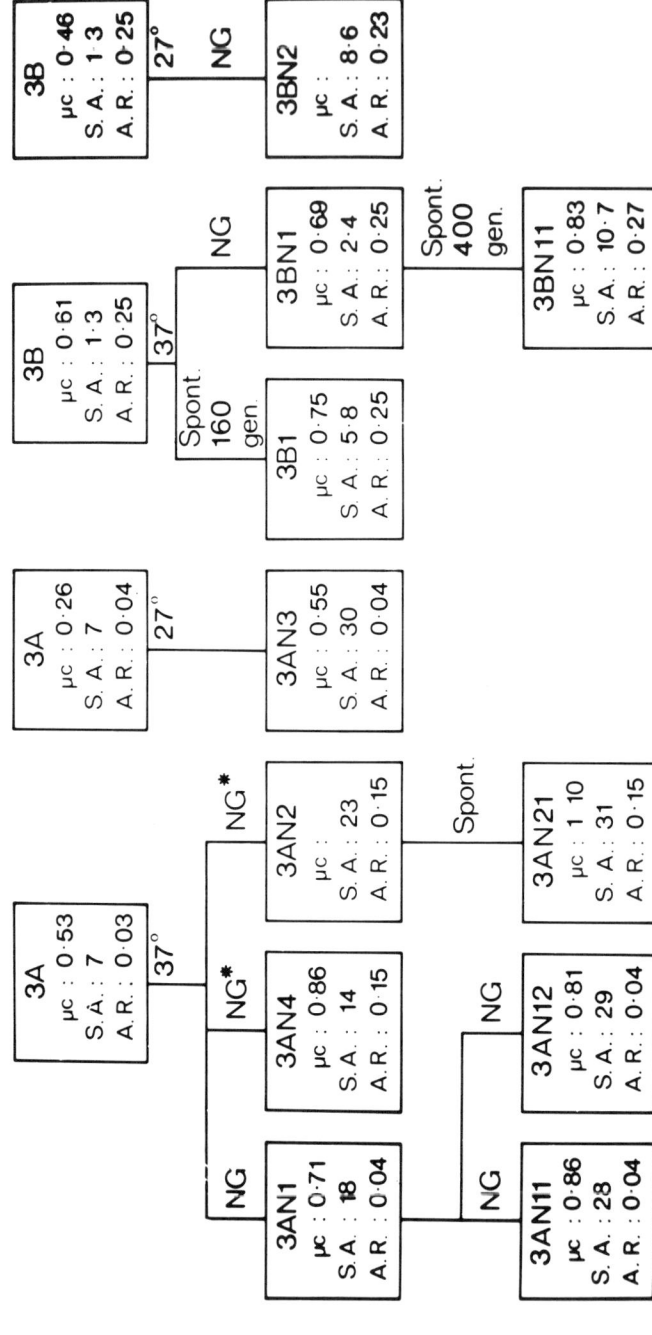

**Figure 7.** Genealogy of further *Klebsiella aerogenes* evolvants. Chemostat cultures of strains A or B, grown on limiting xylitol at 27 or 37°C, were subjected to mild nitrosoguanidine (NG) or severe nitrosoguanidine (NG*) mutagenesis. $\mu_c$, Maximum growth rate of the strain; S.A., the specific activity of ribitol dehydrogenase in cell-free extracts; A.R., the ratio of xylitol dehydrogenase to ribitol dehydrogenase activity.

(strains D, E, and F), and a separate experiment yielded another such strain (G). Further chemostat selection of strain D gave a spontaneous takeover by a strain that superproduces RDH-D.

These results tended to confirm that multiple mutations are required to change enzyme specificity, but subsequent experiments carried out at Imperial College from 1974 to 1979 disproved this hypothesis.

The initial observations were made during chemostat runs designed to test the effect of low dilution rate using strain A', a new prototrophic revertant of the original arginine auxotroph [strain X1 of Wu *et al.* (1968)]. In five such runs, spontaneous takeovers by superproducers of RDH-A were observed, but in another an apparent takeover yielded strains H and J, which proved to have altered ribitol dehydrogenases. Hence, another run was carried out at "normal" diluoion rates (0.2 hr$^{-1}$) and consecutive takeovers were observed in strain K and strain L with altered enzymes, followed by strain L1, which superproduced RDH-L.

Hence, it appears that improved xylitol dehydrogenases *can* arise spontaneously by single point mutations, though gene duplication is a more frequent initial response to the selective pressure. Once gene duplication is established, further gene amplification can readily arise, probably by unequal crossover during recombination.

4.3.1. Properties of Wild-Type and Mutant Ribitol Dehydrogenases

It was clearly necessary to determine the properties and amino acid sequence of wild-type ribitol dehydrogenase so as to compare the properties of the mutant enzymes. Useful purification protocols were developed (Taylor *et al.*, 1974; Rigby *et al.*, 1974; Burleigh *et al.*, 1974) that yielded gram quantities of enzyme. It is a tetramer of 27,000-dalton subunits (Taylor *et al.*, 1974) each containing two thiol groups. It appears to be a "dimer of dimers," since two thiols per tetramer exchange rapidly with disulfides without loss of activity, while two more react slowly, paralleling enzyme inactivation, and the remaining four thiol groups are inaccessible to disulfide exchange (Burleigh *et al.*, 1974). Moreover, precisely two COOH-terminal leucines per tetramer are released by prolonged reaction with carboxypeptidase-A (Dothie *et al.*, 1984).

Table IV shows the results of a comprehensive kinetic study of pure RDH-A, RDH-B (Burleigh *et al.*, 1974), and RDH-D (Dothie *et al.*, 1984). Both of the mutants have threefold to fourfold increase over wild type in specific activity for xylitol under assumed intracellular conditions. This is mediated both by decreases in apparent $K_m$ (RDH-A, 1.20 M; RDH-B, 0.48 M; RDH-D, 0.91 M, at 5 mM xylitol, 1 mM NAD) and increases in $V$ (RDH-A, 20; RDH-B, 36; RDH-D, 48 U/mg). However, it

## Table IV
### Kinetic Constants for Wild-Type and Mutant RDH[a]

| Substrate | Enzyme | $K_P$ (mM) | $K_{NAD}$ (mM) | $K_{P,NAD}$ (mM) | $V_f$ (U/mg) | $V/V_A$[b] 5 mM | 500 mM |
|---|---|---|---|---|---|---|---|
| Ribitol | RDH-A | 8.0 | 0.66 | 10.4 | 166 | — | — |
| Ribitol | RDH-A | 7.2 | 0.37 | 5.9 | 127 | 1.0 | 0.9 |
| Ribitol | RDH-D | 6.7 | 0.18 | 9.8 | 172 | 1.2 | 1.5 |
| Xylitol | RDH-A | 579 | 0.76 | 612 | 20 | — | — |
| Xylitol | RDH-B | 273 | 0.23 | 202 | 36 | 4.5 | 4.3 |
| Xylitol | RDH-D | 394 | 0.29 | 511 | 48 | 3.1 | 3.2 |
| L-Arabitol | RDH-A | 147 | 0.25 | 166 | 16 | — | — |
| L-Arabitol | RDH-B | 170 | 0.21 | 140 | 23 | 1.5 | 1.5 |

[a]The steady-state kinetic analysis was performed on pure proteins according to Burleigh et al. (1974) at pH 7.0: $V = V_f \{1 + K_P/[P] + K_{NAD}/[NAD] + K_{P,NAD}/[P][NAD]\}^{-1}$.
[b]The relative velocity of the mutant enzyme over that of the wild-type enzyme at 1 mM NAD and 5 mM or 500 mM pentitol.

is surprising that the mutant enzymes have essentially unchanged activity for ribitol and for L-arabitol, the other poor substrate for this enzyme. This suggested to Burleigh et al. (1974) that the mutations might operate by shifting equilibria between different quaternary structures of the tetramer rather than by making contact with the three substrates.

This conclusion is supported by a comparison of the properties of the whole family of these enzymes shown in Table III, performed in the main on cell-free extracts of the various strains. All of the enzymes seem to be different! RDH-G, RDH-EB, and RDH-EC have an increased electrophoretic mobility on native gels. RDH-B, RDH-L, and RDH-ED have significantly lower temperature stability, while RDH-D and RDH-F are more stable. There are variable decreases in the apparent $K_m$ for xylitol (large for J and EC, small for D, G, and EB), with little decrease in the apparent $K_m$ for ribitol (except for J, L, and ED), L-arabitol (except ED9), or NAD (except G). The conclusion is that many amino acid replacements give rise to improved xylitol activity without loss in ribitol or L-arabitol activity.

### 4.3.2. The Amino Acid Sequence of RDH-A, RDH-B, and RDH-D

Figure 8 shows the sequence of the wild-type protein (RDH-A) determined from samples of enzyme obtained from strains A, A11, and A211 by a mixture of conventional and mass-spectrometry techniques (Taylor et al., 1974; Morris et al., 1974; Dothie et al., 1984). The latter paper corrects the following errors in the sequence published by Morris *et al.* (1974): Gln-60, Leu-135, Ile-153, Pro-154, Glu-155. Recent analysis suggests that this conclusion may be false.

Also shown in Fig. 8 are the results of partial sequence studies on RDH-D purified from strain D1 (Table III) according to Burleigh *et al.* (1974), which were undertaken in order to locate the mutation(s) responsible for the improved xylitol specificity (Dothie *et al.*, 1984). There was no significant difference in amino acid composition between RDH-A and RDH-D, suggesting that the differences were between the more abundant residues (Asx, Glx, Ala, Val, Leu) or those most prone to analytical error (Ser, Pro, Val, Ile).

An attempt was made to locate the mutation(s) by peptide mapping of elastase digests of RDH-A and RDH-D by paper electrophoresis and chromatography, since this enzyme cleaves at the abundant alanine and valine residues. There were several differences between the two maps, and the relevant peptides were eluted and analyzed. They all proved to arise from minor differences in the elastase cleavages in the separate digests, since they all fitted the RDH-A sequence as shown in Fig. 8.

The search for the mutation(s) was assisted by purifying a small amount of $^{14}$C-labeled RDH-A from strain A7 grown on 400 ml of M9/0.2% xylitol containing 2.75 mCi of [$^{14}$C]-amino acids. A mixture of RDH-D and [$^{14}$C]-RDH-A (10% w/w) was digested with trypsin and chromatographed on DE-52 in 50 mM–1.0 M ammonium bicarbonate. Samples from this column were further digested with chymotrypsin and peptide-mapped on paper, screening for contiguity between radioactivity and ninhydrin or fluoram staining. Figure 8 shows fragments that can be unequivocally identified as unchanged from their position on these maps. Only two major spots were ninhydrin-positive but not radioactive. Both appeared to represent the acidic peptide 198–203 arising from two forms of the tryptic peptide T17 that run in different places on the DE-52 column. Two neutral peptides that probably arose from the tryptic peptide 185–203 were radioactive but ninhydrin-negative. Hence a mutation in this region was indicated.

All of the tryptic peptides from this digest except T10 were purified further by gel filtration and/or paper electrophoresis and subjected to amino acid analysis. All had identical compositions in RDH-D and RDH-A, except that peptide T17 contained one more Pro and less Ala than expected. A thermolysin digest of this tryptic peptide yielded a peptide 188–196 with composition Thr, Pro$_3$, Gly, Val$_3$, Leu. Dansyl Edman sequencing of these peptides simultaneously revealed both the RDH-D (fluorescent DNS derivatives) and RDH-S (radioactive DNS derivatives) sequences:

```
  1         5              10             15             20             25
Met-Lys-His-Ser-Val-Ser-Ser-Met-Asn-Thr-Ser-Leu-Ser-Gly-Lys-Val-Ala-Ala-Ile-Thr-Gly-Ala-Ala-Ser-Gly-
   DT1              DT2                                        DT3
        DE                   DE                                     DE
                                                       DT3-C1
  BT1                BT2                                 BT3
 26            30             35             40             45             50
Ile-Gly-Leu-Glu-Cys-Ala-Arg-Thr-Leu-Leu-Gly-Ala-Gly-Ala-Lys-Val-Val-Leu-Ile-Asp-Arg-Glu-Gly-Glu-Lys-
     DT3                   DT4                  DT5                     DT6
        DE                    DE           DE
        DT3-C2   DT3-C3                          DT5-C1    DT5-C2
              BT3                BT4                 BT5          BT6
 51            55             60             65             70             75
Leu-Asn-Lys-Leu-Val-Ala-Glu-Leu-Gly-Glu-Asn-Ala-Phe-Ala-Leu-Gln-Val-Asp-Leu-Met-Gln-Ala-Asp-Gln-Val-
    DT7                                                              DT8
         DE
                DT8-C1        DT8-C2     DT8-C3    DT8-C4          DT8-C5
  BT7                                                      BT8
 76            80             85             90             95             100
Asp-Asn-Leu-Leu-Gln-Gly-Ile-Leu-Gln-Leu-Thr-Gly-Arg-Leu-Asp-Ile-Phe-His-Ala-Asn-Ala-Gly-Ala-Tyr-Ile-
    DT8                                                    DT9
                                  DE
   DT8-C5    DT8-C6     DT8-C7 DT8-C8      DT9-C2           DT9-C3
                                 BT8
101           105            110            115            120            125
Gly-Gly-Pro-Val-Ala-Glu-Gly-Asp-Pro-Asp-Val-Trp-Asp-Arg-Val-Leu-His-Leu-Asn-Ile-Asn-Ala-Ala-Phe-Arg-
    DT9

   DT9-C4                             DT9-C5
      BT8

126           130            135            140            145            150
Cys-Val-Arg-Ser-Val-Leu-Pro-His-Leu-Ile-Ala-Gln-Lys-Ser-Gly-Asp-Ile-Ile-Phe-Thr-Ala-Val-Ile-Ala-Gly-
   DT11              DT12                                       DT13

                                           DT13-C1          DT13-C2
  BT11            BT12
151           155            160            165            170            175
Val-Val-Pro-Val-Ile-Trp-Glu-Pro-Val-Tyr-Thr-Ala-Ser-Lys-Phe-Ala-Val-Gln-Ala-Phe-Val-His-Thr-Thr-Arg-
   DT13                                                    DT14
           DE                                                   DE
  DT13-C2              DT13-C3    DT13-C4
                                                           BT14
176           180            185            190            195            200
Arg-Gln-Val-Ala-Gln-Tyr-Gly-Val-Arg-Val-Gly-Ala-Val-Leu-Pro-Gly-Pro-Val-Val-Thr-Ala-Leu-Leu-Asp-Asp-
  T15              DT16                                       Pro        DT17
                           DE                      E                     DE
                   DT17-L2:                                   Pro
      BT16                                                              BT17
201           205            210            215            220            225
Trp-Pro-Lys-Ala-Lys-Met-Asp-Glu-Ala-Leu-Ala-Asp-Gly-Ser-Leu-Met-Gln-Pro-Ile-Glu-Val-Ala-Glu-Ser-Val-
    DT17        DT18                                                        DT19
        DE                                                                  E
   DT17-C3           DT19-C1           DT19-C2                DT19-C3
  BT17          BT18
226           230            235            240            245            249
Leu-Phe-Met-Val-Thr-Arg-Ser-Lys-Asn-Val-Thr-Val-Arg-Asp-Ile-Val-Ile-Leu-Pro-Asn-Ser-Val-Asp-Leu.
    DT19           DT20       DT21                    DT22
   T19
     C4   DT19-C5                             DT22-C1      DT22-C2
                              BT20   BT21                 BT22
```

Peptide 188–196
RDH-D:   Val-Leu-Pro-Gly-Pro-Val-Val-Thr-<u>Pro</u>
RDH-A:   Val-Leu-Pro-Gly-Pro-Val-Val-Thr-Ala
Peptide T10
RDH-D:   Val-<u>Gly</u>-Ala-Val-Leu-Pro-Gly-Pro-Val-Val-Thr-Ala-Leu-Leu-Asp-Asp-
RDH-A:   Val-Pro-Ala-Val-Leu-Pro-Gly-Pro-

Hence RDH-D contains a change of Ala-196 to Pro-196 and there is evidence for a "neutral mutation" in RDH-A from strain A7 (see below).

Since it was believed that RDH-D might contain multiple mutations, considerable effort was made to screen the rest of the amino acid sequence. For the smaller tryptic peptides the identical amino acid compositions and contiguity of radioactivity and ninhydrin staining on peptide maps make it almost impossible that there is any sequence difference. For the larger tryptic peptides (e.g., T8), chymotryptic fragments were similarly peptide-mapped, and in no case was there any difference in the ninhydrin staining (RDH-D) and radioactivity (RDH-A). Hence Ala-196 to Pro appears to be the only change in RDH-D, except possibly in peptide T10, which is extremely difficult to isolate (Dothie *et al.*, 1984).

Some effort was made to identify the sequence changes in RDH-B by mapping and analyzing some of the tryptic peptides, as shown in Fig. 8. All of the basic peptides appeared similar on peptide maps, and the amino acid compositions of those peptides that were purified seemed unchanged: in particular, peptide T17 appeared identical to that from RDH-A. Hence, the mutation(s) responsible for the improved xylitol specificity in RDH-B appear to lie within peptides T10, T13, or T19 and are different from those in RDH-D.

## 5. Fluctuating Selective Pressure

The above experiments show that continuous culture on xylitol selects either mutants that superproduce RDH-A or a family of mutants with different single amino acid replacement that yield small improve-

**Figure 8.** Amino acid sequence of ribitol dehydrogenase (RDH)-A, B, and D. The sequence shown is that of RDH-A from Moore *et al.* (1984). Peptides from tryptic digests of RDH-D or RDH-B are labeled DT1, BT1, etc. The DT3-C, etc., represent further digestion by chymotrypsin, and DE peptides were obtained from elastase digests. DT17-L2 was from a thermolysin digest of DT17. Amino acid compositions consistent with the RDH-A sequence are shown as solid lines, and dashed lines indicate identities established by peptide mapping. Dansyl-Edman sequencing of DT17 is shown by arrows.

ments in xylitol specificity. How does this fit the hypothesis for the evolution of new enzyme specificity illustrated in Fig. 1? Gradual movement toward a new enzyme specificity by point mutations was observed, but gene duplication appeared to be a more frequent response to the selective pressure. However, the reversion frequency of these presumed gene duplications was high, and the "silent gene" hypothesis requires that the redundant copy should persist in the population. How would the chemostat population react to a regime of fluctuating growth on xylitol followed by a nonselecting carbon source?

Table V [from Hartley *et al.* (1976)] shows the results of several experiments in which the putative gene-duplicated strain A11 was grown in continuous culture on a nonselective substrate for several generations before a switch back to xylitol selection. Where ribitol or glucose was the nonselective carbon source, the RDH-superproducing phenotype was conserved, but no strains with improved xylitol specificity arose during this phase. When inositol was used as carbon source, the culture rapidly reverted to the ancestral strain A phenotype.

We had observed that synthesis of ribitol dehydrogenase in batch cultures of the constitutive strain A appears to be subject to severe catabolite repression by both glucose and ribitol, but that relatively high levels of the enzyme are produced during growth on inositol. A rationalization of these results is that there is little selective pressure against the strains that superproduce ribitol dehydrogenase during growth on substrates that repress the synthesis of this enzyme: hence, the segregants with a single gene copy remain a minority in the population. Growing on inositol, the gene-duplicated organism would be producing massive quantities of an unnecessary enzyme; one could imagine that the segregant would grow faster. Analysis of the proportion of large versus small colonies on xylitol plates during the final experiment showed that this was the case: large colonies were about 100% to day 8, but fell to almost zero by day 15. Hence, selection against the gene duplication is severe only when the protein is expressed.

## 6. Transfer of the *Klebsiella aerogenes* Ribitol Dehydrogenase Gene into *Escherichia coli* K12

By 1974 an impasse had been reached in this investigation into experimental evolution, since conclusions about the mechanisms of enzyme superproduction demanded studies at the level of RNA and DNA and a tertiary structure of the ribitol dehydrogenase was needed to relate the

## Table V
### Growth of a Gene-Doubled Strain of *Klebsiella aerogenes* on Changing Substrates in Chemostats[a]

| First substrate | Generations | Second substrate | Generations | AR | SA | Strain |
|---|---|---|---|---|---|---|
| — | — | — | — | 0.04 | 18 | A11 |
| Ribitol | 750 (2 UV) | Xylitol | 750 | 0.03 | 26 | A111 |
| Ribitol | 340 (2 UV) | Xylitol | 120 | 0.04 | 25 | A113 |
| Ribitol | 100 (MNNG) | Xylitol | 70 | 0.05 | 33 | A114 |
| Glucose | 700 | Xylitol | 400 | 0.04 | 28 | A112 |
| Inositol | 470 (7 UV) | Xylitol | 120 | 0.03 | 9 | A11S |
| Inositol | 300 | — | — | 0.04 | 5 | A11S |

[a] Chemostats were grown on 0.2% (w/v) substrate at a dilution rate of 0.6 hr$^{-1}$ with weak mutagenesis by UV irradiation or by treatment with $N$-methyl-$N'$-nitro-$N$-nitrosoguanidine (MNNG) as indicated. SA, Ribitol dehydrogenase units/mg protein; AR, ratio of dehydrogenase activities of extracts on 50 mM ribitol or 500 mM xylitol, 1 mM NAD, pH 7.

sequence changes to the altered enzymic properties. Despite several attempts, we were unable to obtain crystals suitable for X-ray crystallography. Knowledge of the genetics of *K. aerogenes* was (and remains) rudimentary, but we felt that an avenue might be opened if we could incorporate the *rbt* operon into *E. coli* K12, which does not grow on pentitols, because it lacks the necessary ribitol and D-arabitol dehydrogenases.

This was achieved by transduction (Rigby *et al.*, 1976) using the specialized transducing phage P1Cm *clr*100, which carries a useful selection marker, chloramphenicol transacetylase, plus a temperature-sensitive repressor mutation (Rosner, 1972). The *K. aerogenes* strain A lysogenic for this phage was selected on broth plates containing chloramphenicol. Temperature induction yielded phage that were used to infect *E. coli* K12 strain CA388 (*gal str tsx*). Selection on plates containing ribitol plus streptomycin gave five colonies, which were all *gal*, *str*, chloramphenicol-resistant, temperature-sensitive, and grew on ribitol, xylitol, and D-arabitol. Further selection on xylitol plates at 42°C gave colonies cured of the phage that would still grow on all three pentitols. Hence, the *rbt* and *dal* operons appear to be closely linked (see also Charnetzky and Mortlock, 1974c; Scangos and Reiner, 1978) and have been incorporated into the *E. coli* genome.

One such *E. coli* strain EA was chosen for further study. Cell-free extracts showed a xylitol/ribitol dehydrogenase activity ratio identical to that of *K. aerogenes* strain A and yielded an enzyme of identical electrophoretic mobility on polyacrylamide gels. The specific RDH activity in the *E. coli* extract was 2.4 U/mg protein, compared with 7.3 U/mg protein for the *K. aerogenes* strain A extract. The *rbt* gene in *E. coli* strain EA was located close to the *his* locus at 40 min on the *E. coli* genome by Hfr time-of-entry mapping.

In retrospect, the optimism with which this interspecies gene transfer was attempted seems unjustified and the success of the experiment is worthy of comment. It implies that the similarity between *E. coli* K12 and *K. aerogenes* 1033, the parental wild type of our strain A, is far greater than would have been guessed. Some strains of *K. pneumoniae* are innately sensitive to coliphage P1 (Streicher *et al.*, 1971), but *K. aerogenes* 1033 is resistant. However, from *K. aerogenes* strain A it was easy to isolate a mutant strain 5A that was sensitive to P1 *vir* and P1Cm *clf*100 but not to P1 *kc* (Rigby *et al.*, 1976). The sensitivity of strain 5A to phage grown on *E. coli* was about 10% of that of the natural *E. coli* host, but when P1Cm *clr*100 was grown on strain 5A the plaque efficiency for the two species was identical. This implies that the structure of their F pili is similar and encourages the concept that genetic exchange between the two species might be common in natural environments.

Moreover, it was easy to incorporate the *K. aerogenes* pentitol operons into the *E. coli* K12 genome and they immediately functioned almost as efficiently as in their natural host on all three pentitols. The *E. coli* K12 definitely lacks pentitol metabolism, since no latent activity could be uncovered by extensive mutagenesis and selection on glycerol plus ribitol (Rigby *et al.*, 1976) and no ribitol permease activity could be detected (M.S. Neuberger, unpublished experiments). But transduction of a small fragment of foreign DNA bestows complete metabolic potential! Not only must transcription and translation of the foreign operons be efficient, but the inserted DNA must either encode permeases that function in the *E. coli* membrane or cause induction of host permeases that would have no function in the absence of the relevant dehydrogenases.

These afterthoughts are made in the light of the discovery by Reiner (1975) that *E. coli* C strains *do* contain genes for ribitol and D-arabitol catabolism and that these are organized into operons of remarkably similar structure to those of *K. aerogenes* (see next chapter), which map close to the *his* operon at about 40 min on the *E. coli* C genome—exactly the same site as in our *E. coli* K12 construct. It appears that the genetic engineering has merely revealed a natural mechanism for a facile interspecies transfer of pentitol metabolism!

## 6.1. Evolution of Ribitol Dehydrogenase in *Escherichia coli*

The *E. coli* strain EA was now a precise analogue of the *K. aerogenes* parent chosen for our chemostat studies, so we could compare the results of selective pressure for growth on xylitol on the same gene in two different organisms. Figure 9 shows the family tree for a series of takeover events in chemostats operated as above. A series of spontaneous takeovers in one chemostat operated at normal dilution rates (0.23–0.8 $hr^{-1}$) produced a series of strains EA1 to EA111 that superproduce eventually massive amounts of unaltered RDH-A, as in Fig. 5. Mild mutagenesis by nitrosoguanidine analogous to that used in Figure 7 immediately yielded a strain EAN1 with improved specificity for xylitol. Another run using UV irradiation yielded a takeover by a strain EA2 producing an improved xylitol dehydrogease EB with decreased temperature stability and increased negative charge, followed by another strain EA21 yielding an even better enzyme EC: the latter retains the increased negative charge, but has normal temperature stability and may contain multiple mutations (Table III).

In another experiment, *E. coli* strain EA was selected on 0.2% xylitol/M9 at very low dilution rate (<0.04 $hr^{-1}$). After 26 generations a gradual takeover by strain EA3 was observed with greatly improved actity toward xylitol (ratio of activities at 1 mM NAD, 500 mM xylitol or 50

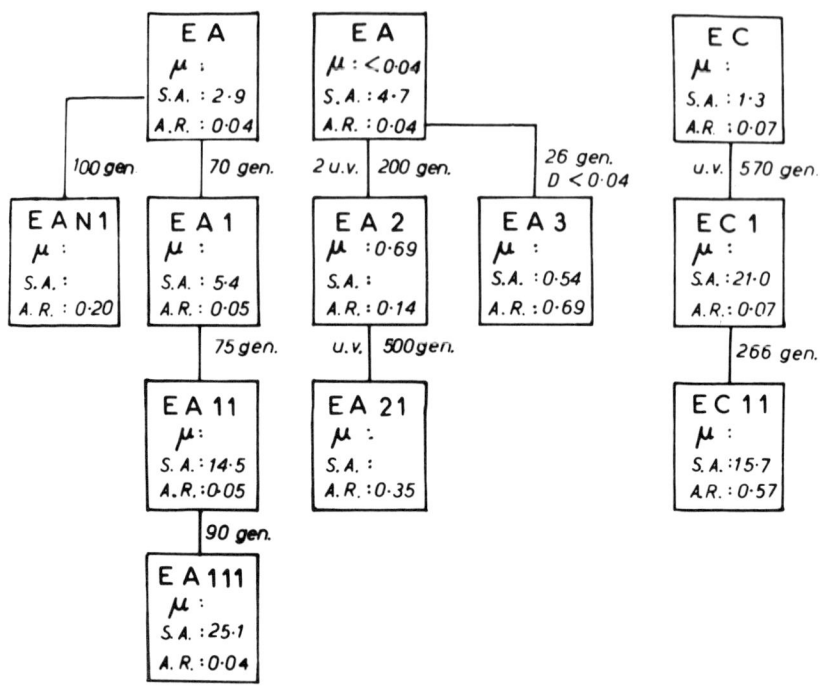

**Figure 9.** Evolution of strains of *E. coli* during chemostat growth on xylitol. EA is the *E. coli* K12–*K. aerogenes* hybrid of Rigby *et al.* (1976). EC is the *E. coli* C strain of Reiner (1975). The evolvant strains indicated had taken over the chemostat after the number of generations shown. The dilution rate $D$ was 0.2–0.8 hr$^{-1}$ except as indicated. EAN1 arose after mild mutagenesis by $N$-methyl-$N'$-nitro-N-nitrosoguanidine (MNNG). Mild UV irradiation of the chemostat was performed as indicated. $\mu$, Maximum specific growth rate (Tempest, 1970); S.A., specific dehydrogenase activity of cell extracts with 50 mM ribitol (U/mg protein); A.R., ratio of the activity with 500 mM xylitol to that with 50 mM ribitol.

mM ribitol was 0.69, compared with 0.04 for RDH-A). Table III shows that the enzyme RDH-ED1 from this strain has decreased temperature stability but normal electrophoretic mobility, and that improved activity for xylitol is in this case at the expense of decreased activity toward ribitol and L-arabitol.

Also included in Fig. 9 and Table III are the results of analogous studies on the strain of *E. coli* C isolated by Reiner (1975) that grows naturally on pentitols. Growth on xylitol plates readily selects strains constitutive for its ribitol dehydrogenase (strain EC). A chemostat selection on 0.2% xylitol at normal dilution rates with UV mutagenesis yielded strain EC1, which makes 16 times the quantity of wild-type enzyme (RDH-

$E_cA$), and this was followed by a spontaneous takeover by a strain producing an altered enzyme, RDH-$E_cB$, with eight times improved specificity for xylitol. This enzyme has unchanged electrophoretic mobility, but slightly decreased temperature stability and slightly reduced activity toward ribitol and L-arabitol.

Our own experiments and those of Reiner (1975) were beginning to suggest that interspecies transfer of these pentitol operons might be a common natural event. If so, the sequence differences between the ribitol dehydrogenases of *E coli* C and *K. aerogenes* might be very small. Using the superproducing strain EC1, we were able to prepare several grams of the pure *E. coli* C enzyme by methods analogous to those of Burleigh *et al.* (1974). Some of the soluble tryptic peptides were purified and sequenced by methods similar to those described for the *K. aerogenes* enzyme, with the results shown in Fig. 8. Of the 85 residues of the sequence in these randomly chosen peptides, there are only four differences from the comparable regions in RDH-A. Hence, it is likely that the overall sequence homology will exceed 90% (Altosaar and Hartley, 1976). Unfortunately, we cannot compare this high homology with a background of species differences between other proteins in *E. coli* C and *K. aerogenes*, and we have already concluded that *E. coli* K12 and *K. aerogenes* are very closely related. However, the result is not inconsistent with the hypothesis that the pentitol operons are prone to interspecies transfer.

## 7. Evolutionary Lessons from the Chemostat Studies

These experiments proved to be a reasonable model for natural microbial evolution of enzyme specificity. They indicate that strains can evolve rapidly to utilize a novel substrate wherever a transport pathway exists and an existing enzyme has a weak activity toward the substrate. Ribitol dehydrogenase is an exceedingly poor xylitol dehydrogenase, but growth can be maintained by pumping in xylitol to very high concentrations. Gene duplication is a frequent response to the selective pressure, followed by further gene amplification to produce very high levels of the poor enzyme. Despite the additional protein synthesis load, such strains can have growth rates approaching those of the wild type on ribitol. It seems likely, therefore, that gene duplications are common in the evolution of new enzyme specificities.

However, these duplications segregate rapidly to single copies when the selective pressure is removed, e.g., by growth on inositol, showing that the excessive protein synthesis load is indeed a selective disadvantage. But when the alternative substrate represses synthesis of the enzyme that is superproduced (e.g., growth on ribitol or glucose in our case) the gene duplication is conserved in the population. Evolution of a system for enzyme induction and repression would therefore seem to be an early requirement for evolution of a divergent enzyme family.

But our studies also give support to advocates of a stepwise acquisition of novel enzyme specificities. There were a very large number of single amino acid changes leading to increased xylitol activity, some of which increased activity ratios for xylitol by over 20 times and yielded growth rates for the organism approaching those on ribitol. It is instructive to note that the new activity was generally acquired without loss of the old activity.

These experiments also speak for the importance of interspecies gene transfer in microbial evolution. The pentitol operons represent an embryonic "pentitol transposon" that can confer on their host the ability to utilize ribitol and D-arabitol and the potential to evolve a pathway for xylitol and possibly also for L-arabitol. This seems to have occurred in *K. aerogenes* and *E. coli* C, almost certainly could occur in *E. coli* K, and might be a widespread phenomenon in other species.

ACKNOWLEDGMENTS. This chapter is dedicated to a generation of chemostat watchers and protein chemists who observed and dissected the events described above: Peter W. J. Rigby and John M. Dothie, who pioneered and developed the whole system; Bruce D. Burleigh and Mark Silcox, who studied the enzymology; Susan S. Taylor and Christopher H. Moore, who broke the back of the protein chemistry with help from Michael J. Smith and Graeme G. Midwinter; Jose Giglio and Eric W. Matthews, who polished off the nasty bits of the sequence of ribitol dehydrogenase; Illimar Altosar, who studied *E. coli* C; and Mary J. Gething, who moved the genes into *E. coli* K12.

## References

Altosaar, I., and Hartley, B. S., 1976, Comparison of ribitol dehydrogenase from *E. coli* C and *K. aerogenes*, in: *Proceedings 10th International Congress of Biochemistry*, p. 200, Abstract 04-6-319.

Awad, W. M., Soto, A. R., Siegel, S., Skiba, W. E., Bernstrom, G. G., and Ochoa, M. S., 1972, The proteolytic enzymes of the K-1 strain of *Streptomyces griseus*, *J. Biol. Chem.* **247**:4144–4145.

Burleigh, B. D., Rigby, P. W. J., and Hartley, B. S., 1974, A comparison of wild-type and mutant ribitol dehydrogenases from *K. aerogenes*, *Biochem. J.* **143**:341–352.

Charnetzky, W. T., and Mortlock, R. P., 1974a, Ribitol catabolic pathway in *Klebsiella aerogenes*, *J. Bacteriol.* **119**:162–169.

Charnetzky, W. T., and Mortlock, R. P., 1974b, D-Arabitol catabolic pathway in *Klebsiella aerogenes, J. Bacteriol.* **119:**170–175.
Charnetzky, W. T., and Mortlock, R. P., 1974c, Close genetic linkage of the determinants of the ribitol and D-arabitol catabolic pathways in *K. aerogenes, J. Bacteriol.* **119:**176–182.
Dothie, J. M., Giglio, J. R., Moore, C. H., Taylor, S. S., and Hartley, B. S., 1984, Ribitol dehydrogenase from *K. aerogenes*: Sequences and properties of wild-type and mutant strains, *Biochem J.* (submitted).
Folk, W. R., and Berg, P., 1971, Duplication of the structural gene for glycyl-transfer RNA synthetase in *Escherichia coli, J. Mol. Biol.* **58:**595–610.
Hartley, B. S., 1966, Enzymes are proteins, *Adv. Sci.* **1966**(May):47–54.
Hartley, B. S., 1974, Enzyme families, in: *Evolution in the Microbial World* (M. J. Carlile and J. J. Skehel, eds.), Elsevier, Amsterdam, pp. 151–182.
Hartley, B. S., and Shotton, D. M., 1971, Pancreatic elastase, in: *The Enzymes*, Vol. 3, 3rd ed. (P. D. Boyer, ed.), Academic Press, New York, pp. 323–373.
Hartley, B. S., Burleigh, B. D., Midwinter, G. G., Moore, C. H., Morris, H. R., Rigby, P. W. J., Smith, M. J., and Taylor, S. S., 1972, Where do new enzymes come from?, in: *Enzymes: Structure and Function* (J. Drenth, R. A. Oosterbaan, and C. Veeger, eds.), North-Holland, Amsterdam, pp. 151–176.
Hartley, B. S., Altosaar, I., Dothie, J. M., and Neuberger, M. S., 1976, Experimental evolution of a xylitol dehydrogenase, in: *Proceedings of the Third John Innes Symposium* (R. Markham and R. W. Horne, eds.), North-Holland, Amsterdam, pp. 191–200.
Herbert, D., Ellsworth, R., and Telling, R. C., 1956, The continuous culture of bacteria: A theoretical and experimental study, *J. Gen. Microbiol.* **14:**601–622.
Herbert, D., Phipps, P. J., and Tempest, D. W., 1965, The chemostat design and instrumentation, *Lab. Practice* **14:**1150–1161.
Hill, C. W., Foulds, J., Soll, L., and Berg, P., 1969, Instability of a missense suppressor resulting from a duplication of genetic material, *J. Mol. Biol.* **39:**563–581.
Horiuchi, T., Horiuchi, S., and Novick, A., 1963, The genetic basis of hypersynthesis of β-galactosidase, *Genetics* **48:**157–169.
Kimura, M., 1969, The rate of molecular evolution considered from the standpoint of molecular genetics, *Proc. Natl. Acad. Sci. USA* **63:**1181–1183.
Kraut, J., Robertus, J. D., Birktoft, J. J., Alden, R. A., Wilcox, P. E., and Powers, J. C., 1971, The aromatic binding site in subtilisin BPN' and its resemblance to chymotrypsin, *Cold Spring Harbor Symp. Quant. Biol.* **36:**117–124.
Lerner, S. A., Wu, T. T., and Lin, E. C. C., 1964, Evolution of catabolic pathway in bacteria, *Science* **146:**1313–1315.
McLachlan, R. D., and Shotton, D. M., 1971, Structural similarities between α-lytic protease of *Myxobacter 495* and elastase, *Nature New Biol.* **229:**202–205.
Morris, H. R., Williams, D. H., Midwinter, G. G., and Hartley, B. S., 1974, A mass-spectrometric sequence study of the enzyme ribitol dehydrogenase from *Klebsiella aerogenes, Biochem. J.* **141:**701–713.
Mortlock, R. P., Fossitt, D. D., and Wood, W. A., 1965, A basis for utilization of unnatural pentoses and pentitols by *Aerobacter aerogenes, Proc. Natl. Acad. Sci. USA* **54:**572–579.
Muller-Hill, B., Fanning, T., Geisler, N., Gho, D., Kania, J., Kathmaan, P., Meissner, H., Schlotmann, M., Schmitz, A., Triesch, I., and Beyruther, K., 1975, The active sites of *lac* repressor, in: *Protein–Ligand Interactions* (H. Sund and G. Blauer, eds.), Walter de Gruyter, Berlin, pp. 211–224.
Pechurkin, N. S., 1969, Continuous cultivation of microorganisms as a means of their autoselection by growth rate in set conditions, in: *Continuous Culture of Microorga-*

*nisms* (I. Malek, K. Beran, Z. Fencl, V. Munk, J. Ricica, and H. Smrckova, eds.), Academic Press, London, pp. 315–322.

Powell, E. O., 1965, Theory of the chemostat, *Lab. Practice* **14**:1145–1149.

Reiner, A. M., 1975, Genes for ribitol and D-arabitol metabolism in *E. coli*: Their loci in C strains and absence in K-12 and B strains, *J. Bacteriol.* **123**:530–536.

Rigby, P. W. J., 1971, An Experimental Approach to Enzyme Evolution, Ph.D. Thesis, University of Cambridge.

Rigby, P. W. J., Burleigh, B. D., and Hartley, B. S., 1974, Gene duplication in experimental enzyme evolution, *Nature* **251**:200–204.

Rigby, P. W. J., Gething, M. J., and Hartley, B. S., 1976, Construction of intergeneric hybrids using bacteriophage P1CM: Transfer of the *K. aerogenes* ribitol dehydrogenase gene to *E. coli, J. Bacteriol.* **125**:728–738.

Rosner, J. L., 1972, Formation, induction and curing of bacteriophage P1 lysogens, *Virology* **48**:679–689.

Russell, R. L., Abelson, J. N., Landy, A., Gefter, M. L., Brenner, S., and Smith, J. D., 1970, Duplicate genes for tyrosine tRNA in *E. coli, J. Mol. Biol.* **47**:1–13.

Scangos, G. A., and Reiner, A. M., 1978, Ribitol and D-arabitol catabolism in *Escherichia coli, J. Bacteriol.* **134**:492–500.

Streicher, S. L., Bender, R. A., and Magasanik, B., 1971, Transduction of the nitrogen-fixation genes in *Klebsiella pneumoniae, J. Bacteriol.* **121**:320–331.

Taylor, S. S., Rigby, P. W. J., and Hartley, B. S., 1974, Ribitol dehydrogenase from *K. aerogenes*: Purification and subunit structure, *Biochem. J.* **141**:693–700.

Tempest, D. W., 1970, The continuous cultivation of micro-organisms: 1. Theory of the chemostat, in: *Methods in Microbiology*, Vol. 2 (J. R. Norris and D. W. Ribbons, eds.), Academic Press, London, pp. 259–276.

Wright, C. S., Alden, R. A., and Kraut, J., 1969, Structure of subtilisin BPN' at 2.5 Å resolution, *Nature* **221**:235–242.

Wu, T. T., Lin, E. C. C., and Tanaka, S., 1968, Mutants of *Aerobacter aerogenes* capable of using xylitol as a novel carbon source, *J. Bacteriol.* **96**:447–456.

Zabin, I., and Fowler, A. V., 1980, β-Galactosidase, the lactosepermease protein, and thiogalactoside transacetylase, in: *The Operon* (J. H. Miller and W. S. Reznikoff, eds.), Cold Spring Harbor Laboratory, New York, pp. 89–122.

CHAPTER 3

# The Structure and Control of the Pentitol Operons

## B. S. HARTLEY

## 1. Introduction

This chapter discusses evolutionary lessons from the structure of the pentitol operons. The studies described in the last chapter, together with the genetic mapping of Charnetzky and Mortlock (1974) and the parallel work of Reiner (1975; Scangos and Reiner, 1978a) in *E. coli* C, had made clear that the *rbt* and *dal* operons were probably contiguous, with a gene order *rbtK-rbtD-rbtC-dalB-rbtB-dalC-dalD-dalK-*, where *rbtK* encodes D-ribulokinase (DRK); *dalK*, D-xylulokinase (DXK); *rbtD*, ribitol dehydrogenase (RDH); and *dalD*, D-arabitol dehydrogenase (ArDH); and *rbtB*, *rbtC*, *dalB*, and *dalC* are control loci for the appropriate operons. Detailed study of the structure of this region might elucidate (1) the nature of events leading to the enzyme superproduction discussed in the previous chapter, (2) the control of the operons, and (3) potential homologies between the corresponding *rbt* and *dal* genes: their arrangement invites speculation about an ancestry arising from an invert gene duplication.

### 1.1. Construction of λp *rbt*

When this work was begun in 1974 recombinant DNA techniques were not as well developed as they are today, so we chose to use the technique of Shimada *et al.* (1972, 1973) to try to incorporate λ prophage

---

*B. S. HARTLEY* • Department of Biochemistry, Imperial College of Science and Technology, London SW7 2AZ, England.

into secondary attachment sites within our *Escherichia coli* strain EA, which contains the pentitol operons of *Klebsiella aerogenes* (Neuberger and Hartley, 1979). In strains deleted for the normal λ attachment site (*att* λ), phage λ will lysogenize at 0.5% of normal efficiency in sites scattered throughout the genome. If one of these were close to the *rbt* operon, we could use the very powerful selection of growth of a lysogen on ribitol to isolate a transducing phage that had picked up this operon by illegitimate excision.

Hence the *rbt* and *dal* operons were transduced by phage P1 from *E. coli* EA into an *att* λ-deleted strain (*E. coli* RW592) to yield strain NC596: *rbt*101 *rbtD*$^+$ *rbtK*$^+$ *dal*$^+$ *thi* Δ(*gal att* λ *bio*), where *rbt*101 denotes the constitutive RDH control mutation and *rbtD*$^+$, *rbtK*$^+$, and *dal*$^+$ denote the wild-type *K. aerogenes* genes transferred from strain EA. Phage λ627 was used as the parental phage, since it is thermoinducible and will not lyse a suppressor-free host. Also, it contains large deletions that increase the likelihood of a plaque-forming transducing derivative. Lysogens of NC596 and λ627 were selected by immunity to a virulent λ phage and then temperature-induced for phage production. The lysate containing $10^{11}$ pfu/ml was infected into *E. coli* NC4 (*supF nalA tonA*), which were then plated onto minimal ribitol medium containing nalidixic acid. One *rbt*$^+$ *dal*$^-$ colony was found in $10^{13}$ phage screened in this way. Phage induced from this colony gave 3% *rbt* transductants when infected into a standard *E. coli* background (CSH62: *HfrH thi*) to give strain NC621, which was used as a subsequent source of λp *rbt*: all phage harvested from this strain proved to transduce *rbt*$^+$ into wild-type *E. coli* K12 strains.

### 1.2. Construction of λp *rbt dal*

The original event leading to λp *rbt* clearly excluded the adjacent *dal* operon, but this was easily remedied by lysogenizing λp *rbt* back into a *rbt*$^+$ *dal*$^+$ Δ*att*λ host. Then *rec*-mediated recombination yielded lysogens within the *rbt* region that could be selected for the ability to transduce the *dal* phenotype into wild-type *E. coli*. One such transductant yielded colonies that stained for both RDH and ArDH activity and allowed growth on ribitol or D-arabitol. The phage from this strain were infected into CSH62 to give NC623 as a standard source of λp *rbt dal*.

Table I shows the activity of lysogens of λp *rbt* and λp *rbt dal* compared with previously described *K. aerogenes* and *E. coli* strains. As expected, RDH and DRK are constitutive in both λp *rbt* and λp *rbt dal* lysogens, and ArDH and DXK are absent in λp *rbt* lysogens and inducible in λp *rbt dal* lysogens.

The *rbt* operon in λp *rbt* and λp *rbt dal* therefore carries the original

## Table I
### Specific Activities of Pentitol Operon Enzymes in Cell-Free Extracts of Various Strains[a]

| | Constitutive | | Inducible | | Constitutive | |
|---|---|---|---|---|---|---|
| | RDH | DXK | ArDH | DXK | ArDH | DXK |
| K. aerogenes FG5 | 0 | 0.06 | 0.38 (0.34) | — | — | — |
| K. aerogenes A' | 2.0 (0.046) | 188 | — | — | — | — |
| E. coli EA | 1.3 (0.048) | 86 | 0.25 (0.35) | — | — | — |
| E. coli CSH62 | 0 | 0 | 0 | 0 | — | — |
| E. coli CSH62 (λp rbt) | 1.4 (0.042) | 116 | 0 | 0 | — | — |
| E. coli CSH62 (λp rbt dal) | 1.4 (0.045) | 79 | 0.21 (0.35) | 0.30 | 0.08 (0.35) | 0.06 |
| E. coli CSH62 (λp rbt dal201) | — | — | — | — | 0.96 (0.37) | 0.70 |
| E. coli CSH62 (λp rbt dal202) | — | — | — | — | 1.10 (0.36) | 0.80 |
| E. coli L250 (λp rbt dal201)[2] | — | — | — | — | 1.70 (0.36) | — |

[a]Specific enzyme activities are expressed as units/mg soluble protein, measured and defined as in Neuberger and Hartley (1979), except that DRK units are times $10^{-3}$. Activity ratios (shown in parentheses) are 500 mM xylitol/50 mM ribitol for RDH and 25 mM D-mannitol/50 mM D-arabitol for ArDH. Cell extracts were made after overnight growth on 2% casamino acids, supplemented with 0.2% D-arabitol where indicated. Dashes indicate not determined.

RDH-constitutive control mutation ($rbt^c = rbt101$), which, from the data in Table I, is presumably in a repressor controlling both *rbtD* and *rbtK*. An analogous ArDH-constitutive control mutation was selected by infecting λp *rbt dal* into *E. coli* strain L250 (*mtlA mtlD*), which lacks D-mannitol phosphotransferase and D-mannitol phosphate dehydrogenase activity and so cannot grow on D-mannitol. D-Arabitol dehydrogenase has a side specificity for mannitol, which it converts to fructose (Lin, 1961), so a lysogen of λp *rbt dal* in strain L250 might grow on D-mannitol if there was constitutive expression of *dalD*. Such lysogens gave rise to mannitol derivatives at a frequency of $10^{-5}$, whereas the reversion frequency of L250 was less than $10^{-8}$. Two such colonies were selected, and phage harvested from these were infected into strain CSH62 for comparison with the other strains shown in Table I, giving lysogens of λp *rbt dal*201 and λp *rbt dal*202. Both show constitutive ArDH and DRK activity, suggesting that both mutations are in a repressor controlling *dalD* and *dalK*. Some strains gave large colonies on these minimal D-mannitol plates and proved to synthesize both RDH and ArDH at about twice the level of the monolysogen. These proved to be dilysogens constitutive for both operons: one such strain, NC260, was used for purification of D-arabitol dehydrogenase (Neuberger *et al.*, 1979) as described below.

## 2. The Structure of λp *rbt* and λp *rbt dal*

### 2.1. Genetic Analysis

Figure 1 [from Neuberger and Hartley (1979)] indicates the genotype of the λ627 strain used to incorporate the *K. aerogenes rbt* operon from *E. coli* K12 strain EA. Attachment of phage λ into its normal site in *E. coli* occurs through the *att* sequence, so it was reasonable to assume that this might also have been used in the secondary-site attachment. Since both λp *rbt* and λp *rbt dal* could form plaques and give rise to thermoinducible lysogens without a helper phage, genes *A–J* and *cI–R* must be present. Tests for the *red*, *gam*, and *xis* functions were positive in both λp *rbt* and λp *rbt dal*, and the *int* function was present in λp *rbt*. Hence, there was no loss of λ functions when the *rbt* and *dal* genes were incorporated into this phage, so it is almost certainly of the λp *gal* class. However, λp *rbt dal* has lost the *int* function during the recombination and asymmetric excision of λp *rbt* with the pentitol operons present in the original *E. coli* construct. It should properly be classed as a λ *bio* type.

As discussed above, the levels, activity ratios, and inducibility of the four pentitol operon enzymes indicate that λp *rbt* contains *rbtD* and *rbtK* only, while λp *rbt dal* has all four genes and control properties identical to the original *K. aerogenes* strain A.

## 2.2. Physical Analyses

The size of λ phages can conveniently be estimated by the kinetics of sodium pyrophosphate inactivation (Parkinson and Huskey, 1971). This indicated that while the original λ627 was 86% of the length of wild-type λ, λp *rbt* was 91.5% and λp *rbt dal* was >100%.

Restriction endonuclease mapping (Fig. 1) confirmed that the *rbt* operon was close to the *att* site in λp *rbt*, within an insert of about 6 kb. The λp *rbt dal* appeared to contain all of this insert plus an additional 4.4 kb of foreign DNA, which presumably contains the *dal* operon. The mapping confirmed that a small piece of the λ DNA between the *att* site and the *xis* gene has been deleted during the formation of λp *rbt dal*, consistent with excision by illegitimate recombination between a λ sequence within *int* and a sequence of *K. aerogenes* DNA (or possibly *E. coli* DNA) just beyond the *dal* operon (Neuberger and Hartley, 1979).

## 2.3. How Did λp *rbt* and λp *rbt dal* Arise?

The history of these events has some relevance to our evolutionary story, since the gene transfers were achieved by "biological events" rather than by *in vitro* recombinant DNA technology. Figure 2 shows the assumed mechanism (Loviny *et al.*, 1981).

Phage λ inserts at its normal attachment site in *E. coli* by a single reciprocal recombination between a sequence BOB' in the bacterial *att* site and POP' in the phage *att* site (Campbell, 1962). The phage *int* gene and at least two *E. coli* proteins are involved (Miller and Friedman, 1967; Williams *et al.*, 1977). There is a 15-bp identity between the two O sites, but the B, B', P, and P' sequences are nonhomologous and functionally distinct (Guerrini, 1969). Assuming that the mechanism is similar for secondary *att* sites in the *E. coli* genome, we would expect insertion of λ at a ΔOΔ' sequence (Shimada *et al.*, 1973) between *rbtD* and *dalD* as shown in Fig. 3. This sequence should lie in the 0.31-kb *BstA–Bst* λ3 fragment shown in Fig. 1. This fragment was therefore excised from plasmid pRD101 (C. L. Koh, unpublished work), which contains the *Bst*1–*Bst*3 region (see Fig. 1) cloned into the *Bst*I site of plasmid pBR313 (Bolivar *et al.*, 1977). After end-labeling with [$\gamma$-$^{32}$P]-ATP and *Hin*fI digestion, the two labeled fragments were sequenced by the method of Maxam and Gilbert (1977). A sequence ΔO*P should lie at the junction of the *K. aerogenes* DNA

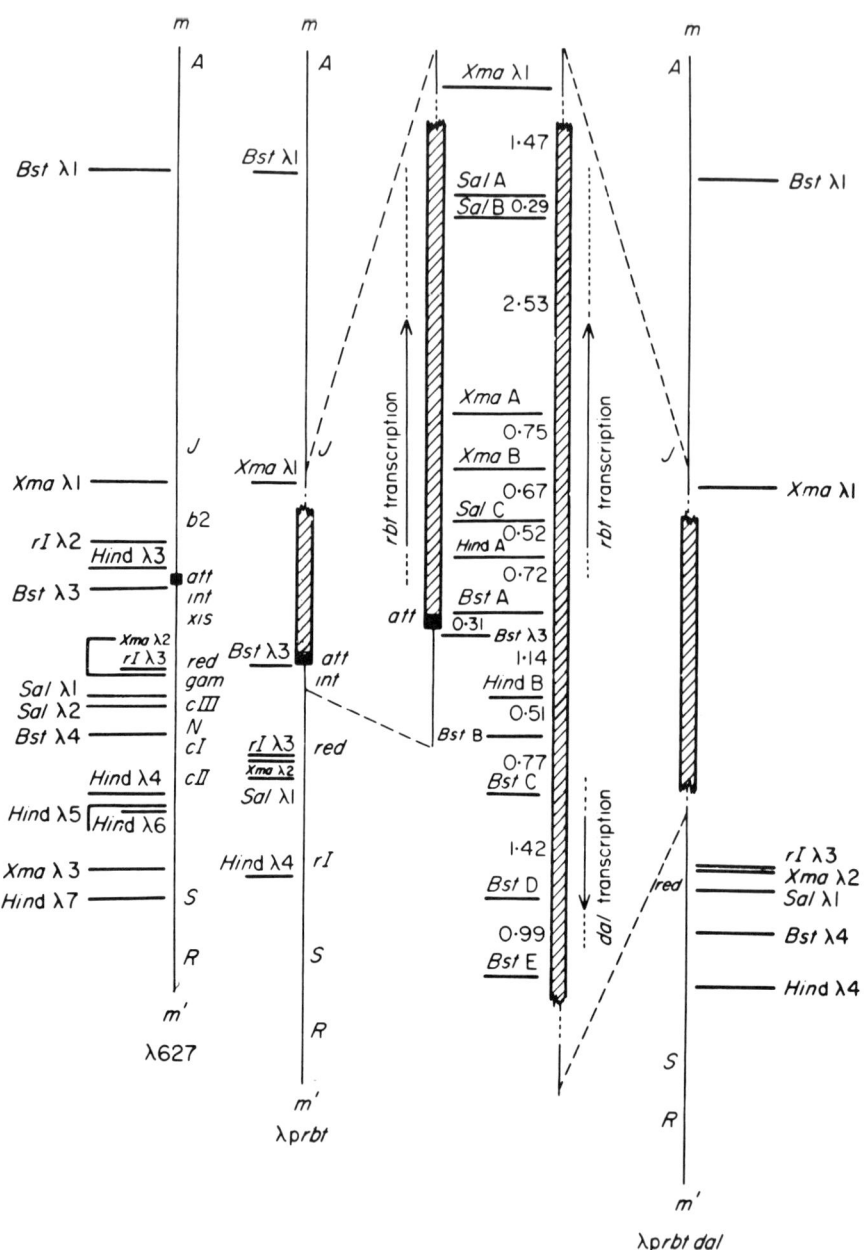

and the known P' sequence in λ; this was found, as shown in Fig. 2 (Loviny et al., 1981).

The whole sequence of the original secondary *att* site $\Delta O^*\Delta'$ should lie in the 1.65-kb *BstA–BstB* fragment of λp *rbt dal* (Fig. 1). This fragment was excised, end-labeled, digested with *Hin*dIII, and sequenced as above. Figure 3 shows that there is no homology between the $\Delta$ or $\Delta'$ sequences and the P or P' sequences of the phage or the B or B' sequences of the normal *E. coli att* site. The insertion of the phage appears to be due solely to a strong homology between the 15-bp O* sequence in the secondary site and the O sequence in the λ *att* site. A similar homology has been found in other λ secondary sites, as indicated. That in the *rbtD–dalD* region shows the greatest homology to the primary *att* sequence, so it is not surprising that the phage inserted here at reasonable frequency. The moral is that such events may not be rare in microbial evolution.

## 3. Bipolar Transcription of the Pentitol Operons

The phage λp *rbt dal* is a useful tool to test the hypothesis that the two operons are encoded on opposite strands of the DNA (Neuberger and Hartley, 1979). The left (L) and right (R) strands of λ phage can be

---

**Figure 1.** Physical structures of λ p *rbt* and λ p *rbt dal*. The phages contain both λ DNA (solid line) and bacterial DNA (hatched rectangle). The approximate positions of the *rbt* and *dal* structural gene transcriptions are indicated. Target sites for restriction endonucleases *Eco*RI, *Bst*I, *Hin*dIII, *Sal*I, and *Xma*I are designated *rI*, *Bst*, *Hind*, *Sal*, and *Xma*, respectively. This designation is followed by a letter (A, B, C, etc.) for sites in the bacterial substitution, sites on λ DNA being numbered according to the position of the corresponding site on wild-type λ, numbering from the left-hand end. The left and right ends of the phage DNAs are designated *m* and *m'*, respectively. The restriction map of λ627 gives the locations of cleavage sites for the five endonucleases; the approximate positions of various relevant λ genes are also indicated. Several $\lambda^+$ restriction sites are missing from λ627. The λ cleavage maps are taken from Thomas and Davis (1975) for *Eco*RI; Haggerty and Schleif (1976) for *Bst*I [as *Bst*I is an isoschizomer of *Bam*HI (Catterall and Welker, 1977)]; Robinson and Landy (1977) for *Hin*dIII; Kamp *et al*. (1977) for *Sal*I; and Lindahl *et al*. (1977) for *Xma*I. The length of wild-type λ DNA is taken as 46.5 kb and all length measurements are given in kb. The structures of λp *rbt* and λp *rbt dal* are given with only the location of the λ target site proximal to each side of the bacterial substitution for each endonuclease being indicated. A magnification of the bacterial substitutions presents the locations of restriction sites in the *K. aerogenes* DNA. All data obtained were consistent with these maps and all appropriate fragments were observed in single and double digests. (Both *shin*dIIIA–*ssal*IC and *sbst*IB–*shin*dIIIB comigrate with *shin*dIII $\lambda_4$–*shin*dIII $\lambda_5$, but the existence of *sbst*IB–*shin*dIIIB was confirmed from the restriction map of plasmid pRD351 and that of *shin*dIIA–*ssal*IC from gel transfer hybridization.)

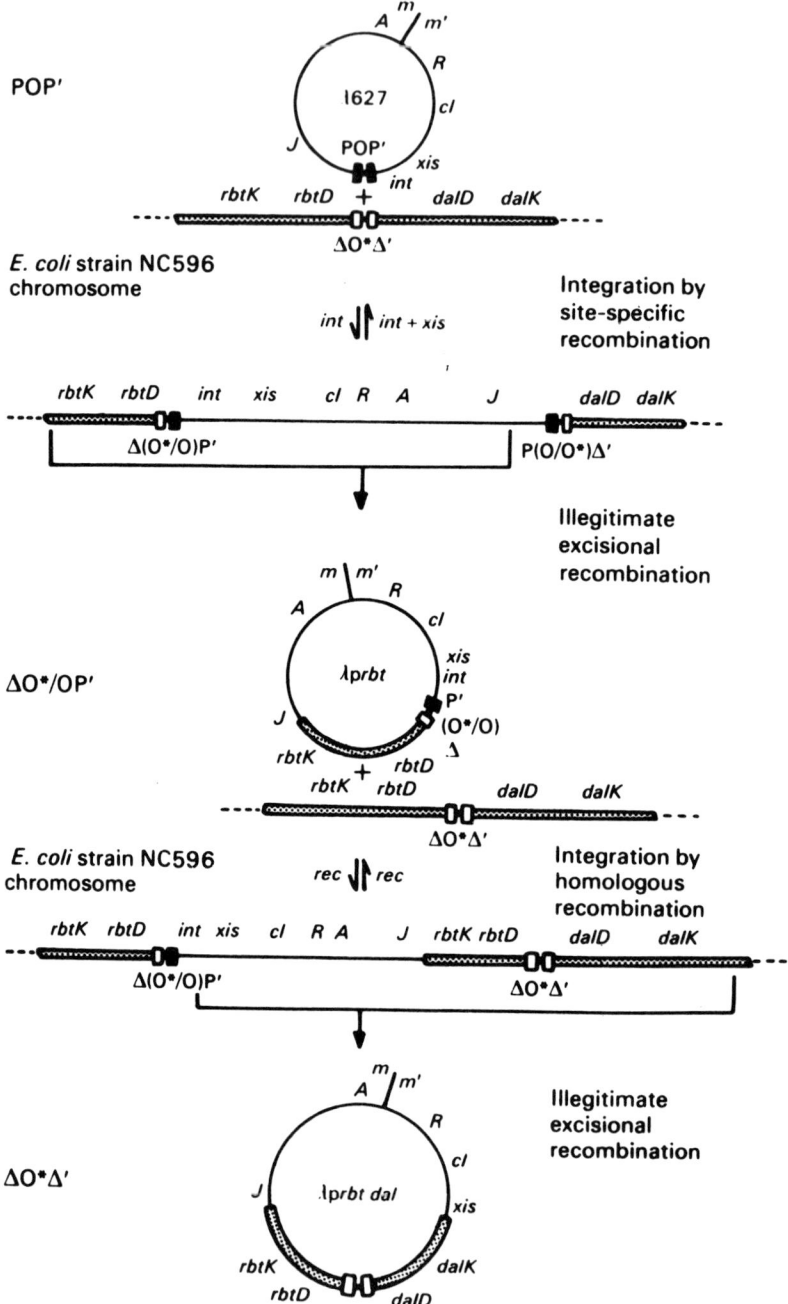

separated by centrifugation in CsCl in the presence of poly(U,G), which binds preferentially to the L strand. The two strands of λp *rbt dal* were also separated in this way, and late λ627 mRNA hybridized predominantly to the first peak eluting from the CsCl density gradient, confirming that it was the R strand. Using mRNA from cultures of the *rbt*-inducible *K. aerogenes* strain FG5, we detected no hybridization with mRNA from cells grown in the absence of ribitol, but there was clear hybridization to the L strand only with mRNA from ribitol-induced cultures. Cultures of a ribitol dehydrogenase superproducer, *K. aerogenes* strain A114, gave strong hybridization to the L strand but none to the R strand. Hence the *rbt* operon is encoded by the λp *rbt* L strand.

In a similar experiment with the separated strands of λp *rbt dal* (Fig. 4), the mRNA from the RDH-constitutive strain A114 again bound to the L strand and λ627 mRNA to the R strand. Strain FG5 mRNA from uninduced cells bound to neither strand without induction, but hybridized to the L strand if cultures were induced with ribitol and to the R strand if D-arabitol was the inducer. Hence the two operons are indeed transcribed in a bipolar fashion from opposite strands of the DNA.

## 4. The Pentitol Operon Enzymes

Extensive studies of wild-type and mutant ribitol dehydrogenases were discussed in the last chapter, but is is now necessary to investigate the other three enzymes.

### 4.1. D-Arabitol Dehydrogenase

D-Arabitol dehydrogenase was purified from a 60-liter culture of *E. coli* NC260, the dilysogen for λp *rbt dal*201, which is constitutive for both operons (Neuberger *et al.*, 1979). Methods analogous to those of Burleigh

---

**Figure 2.** Model for the formation of specialized transducing phages λp *rbt* and λp *rbt dal*. Phage λ627 was infected into the *E. coli* K12–*K. aerogenes* hybrid strain NC596 [*Hfr*H *rbt*101 *rbtD*⁺*dal*⁺]*thi* Δ(*gal att* λ-*bio*)] and a population of secondary site lysogens induced for phage production; λp *rbt* was isolated from this phage lysate. It is envisaged that, in the secondary site lysogen that gave rise to λp *rbt*, phage λ627 had integrated into the NC596 chromosome by *int*-mediated site-specific recombination at a secondary *att* located between *rbtD* and *dalD*. Phage λp *rbt dal* was isolated from a phage lysate induced from a population of NC596 (λp *rbt*) lysogens and it is envisaged that in the lysogen that gave rise to λp *rbt dal*, phage λp *rbt* had integrated into the NC596 chromosome in the region of the pentitol operons by rec-mediated homologous recombination.

```
                -25          -20          -15          -10           -5            0            5           10          15          20          25
                 .            .            .            .            .            .            .           P'            .            .            .
λrbt         G C C G G C G A T G C T G G A C G C G G T T T T T A T A C T A A G T T G G C A T T A T A A A A A A G
                              Δ
λrbt dal     G C C G G C G A T G C T G G A C G C G G T T T T T C G A T T A T T G A C G C C G G C G A A G C C A G
                              Δ                                                                        Δ'

trp C                                              T A A T G T T T A T A A A T G
bfe                                                G C T A T T T T A C C A C G A
gal T                                              C T T T G T T T T C A A A A A
Primary att (gal-bio)                              G C T T T T T A T A C T A A
```

**Figure 3.** Secondary attachment sites for λ phage. Only 5′–3′ strands are shown. The box compares the "O-regions" of various secondary sites with those in λp *rbt* and λp *rbt dal*: *trpC* (Christie and Platt, 1979), *bfe* (Csordas-Toth *et al.*, 1979), *galT* (Bidwell and Landy, 1979), and primary *att* (Landy and Ross, 1977). Δ and Δ′ are the flanking *K. aerogenes* sequences and P′ is one of the flanking λ sequences.

*et al.* (1974) yielded 280 mg (53% yield, 17.7 × purification) of enzyme giving a single band on SDS–polyacrylamide gels equivalent to molecular weight 46,500. The specific activity was 37.1 U/mg and the sample contained less than 0.01% (w/w) of ribitol dehydrogenase, the major impurity. The enzyme retains >95% activity when stored at −20°C in 5 mM NAD–50% (w/v) glycerol, but has a half-life at 45°C of only 20 min.

The protein is clearly a monomer, since gel filtration against known molecular weight markers gave a molecular weight of 43,000 and sedimentation equilibrium gave 44,000. Some minor bands equivalent to dimers, tetramers, and hexamers were observed during electrophoresis of the native protein on polyacrylamide gels, but the activity was only in the monomer and only this band was seen on gels where preelectrophoresis with dithiothreitol had removed oxidizing agents. Hence this enzyme is clearly active as a monomer of about 46,000 daltons.

The steady-state kinetic parameters for D-arabitol dehydrogenase against D-arabitol and D-mannitol are compared in Table II with those of ribitol dehydrogenase against ribitol and xylitol (Neuberger *et al.*, 1979): the enzyme had no activity against ribitol, xylitol, or D-sorbitol. One observes for D-arabitol dehydrogenase that the maximum velocity is the same for D-arabitol and D-mannitol. Both contain the same stereochemistry at the first five carbon atoms, and it is likely that they bind in an identical way to the catalytic site (the bold letter indicates the position of the oxidation):

$$
\begin{array}{cc}
\text{CH}_2\text{OH} & \text{CH}_2\text{OH} \\
| & | \\
\text{HO}-\mathbf{C}-\text{H} & \text{HO}-\mathbf{C}-\text{H} \\
| & | \\
\text{HO}-\text{C}-\text{H} & \text{HO}-\text{C}-\text{H} \\
| & | \\
\text{H}-\text{C}-\text{OH} & \text{H}-\text{C}-\text{OH} \\
| & | \\
\text{CH}_2\text{OH} & \text{H}-\text{C}-\text{OH} \\
 & | \\
 & \text{CH}_2\text{OH} \\
\text{D-Arabitol} & \text{D-Mannitol}
\end{array}
$$

In contrast, there are big changes in both the Michaelis constant ($K_{app}$) and the maximum velocity ($V_{app}$) of ribitol dehydrogenase for ribitol, xylitol, and L-arabitol, consistent with the speculation that they bind to different conformations of the tetramer (Burleigh *et al.*, 1974):

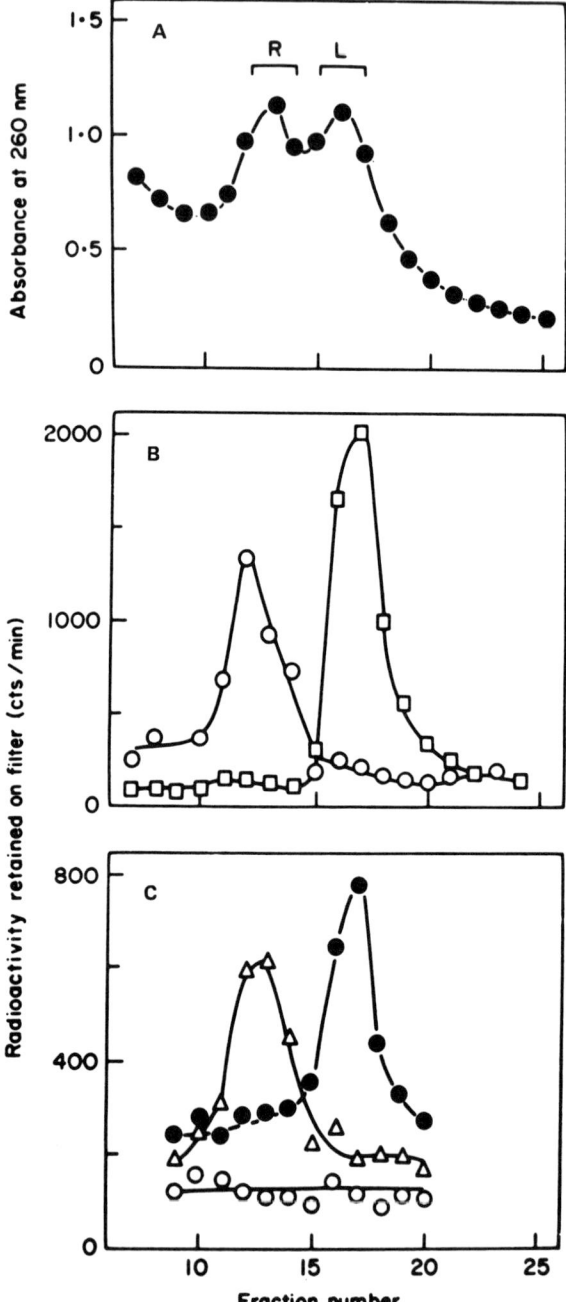

```
      CH₂OH              CH₂OH              CH₂OH
       |                  |                  |
    H—C—OH             H—C—OH             H—C—OH
       |                  |                  |
    H—C—OH             HO—C—H             H—C—OH
       |                  |                  |
    H—C—OH             H—C—OH             HO—C—H
       |                  |                  |
      CH₂OH              CH₂OH              CH₂OH

      Ribitol            Xylitol           L-Arabitol
```

There are big differences in the amino acid compositions of D-arabitol dehydrogenase and ribitol dehydrogenase, and antibodies to the former showed no cross reaction with the latter (Neuberger et al., 1979). The $NH_2$-terminal sequence of purified ArDH was determined by sequenator analysis, and several peptides from an elastase digest were sequenced by "mixture-analysis" mass spectrometry as shown in Fig. 5 (J. E. Walker, A. Dell, M. S. Neuberger, and B. S. Hartley, unpublished results). This technique rapidly generates short runs of sequence that are invaluable in establishing reading frames in DNA sequences of genes (see below). There is no obvious homology between these and sequences in ribitol dehydrogenase, though the quantity and quality of the sequences make this a weak conclusion.

### 4.2. D-Ribulokinase

D-Ribulokinase was purified from *E. coli* strain NC629, which superproduces this enzyme (Neuberger et al., 1981). Growth of a λp $rbt^c$ lysogen on xylitol rapidly selects strains with multiple prophage copies that synthesize large amounts of both ribitol dehydrogenase and D-ribulokinase; these can be recognized as major bands on SDS–polyacrylamide gel electrophoresis of crude cell-free extracts. Several purification steps

**Figure 4.** [$^3$H]-RNA hybridization to separated λp *rbt dal* DNA strands. (A) Elution profile of λp *rbt dal* strand separation centrifugation. (B) Samples (40 μl) of each separated DNA strand fraction were hybridized with $3.2 \times 10^4$ counts/min of A114 RNA (□, specific activity $7.5 \times 10^4$ counts/min per μg), and 10-μl samples of each fraction were hybridized with $3.2 \times 10^4$ counts/min of late λ mRNA (○, specific activity $7.4 \times 10^4$ counts/min per μg). (C) Samples (20 μl) of each fraction were hybridized with $2.1 \times 10^6$ counts/min of ribitol-induced FG5 RNA (●, specific activity $9.0 \times 10^5$ counts/min per μg), with $8.9 \times 10^5$ counts/min of D-arabitol-induced FG5 RNA (△, specific activity $2.8 \times 10^5$ counts/min per μg), and with $0.5 \times 10^5$ counts/min of uninduced FG5 RNA (○, specific activity $5 \times 10^4$ counts/min per μg).

## Table II
### Kinetic Constants for D-Arabitol Dehydrogenase and Ribitol Dehydrogenase[a]

| Enzyme | Substrate | $K_{app}$ (mM) | $V_{app}$ (U/mg) | $K_P$ (mM) | $K_{NAD}$ (mM) | $K_{P,NAD}$ (mM)$^2$ | $V_f$ (U/mg) | $K$ (sec$^{-1}$) |
|---|---|---|---|---|---|---|---|---|
| ArDH | D-Arabitol | 24 | 52 | 20 | 0.04 | 4.4 | 55 | 40 |
| ArDH | D-Mannitol | 70 | 52 | — | — | — | — | — |
| RDH | Ribitol | 11 | 93 | 8 | 0.66 | 10.4 | 166 | 75 |
| RDH | Xylitol | 1000 | 10.5 | 579 | 0.76 | 612 | 20 | 36 |
| RDH | L-Arabitol | 267 | 12 | 147 | 0.25 | 166 | 16 | 28 |

[a] Steady-state constants are determined according to Cleland (1963). $K_{app}$ and $V_{app}$ are at 0.83 mM NAD. $K_P$ and $K_{P,NAD}$ are the constants at saturating NAD, and $K_{NAD}$ at saturating pentitol. $V_f$ and $K$ are for the forward reaction.

N-terminal sequence:

```
        1                                        10                       20
Met-Asn-Asn-Gln-Phe-Thr-Trp-Leu-His-Ile-Gly-Leu-Gly(Ser)Phe-His-Arg(Ala)---(Glx)---Tyr-Leu-
```

Elastase peptides:

| | | | | |
|---|---|---|---|---|
| A1: | Thr-Leu-Pro-Tyr-Gln-Tyr- | B1: | Leu-Leu-Gly- | C1: Gly-Leu-Gln-Pro- |
| A2: | Gly-Leu-Gln-Pro-Leu- | B2: | Thr-Leu-Thr-Asp-Val-Leu- | C2: Asp-Phe-Leu- |
| A3: | (Glx-Ala-Ala-Met-Gly-Ala) | B3: | Gln-Lys-Leu-...,Asn-Pro-Tyr-Leu- | |
| D1: | Met-Leu-Ala-Pro- | E1: | Asn-Pro-Gln-Thr-Lys-Val-Leu- | F1: Tyr-Thr-Leu-Leu- |
| D2: | Met-Leu-Ala-OMe | E2: | Leu-Phe-Gly-Asp-Leu-Ala- | F2: Met-Leu-Pro-Ala-OMe |
| D3: | (Leu-Thr-Thr-Leu) | E3: | Ala-Asp-Leu-Pro-Ala- | |
| G1: | Tyr-Ala-Leu-Ala-OMe | H1: | (Lys-Leu-Leu-Glu-Leu-Ala-OMe) | I1: Lys-Leu-Leu-Pro |
| J1: | Ala....Asn-Pro-Tyr-Leu- | K1: | Ala...Trp-Tyr-Leu-OMe | L1: Leu-Leu-Asn-Cys- |
| J2: | Tyr-Thr-Leu-Leu- | K2: | Ser-Leu-Gln-Lys- | |
| K1: | Leu-Tyr-Gly- | L1: | (Leu-Ala-Gly-Asn-Gln-Leu-Gly-) | M1: Gln-Lys-Leu-Leu- |
| N1: | Thr-Ser-Leu-Gln-Lys-Leu- | O1: | Gly-Tyr-Tyr- | P1: Phe-Met-Glu-Gln- | Q1: Gly-Lys-Gln- |

Figure 5. Partial amino acid sequence of D-arabitol dehydrogenase. The NH₂-terminal sequence was determined by conventional sequenator analysis (J. E. Walker, unpublished results); parentheses indicate doubtful assignment and dots indicate no assignment possible. The elastase peptides (M. S. Neuberger and A. Dell, unpublished results) were separated by Dowex ion-exchange chromatography and fractions from this column were analyzed by electron impact mass-spectrometry "mixture analysis" after acetylation and permethylation (Morris et al., 1974). This technique does not distinguish Leu and Ile, but yields NH₂-terminal sequences and COOH-terminal sequences (e.g., Met-Leu-Ala-OMe) from impure mixtures. Sequences shown in parentheses were of doubtful significance. Sequences underlined were recognized in the nucleotide sequence of *dalD*.

were necessary: DEAE–cellulose chromatography, gel filtration, hydroxyapatite chromatrography, and ammonium sulfate precipitation gave 81 mg of pure D-ribulokinase from 300 g (wet weight) of cells.

The enzyme was homogeneous by the criteria of native and SDS–polyacrylamide gel electrophoresis with a subunit molecular weight of 60,000 and a native molecular weight of 112,000 by calibrated gel filtration. Its $NH_2$-terminal sequence was

$\overset{①}{\text{Met}}$-His-Asn-Asn-Thr-Glx-Asn-Ile-Ile-Gly-Val-$\overset{⑫}{\text{Asp}}$-

### 4.3. D-Xylulokinase

D-Xylulokinase was purified from *E. coli* strain NC260, which harbors multiple copies of a λp $rbt^c$ $dal^c$ prophage that is constitutive for expression of both operons (Neuberger *et al.*, 1979). This allows growth in the absence of D-arabitol or D-xylose, which both induce the *E. coli* D-xylulokinase encoded in the D-xylose operon: separation of the two enzymes is difficult (Wilson and Mortlock, 1973). Cell-free extracts were purified by DEAE–cellulose chromatography, gel filtration, chromatography, and ω-aminobutyl agarose and then on hydroxyapatite, followed by ammonium sulfate precipitation. The final yield of 5.1 mg of pure enzyme from 300 g (wet weight) of cells probably represents a poor recovery, since the enzyme is cold-labile (Wilson and Mortlock, 1973).

The enzyme was homogeneous in native or SDS–polyacrylamide gel electrophoresis, with subunit molecular weight of 54,000 and a native molecular weight of 110,000 by calibrated gel electrophoresis.

Its $NH_2$-terminal sequence was

$\overset{①}{\text{Met}}$-Tyr-Leu-Gly-Ile-Asp-Leu-Gly-Thr-$\overset{⑩}{\text{Ser}}$-Glu-Val-Lys-Ala-Leu-Val-Ile-
$\overset{⑳}{\text{Asp}}$-Glu-Asn-His-$\overset{㉔}{\text{Glu}}$-Val-Ile-

This shows no significant homology with the $NH_2$-terminus of D-ribulokinase.

## 5. Substrate Specificity of the Pentitol Operon Enzymes

The conformations of the pentitols and pentuloses shown in Fig. 6 do much to explain the pattern of side specificities shown by the enzymes of the pentitol operons. Xylitol and L-arabitol each have one hydroxyl different from ribitol, accounting for their activity with ribitol dehydrogenase, whereas D-arabitol differs at the critical C-2 hydroxyl. It is also

# THE STRUCTURE AND CONTROL OF THE PENTITOL OPERONS 71

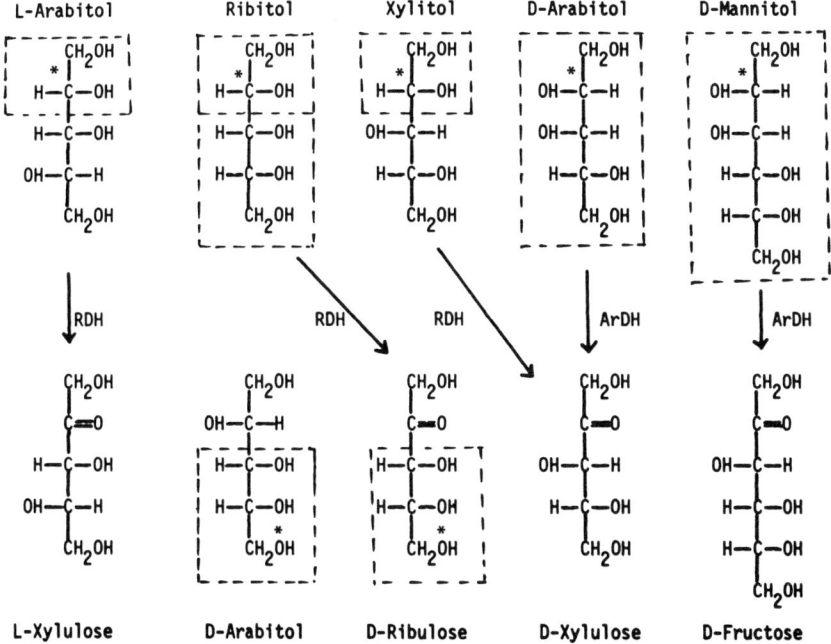

Figure 6. Conformations of substrates for the pentitol operon enzymes. Boxes indicate common structures probably responsible for the side specificity of ribitol dehydrogenase for xylitol and L-arabitol, of D-arabitol dehydrogenase for D-mannitol, and of D-ribulokinase for ribitol and D-arabitol. The positions of dehydrogenation or phosphorylation are indicated by asterisks.

not surprising that D-mannitol fits the catalytic site of D-arabitol dehydrogenase. Figure 6 also shows the side specificity of D-ribulokinase for ribitol and D-arabitol, which have the same C-3 and C-4 hydroxyl conformations as D-ribulose next to the C-5 hydroxyl that is phosphorylated. The $K_{app}$ at 0.5 mM ATP, 38°C, pH 7.9, is 220 mM for ribitol and 140 mM for D-arabitol, compared with 0.4 mM for D-ribulose, and $V_{app}$ is 12 μmole/min per mg for ribitol, 6.6 μmole/min per mg for D-arabitol, and 71 μmole/min per mg for D-ribulose. D-Xylulokinase has a $K_{app}$ of 0.8 mM and a $V_{app}$ of 150 μmole/min per mg for D-xylulose, but showed no detectable activity toward any other substrate tested. Nor could any significant D-ribulokinase activity be detected (>0.1 μmole/min on 100 mM substrate) on D-xylulose, xylitol, L-arabitol, galactitol, D-mannitol, D-sorbitol, erythritol, glycerol, D-ribose, D-arabinose, D-fructose, or D-glucose.

These negative results cast a small shadow over the interesting speculations of Scangos and Reiner (1978b) concerning "lethal synthesis" of pentitol or hexitol phosphates in *E. coli* C, which contains pentitol operons analogous to those of *K. aerogenes* (see Chapter 1). They observed that D-arabitol was toxic to strains that produce D-xylulokinase constitutively but lack D-arabitol dehydrogenase activity. Mutants that became D-arabitol resistant were found to have lost the D-xylulokinase activity, so they suggested that this enzyme might convert D-arabitol to D-arabitol 5-phosphate, which was a metabolic inhibitor. By analogy, they proposed that D-ribulokinase might convert galactitol, L-arabitol, and D-arabitol into similarly lethal hexitol or pentitol phosphates.

Neuberger *et al.* (1981) showed that galactitol, L-arabitol and D-arabitol were toxic to a λp *rbt* lysogen of *E. coli* K12, which expresses the *K. aerogenes* ribitol dehydrogenase and D-ribulokinase constitutively. The *E. coli* K12 host was resistant to these compounds. However, they note that in some cases lethal products other than pentitol phosphates might be responsible, e.g., L-xylulose produced from L-arabitol by ribitol dehydrogenase. The lethal synthesis might represent a very low-level activity, difficult to reveal by normal enzyme assays.

## 6. *rbt* Messenger RNA

The massive superproduction of ribitol dehydrogenase elicited by the experimental evolution of *K. aerogenes* strains in continuous culture on xylitol gives rise to obvious questions about the evolution of the *rbt* mRNA. Is it transcribed more abundantly from a single-copy gene, e.g., by up-promoter mutations, or by removal of catabolite repression? Does it translate more protein from a single mRNA strand, e.g., by increased initiation rate or by increased mRNA stability? Or is the enzyme superproduction related entirely to a gene dosage effect arising from gene amplification?

### 6.1. A Switch in *rbt* mRNA Translation in Mid Log Phase

To answer some of these questions, Bahramian and Hartley (1980) devised a method for radioimmune labeling of polysomes engaged in synthesis of ribitol dehydrogenase. It had previously been shown that antibodies raised to a pure protein can react with nascent chains of that protein attached to polysomes (e.g., Cowie *et al.*, 1961; Williamson and Askonas, 1967; Hamlin and Zabin, 1972), and Schechter (1974) used this property to purify specific eukaryotic polysomes by a double antibody precipitation technique. That method has not found favor, because of a lack of specificity in the antibody binding (see below) and has been used mainly with eukaryotic polysomes because of the known instability of prokaryotic mRNA. However, rapid and delicate cell lysis can yield undegraded prokaryotic polysomes (Godson and Sinsheimer, 1967), and modifications of this technique yielded highly reproducible polysome profiles from cultures of *K. aerogenes* strain A111, which superproduces ribitol dehydrogenase. Homospecific antibodies to the pure enzyme were raised in rabbits, purified by affinity chromatography on ribitol dehydrogenase covalently bound to Sepharose, and labeled with $^{125}I_2$. When these were mixed with pelleted polysomes before sucrose density gradient centrifugation, profiles such as those in Fig. 7A were observed. There is clearly more antibody bound to the polysomes from the constitutive RDH-superproducing strain A111, but a huge background of nonspecific binding is revealed in the control of the RDH-inducible strain FG5 grown in the absence of inducer. This was eliminated by adding heparin to the mixture of [$^{125}I$]antibody and polysomes before density gradient centrifugation. The profiles in Fig. 7B show that the method *can* reveal the size and quantity of polysomes engaged in synthesis of ribitol dehydrogenase.

It was surprising to discover that there was a sudden and dramatic change in these profiles at a specific point in the batch cultures of strain A111, regardless of whether these were grown on rich medium or on minimal xylitol medium. The pattern in Fig. 7C(a) is typical of cells harvested at a low culture density ($A_{650} < 0.30$), while the pattern in Fig. 7C(b) is established at $A_{650} = 0.56$ and persists throughout exponential phase and into stationary phase. It should be noted that this switch occurs in midexponential phase after about 3–4 exponential doublings of the original stationary phase inoculum, and that there is no change in the growth curve at this point.

At the switch point there is a sudden drop in the proportion of large polysomes (8–50 ribosomes/mRNA) and a rise in the proportion of small polysomes (1–8 ribosomes/mRNA) plus a 20% increase in the amount of ribosomal subunits. The distribution of polysomes carrying nascent RDH changes in the opposite sense: predominantly small polysomes at $A_{650} <$ 0.3; above an $A_{650}$ of 0.56 a dome-shaped profile is observed extending to about 50 ribosomes/mRNA and peaking at about 15 ribosomes/mRNA. On rich medium the profiles are similar, except that above the switch point the dome peaks at about 7 ribosomes/mRNA and extends only to about 25 ribosomes/mRNA.

This strange phenomenon prompted Bahramian and Hartley (1980) to examine the level of ribitol dehydrogenase in cell-free extracts of various strains during batch culture of cells grown on 2% casamino acids. It was known that synthesis of ribitol dehydrogenase is strongly repressed

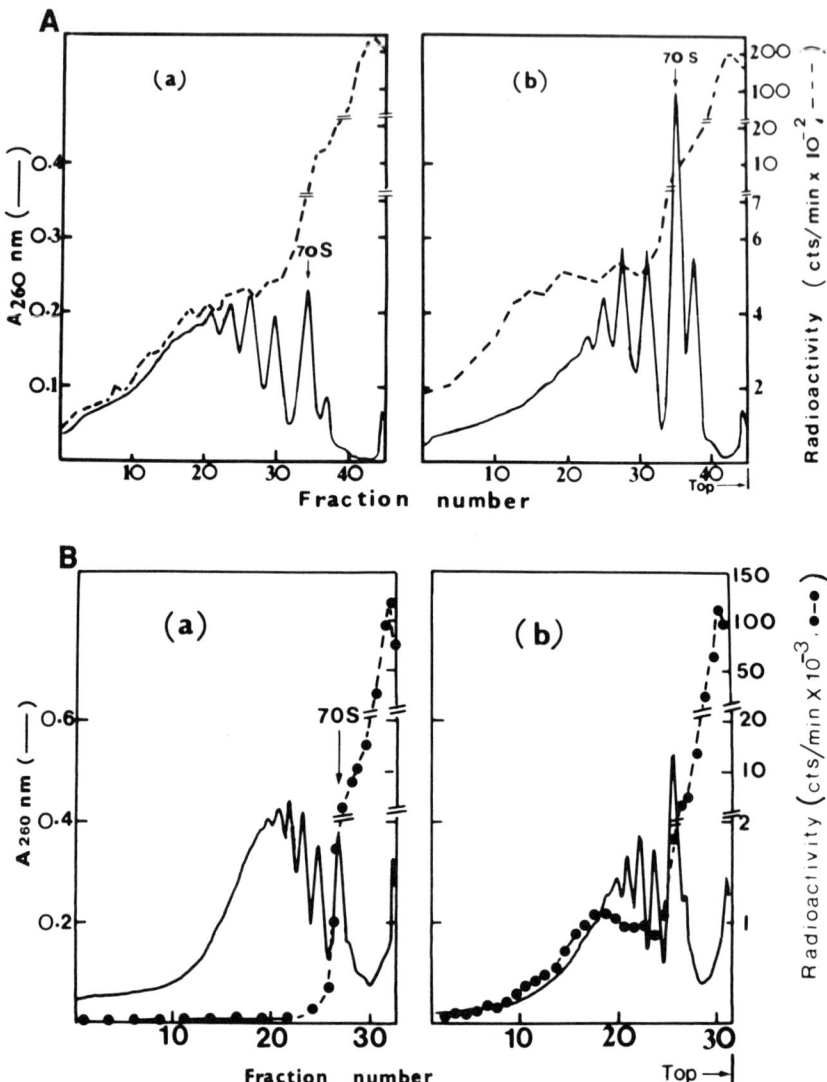

during growth on glucose and to a lesser extent on pentitols, and that addition of cAMP to the culture releases this catabolite repression (Neuberger and Hartley, 1979). However, catabolite repression is minimal during growth on casamino acids. Figure 8 demonstrates that a sudden switch in the level of ribitol dehydrogenase in the cell occurs at exactly

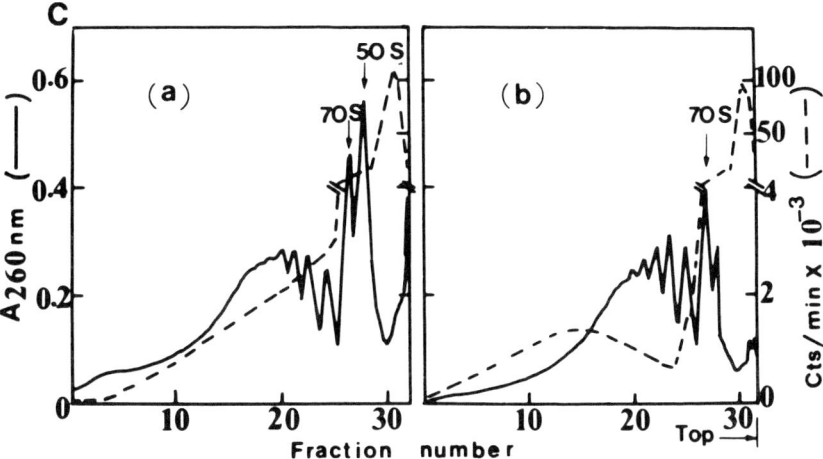

Figure 7. (A) Interaction of purified $^{125}$I-labeled anti-ribitol dehydrogenase with purified polysomes from *K. aerogenes* strains FG5(A) and A111(B). Five micrograms of iodinated antibody was incubated on ice with 0.8 mg polysomes in 0.5 ml for 40 min, then centrifuged for 3.5 hr. (B) Specific interactions of polysomes involved in ribitol dehydrogenase synthesis with $^{125}$I-labeled anti-ribitol dehydrogenase in the presence of heparin. Polysomes (0.8 mg) in 1.8 ml buffer containing 10 U/ml of sodium heparin and 0.1 M sucrose were incubated for 45 min on ice with 30 µg of $^{125}$I-labeled anti-ribitol dehydrogenase (specific activity 7000 counts/min per µg), then centrifuged for 3 hr. (a) Purified polysomes from *K. aerogenes* FG5 strain grown on tryptone-yeast extract medium up to culture $A_{650} = 0.6$; (b) purified polysomes from *K. aerogenes* A111 strain grown on tryptone-yeast extract up to culture $A_{650} = 0.75$. (C) Size distribution of ribitol dehydrogenase-synthesizing polysomes before and after the "critical culture density." A 1-liter mineral salts medium/xylitol culture of *K. aerogenes* A111 was grown at 37°C. One half of this culture was harvested at $A_{650} = 0.30$, the other half at (b) $A_{650} = 0.56$. The cells were lysed and the polysomes purified. About 0.8 mg of each polysome sample was incubated with excess (30 µg) of $^{125}$I-labeled anti-ribitol dehydrogenase (specific activity 8000 counts/min per µg) in 1.85 ml for 45 min on ice; the sample was then layered carefully onto 56 ml of a linear 0.5–1.5 M sucrose gradient, and the polysomes sedimented at 24,000 rev/min for 3 hr at 2°C.

the same switch point as was observed in the polysome profiles, whatever strain of *K. aerogenes* is used. In strain A111, for example, the RDH level remains constant at about 0.5% of the total soluble protein up to a culture density of $A_{650} = 0.6$, rising continuously thereafter to 2.5% at $A_{650} = 6.0$.

These results suggest that there is a dramatic switch in the whole pattern of protein synthesis of the cells in mid log phase of growth, without any discernible effect on the growth rate of the organism. Bahramian and Hartley (1980) rationalize this by noting that the pattern of proteins in

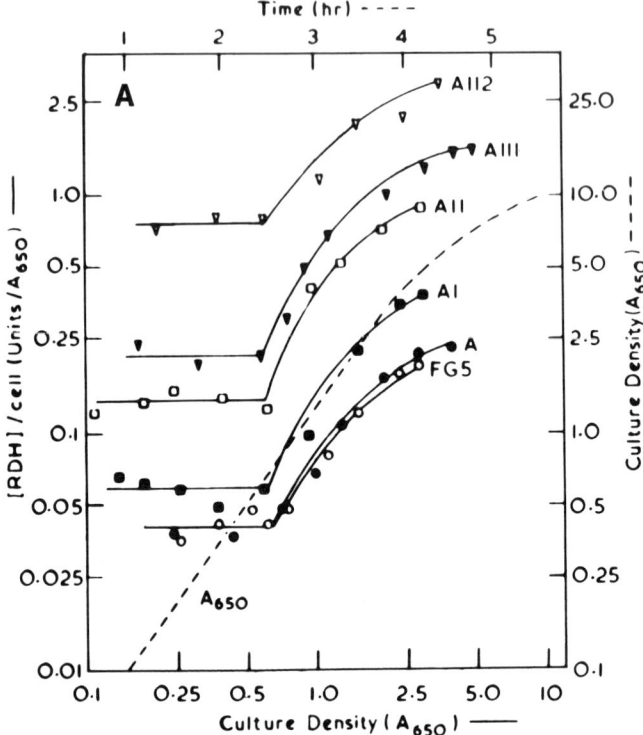

**Figure 8.** Variations in specific activity of ribitol dehydrogenase during growth of various strains of *K. aerogenes*. Ribitol dehydrogenase activity was assayed in extracts of cells grown on mineral salts medium supplemented with casamino acids and harvested at various culture densities. (A) The activity of ribitol dehydrogenase per unit of culture absorbance ($A_{650}$); (○) the ribitol dehydrogenase-inducible strain FG5 grown in medium supplemented with 0.2% (w/v) ribitol; (●) the ribitol dehydrogenase-constitutive strain A. (■) Strain A1, (□) A11 (□), (▼) A111, and (△) A112 are strains that superproduce ribitol dehydrogenase, selected by continuous culture on xylitol. (B) The specific activity of ribitol dehydrogenase in extracts of strain A111, expressed as either (○) ribitol dehydrogenase units/culture $A_{650}$ or (×) ribitol dehydrogenase/mg soluble protein. (C) The growth rates of (▼) strain A111 on mineral salts medium/casamino acids and (○) strain FG5 on mineral salts medium supplemented with casamino acids and 0.2% (w/v) ribitol.

resting stationary phase cells is different from that in exponential phase cells. Koch (1971) has observed that starvation is the natural lot of most microorganisms for most of their natural existence. Protein synthesis must be trimmed down to the minimum necessary for survival, but the ability to switch into rapid growth on encountering a feast of substrate would be

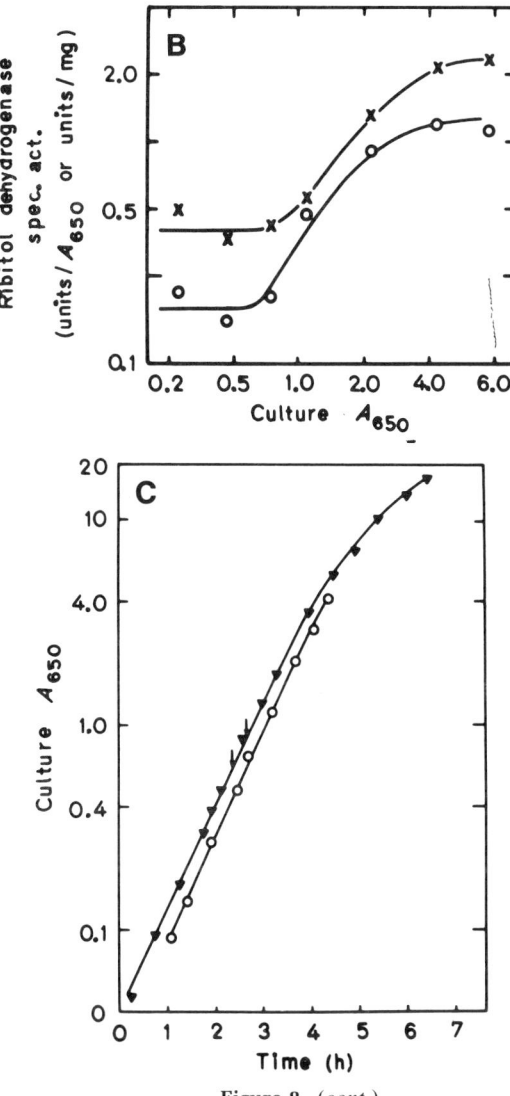

Figure 8. (cont.)

a selective advantage. The adjustment from "starvation mode" to "feast mode" appears to be a sudden rather than a gradual event and in this case is *not* associated with a lag phase.

Shortage in stationary phase cells of initiation factors for protein synthesis might be one explanation. In that case competition between different ribosome-binding sequences for the scarce initiation factor would govern the pattern of proteins initially made. Poorly translated proteins such as repressors would be in short supply. However, synthesis of the rate-limiting initiation factor itself would be a priority requirement and the level of this would rise to a point where repressors are made in adequate amounts to shut down transcription of their relevant operons. At that point the pool of free initiation factors would suddenly rise and the cell would switch rapidly to the familiar pattern of transcriptional control.

The polysome patterns observed are consistent with this hypothesis. Before the switch point the *rbt* RNA must compete with a mass of other mRNAs engaged in adjusting enzyme levels to the appropriate "feast norm." There will be few ribosomes per mRNA because these are in short supply. When stable transcriptional control is established, the level of competing mRNAs will drop and the size of the average polysome will decrease, but the number of ribosomes per *rbt* mRNA will increase because this mRNA continues to be abundant. One should treat this hypothesis with caution because the supporting evidence is slim, but the phenomenon itself is intriguing.

### 6.2. Purification and Properties of *rbt* mRNA

Since the rabbit anti-RDH specifically binds in the presence of heparin to polysomes synthesizing ribitol dehydrogenase, it can be used to purify *rbt* mRNA by double antibody precipitation (Bahramian and Hartley, 1982). Sheep antibodies monospecific for rabbit anti-RDH were purified by affinity chromatography on columns of the latter protein attached to Sepharose. Washed polysomes were carefully prepared from 1-liter cultures of *K. aerogenes* strain A111 grown to an $A_{650}$ of 0.6–0.7 on minimal xylitol medium and incubated at 4°C in the presence of heparin first with rabbit anti-RDH and then with sheep anti-rabbit Ig. The ratios of polysomes and the two antibodies are important to achieve complete precipitation. The precipitate that collected overnight at 4°C was washed and used to isolate effectively pure *rbt* mRNA. No precipitate formed with a control of the RDH-inducible *K. aerogenes* strain FG5 grown in the absence of inducer.

These *rbt* polysomes formed 2.4% of the total polysome population, indicating that *rbt* mRNA is at least that proportion of total mRNA. The RNA was extracted with phenol from the precipitate of *rbt* polysomes and used in a crude *E. coli* cell-free translation system with [$^{35}$S]-methionine plus unlabeled amino acids. Radioautography revealed a major band on SDS–polyacrylamide gel electrophoresis with identical mobility to the 27,000 subunit of ribitol dehydrogenase. Peptide maps of this radioactive product revealed characteristic RDH peptides. The relative translation efficiency of the *rbt* mRNA was about 40 times that of controls of MS2 RNA or Qβ RNA used in the same experiment. Moreover, the *rbt* mRNA directed linear incorporation of radioactive methionine into protein for over 20 min, whereas the endogenous *E. coli* mRNA had a half-life of about 7.5 min.

Hence, the *rbt* mRNA extracted in this way appears to be pure and unusually efficient and stable for a bacterial mRNA in the cell-free translation system. When the extracted mRNA was subjected to agarose gel electrophoresis in the presence of methyl mercury hydroxide (Bailey and Davidson, 1976) and then hybridized with a nick-translated probe of [$^{32}$P]-λp *rbt*, a major band corresponding to about 1500 nucleotides was observed, plus a streak of fragments from 400 to 1400 nucleotides. Since 741 nucleotides are required to encode a complete ribitol dehydrogenase subunit and about 1700 for D-ribulokinase, it is clear that most of the purified *rbt* mRNA was too small to encode both proteins. Indeed, no product equivalent to a D-ribulokinase subunit of 60,000 daltons could be detected in the cell-free translation system. This implies that the majority of *rbt* transcripts in the cell are incomplete, translation of the *rbtD* gene having taken place before the *rbtK* gene is fully transcribed.

Godson (1976) has reported that most of the *E. coli* polysomes in early log-phase cells are attached to a DNA/membrane complex and can be released by adding deoxycholate plus DNAse. Bahramian and Hartley (1982) confirmed this result, but found that only 50% of the *rbt* polysomes were DNA/membrane bound, compared with 75% of total polysomes. If this result is taken together with the absence of full-length transcripts revealed above, it suggests that there may be premature transcription termination before or within the *rbtK* gene.

### 6.3. Is *rbt* mRNA Superstable?

This would be one explanation for the superproduction of ribitol dehydrogenase. Two experiments suggested that *rbt* mRNA might be unusually stable. The translation efficiency and *in vitro* half-life of the isolated *rbt* mRNA in the cell-free translation system were greater than those of MS2 or Qβ RNA, both of which are judged to be stable. Moreover, the relative size of *rbt* polysomes exceeds that of most polysomes, consistent with *in vivo* stability. To test this hypothesis, profiles of *rbt* polysomes and total polysomes were examined at various times after addition of rifampicin to log-phase cultures of *K. aerogenes* strain A11 (Bahramian and Hartley, 1982). Rifampicin inhibits DNA-dependent RNA polymerase and hence blocks transcription. The decay in *rbt* polysomes was slower than that in total polysomes when low amounts of rifampicin were used, but amounts of inhibitor adequate to block RNA synthesis caused complete immediate cessation of both growth and RDH production. This suggests that the *rbt* promoter may have high affinity for the RNA polymerase but that the *rbt* mRNA is *not* superstable.

This was confirmed by Bahramian *et al.* (1982) by adding [$^3$H]uridine to exponential cultures of strain A111 in a "pulse-chase" fashion and extracting the RNA rapidly at intervals thereafter. This RNA was hybridized to nitrocellulose filters containing the L strand of λp *rbt*, which encodes the *rbt* operon (see above). The half-life of the [$^{14}$C]-*rbt* mRNA appeared to be less than 3 min.

In a different experiment, a "pulse-chain" experiment after addition of rifampicin to cultures of *K. aerogenes* was performed in order to observe rates of uptake of [$^{14}$C]amino acids into newly synthesized ribitol dehydrogenase. The [$^{14}$C]ribitol dehydrogenase in cell extracts at various times after the addition of rifampicin was precipitated with rabbit anti-RDH plus sheep anti-rabbit Ig. In the RDH-superproducing strain A111, 4.6% of the total pulse-labeled protein was ribitol dehydrogenase at the time of addition of rifampicin, and protein synthesis ceased 20 min later, at which time 13% of the [$^{14}$C]protein was ribitol dehydrogenase. In fully induced cultures of the RDH-inducible strain FG5, 4.6% of the newly synthesized protein was initially ribitol dehydrogenase, and the proportion was the same when protein synthesis ceased 20 min after rifampicin was added. These results confirm that *rbt* mRNA is not superstable but that the *rbt* promoter is more efficient than the average promoter.

## 7. DNA Sequencing of the Pentitol Operons

A major effort was made to determine the DNA sequence of the relevant regions of the two pentitol operons. Figure 9 shows the restriction endonuclease map of the *K. aerogenes* DNA insert within the *K. aerogenes* chromosome. A comparison of this map with that of λp *rbt* and λp *rbt dal* (Fig. 3) shows that the *Bst*L and *Hin*R sites are absent in λp *rbt dal*, but all of the other sites indicated are present. Hence, effectively all of the DNA picked up from *E. coli* strain EA during the formation of λp *rbt dal* must have arisen from the *K. aerogenes* DNA that had been originally transduced into it from *K. aerogenes* strain A. This makes λp *rbt dal* an excellent tool for investigating the structure of the two operons.

The sequencing was conducted by cloning appropriate fragments into plasmid pBR322. The origin of the *rbt* mRNA had been localized between *Sal*C and *Hin*dA by hybridizing restriction fragments with pulse-labeled mRNA (Neuberger and Hartley, 1979). Hence the complete *rbtD* sequence

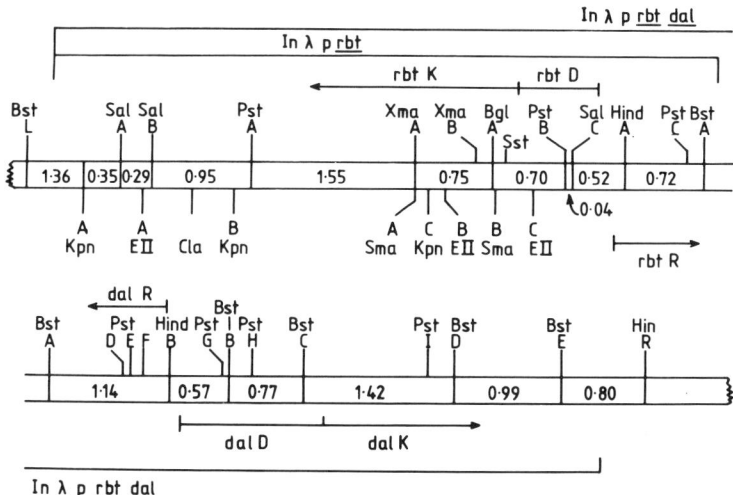

**Figure 9.** Restriction endonuclease map of *K. aerogenes* DNA insert within the *K. aerogenes* chromosome.

was obtained from clones containing the 0.52-kb *Hind*A–*Sal*C fragment or the 0.70-kb *Pst*B–*Bgl*A fragment overlapped by sequences from the 4-kb *Sal*C–*Sal*B fragment (Loviny *et al.*, 1984). The sequence of this latter large fragment is still incomplete, so the sequence encoding D-ribulokinase is still unknown.

The region encoding the constitutive *rbt* repressor (*rbtR$^c$*) was obtained from clones of the *Sal*C–*Bst*A fragment and the *Hind*A–*Bst*A fragment of λp *rbt dal*. Since this region encodes a defective repressor [the original X1 mutation of Wu *et al.* (1968)], the wild-type *rbtR* gene was cloned from the inducible *K. aerogenes* strain FG5 by inserting *Bst*I fragments of *K. aerogenes* DNA into the *Bam* site within the *tet* gene of pBR322 and transforming into *E. coli* K12 strain DC10 (*trpR gal hsdR hsdM$^+$*), which lacks pentitol operons. Selection on minimal ribitol plates containing ampicillin yielded two colonies (pJCW1 and pJCW2) that were both tetracycline-sensitive, as expected. Plasmid pJCW1 was shown to contain a 7.3-kb insert of *K. aerogenes* DNA whose restriction endonuclease map (Fig. 10) clearly corresponds to the fragment *Bst*L–*Bst*A in Fig. 9. The *E. coli* DC10/pJCW1 transformant grew on ribitol but not on xylitol or D-arabitol, implying that the plasmid carries *all* the information necessary for inducible growth on ribitol. In contrast, a derivative of *E. coli* K12 strain EA (Sc25: F$^-$ *supE xyl$^+$ mtl$^-$ rbt$^c$ dal$^c$*), which carries the constitutive *rbt* and *dal* operons of *K. aerogenes*, grew well on all three pentitols. Enzyme activities in cell-free extracts of strain DC10/pJCW1 showed high ribitol dehydrogenase and D-ribulokinase activity after growth on ribitol but low levels on casamino acids. The plasmid appeared unstable in the absence of ribitol or ampicillin as selective agents. Hence pJWC1 encodes a functional *rbt* repressor.

The *dal* repressor region was sequenced from fragments spanning the *SalC–BstC* region cloned into plasmid pBR322. Table III shows the structure of these plasmids and their properties when transformed into *E. coli* SC24 ($F^-$ *SupE mtlA mtlD rbt$^c$ dal$^c$*), a spontaneous derivative of *E. coli* SC20 ($F^-$ *SupE mtlA mtlD rbt$^c$ dal$^+$*). Strain SC24 will grow on mannitol by virtue of its oxidation to D-mannose by the constitutive D-arabitol dehydrogenase, but this growth is abolished by a plasmid that expresses a functional *dal* repressor. Table III also includes the activities of the two dehydrogenases in cell-free extracts of the transformants. It is clear that a functional *dal* repressor is encoded in the *BstA–HindB* fragment.

Neuberger and Hartley (1979) had shown that *dal* mRNA hybridized strongly to the *BstC–BstD* fragment and this was cloned as plasmid pRD252, whereas pRD256 contains *BstB–BstC* and pRD257 contains *BstD–BstE*.

DNA sequencing was mainly by the chain termination method of Sanger *et al.* (1977), using a "universal primer" and the M13 cloning vectors developed by Messing *et al.* (1981) and Messing and Vieira (1982), though the chemical method of Maxam and Gilbert (1977) was occasionally employed. Relevant features of the sequencing strategies are described in relation to the results discussed below.

### 7.1. The Ribitol Dehydrogenase Gene

This is contained within the 1.2-kb *HindA–BglA* fragment. Figure 10 shows the DNA sequence and the coding region. The sequence is identical to that described in the previous chapter, except that Ser–Ser in the DNA sequence replaces Ala–Val at positions 146 and 147 in the protein sequence, and residue 212 is encoded Asn and not Asp as in the protein sequence. The latter difference is almost certainly an error in the protein sequence determination, since this is an Asn–Gly sequence, which deamidates very readily during peptide purifications. However, we have carefully checked the evidence in both the DNA and protein sequencing and conclude that the Ser-146, Ser-147 change from Ala-144, Val-147 in the protein is a genuine difference.

We believe that the protein sequence is that of genuine wild-type *K. aerogenes*, since the protein sequencing was performed on several samples from different RDH-superproducing strains. Moreover, the sequence in the "specificity mutant" RDH-D was the same in this region as RDH-

## Table III
### In Vivo Assays for dal Repressor Activity

| Host strain | Plasmid | Insert | Growth on mannitol | [ArDH] (U/mg) | [RDH] (U/mg) |
|---|---|---|---|---|---|
| SC20 | — | — | — | 0 | 1.7 |
| SC24 | — | — | + | 0.9 | 1.8 |
| SC24 | pBR322 | — | + | 0.9 | 1.5 |
| SC24 | pRD202 | HindA–BstA | + | 1.1 | 1.7 |
| SC24 | pRD211 | SalC–BstA | + | 0.9 | 2.1 |
| SC24 | pRD255 | BstA–BstB | — | 0 | 1.8 |
| SC24 | pRD259 | HindA–HindB | — | 0 | 1.3 |
| SC24 | pRD262 | BstA–HindB | — | 0 | 1.6 |
| SC24 | pRD261 | PstC–PstD | + | 0.9 | 1.9 |

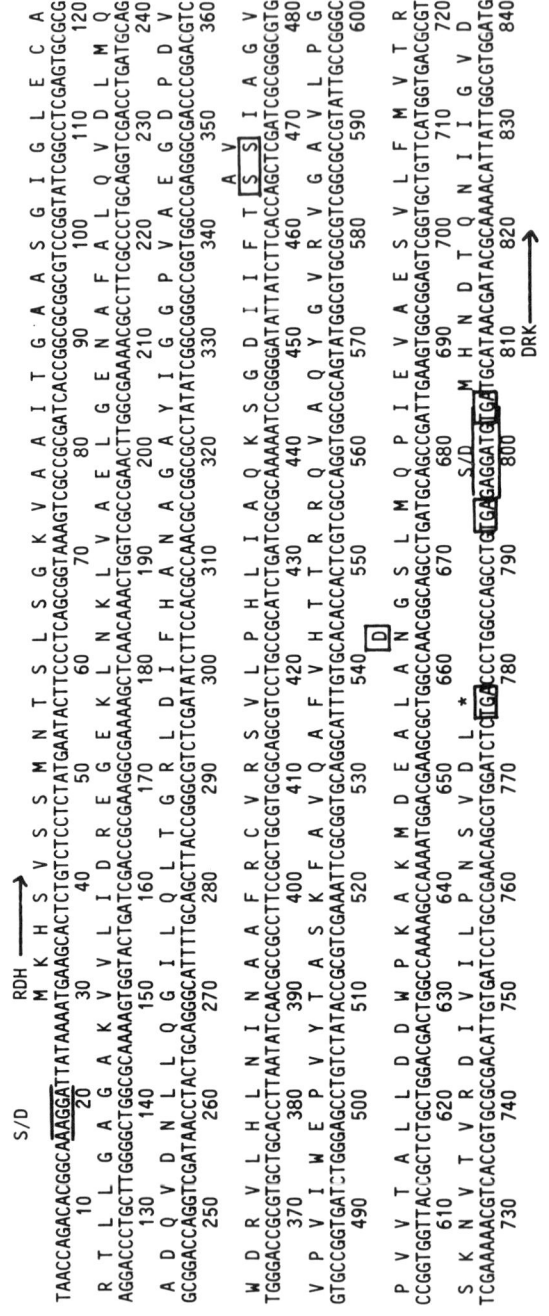

**Figure 10.** The *rbtD* gene and *rbtD–rbtK* intercistronic region. Ribosome binding sites (Shine–Dalgarno, S/D) and stop codons (TGA) are illustrated, together with differences from the observed protein sequence (Dothie *et al.*, 1984), which has Ala–Val at residues 146 and 147 and Asp at residue 212.

A. It seems improbable that the difference arose during the extensive chemostat evolution.

It may, however, have arisen during the extensive genetic manipulations involved in transferring the DNA to M13 phage. The gene was moved first from *K. aerogenes* strain A into *E. coli* K12 by P1 transduction—a rare event. Thereafter it was incorporated into λp *rbt*—another rare event; and eventually into λp *rbt dal* by illegitimate recombination. The sequencing of the 146–147 region was from fragments derived from both λp *rbt* and λp *rbt dal*: both encoded Ser–Ser at this position. Hence, it is unlikely that the sequence difference is an artifact of M13 cloning. There is a strong inference that the mutation must be present in *E. coli* strain EA, since the mechanism envisaged for formation of λp *rbt dal* involves a second integration of the chromosomal *rbt* region into the phage. If so, this could be an entirely adventitious mutation, but it might represent a genuine selective pressure on surface residues in this protein in the different cellular environments in the two organisms. There is some reason to favor the latter argument, since several mutations are required to move from an Ala–Val codon GCXGUX to the observed Ser–Ser codons AGCTCG.

### 7.2. The D-Arabitol Dehydrogenase Gene

Figure 11 shows the protein sequence derived from its DNA sequence, together with the peptide sequences previously determined (Fig. 3). A search for amino acid sequence homologies between this sequence and that of ribitol dehydrogenase revealed no significant homology, and a comparison of potential secondary structures based on the Chou and Fasman (1974) algorithm was equally unilluminating. Hence, these two proteins of similar function have quite different primary and tertiary structures.

### 7.3. The *rbt* Repressor Protein

This sequence was established from fragments cloned either from λp *rbt dal*, giving the sequence of the original constitutive mutation of Wu *et al.* (1968), or from plasmid pJCS1, giving the native wild-type repressor. Figure 12 shows the DNA sequence and the amino acid sequence of the wild-type repressor. The only change in the sequence of the constitutive gene was an insertion of an extra base pair at position 829 in the *Sal*C–*Bst*A fragment, which causes a frameshift resulting in chain termination at residue 133, -Trp-Val-Val-Ala-Thr-Gly-Ala-Thr-Asn-Pro-Asp- in wild type, and -Trp-Gly-Gly-Arg-Asp-Arg-Gly-Asp-Gln-Ser-Gly-COOH in the mutant.

MET ASN ASN GLN PHE THR TRP LEU HIS ILE GLY LEU GLY SER PHE HIS ARG ALA HIS HIS ALA TRP TYR LEU HIS HIS LEU ILE ALA SER
GLY ASP ASN HIS TRP ARG ILE SER ALA GLY ASN ILE ARG ASN ASP ALA GLU GLN VAL VAL GLN ALA LEU ALA GLN GLY ARG TYR
VAL LEU GLU THR VAL SER PRO GLU GLY GLU ARG GLU TYR GLU ILE THR SER ILE GLN LYS LEU LEU PRO TRP GLN ALA GLY LEU GLN
PRO LEU ILE ASN GLU GLY ALA ASN PRO GLN THR LYS VAL ILE ALA PHE THR VAL GLU VAL GLY TYR THR ARG HIS ARG
LEU GLU THR SER ASN PRO ASP LEU GLN ALA ASP LEU GLN GLY CYS LYS THR ILE TYR GLY THR LEU ASP ALA ASP PRO GLU LYS ARG
MET ALA ASP ASN ALA GLY PRO LEU THR LEU LEU ASN CYS ASP ASN VAL ARG HIS ASN GLY ARG PHE HIS ASP GLY MET VAL GLU PHE
LEU GLN LEU THE GLY LYS GLN ALA VAL ILE ASP TRP MET ALA ALA ASN THR THR CYS PRO ASN THR MET VAL ASP ARG VAL THR PRO ARG
PRO ALA ALA ASP LEU PRO ALA ARG ILE LYS ARG GLN ALA GLY ILE ASP ASP LYS VAL GLU MET GLY THR PHE ILE GLN TRP PRO VAL
VAL GLU ASN ASN PHE ARG ASP VAL ARG PRO ASN LEU GLU ALA VAL GLY VAL GLU MET VAL GLU SER ALA SER PRO TYR GLU GLU ALA LYS
ILE ARG ILE LEU ASN ALA SER HIS SER CYS ILE ALA TRP ALA GLY THR LEU ILE GLY GLN GLN TYR ILE HIS GLU SER THR LEU THR ASP
VAL ILE TYR ALA ILE ALA ASP ARG PHE THR ASN PRO TYR ILE GLN ASP THR ASN GLN ARG ALA VAL SER ALA ALA THR THR ALA TRP CYS ARG PRO THR GLY TYR
GLY LEU LYS ARG GLU PHE THR ASN PRO TYR ILE GLN ASP THR ASN GLN ARG ALA VAL ALA ALA ASP GLY PHE SER LYS ILE PRO ALA MET ILE ALA
PRO THR LEU GLN GLU CYS TYR GLN ARG GLY VAL ARG PRO GLU ALA THR ALA LEU PHE PHE VAL PHE MET GLU GLN TRP
HIS LYS GLY THR LEU PRO TYR GLN TYR GLN ASP GLY ILE LEU ASP ALA GLN ALA LEU HIS GLU MET PHE GLU ALA GLN ASP PRO VAL ALA
VAL PHE ALA ARG ASP LYS ALA LEU PHE GLY ASP LEU ALA ASN ASN ALA ASP PHE LEU ALA LEU MET ARG GLU LYS VAL ALA ALA VAL TYR
THR LEU ILE ASN

Elastase peptides are boxed.
MW = 49,800
454 Amino acids

**Figure 11.** Amino acid sequence of *Klebsiella aerogenes* arabitol dehydrogenase, showing elastase peptides.

It is clear that this much truncated peptide (the wild type has 270 residues) could not function as a repressor.

There is a significant homology between the $NH_2$ terminus of the *rbt* repressor protein and that of the *lac* repressor (Fig. 13), a region that is known to be important for DNA binding from mutational evidence (Muller-Hill, 1975) and from tertiary structure evidence of other DNA-binding proteins Cro (Anderson *et al.*, 1981), CRP (McKay and Steitz, 1981), and λ repressor (Pabo and Lewis, 1982). Of the first 70 residues, 41% are identical. Moreover, a secondary structure comparison based on the Chou and Fasman (1974) method reveals that a "helix-turn-helix" structure can be built as proposed by Weber *et al.* (1982) for the *lac* repressor. Also, there is strong homology between the *rbt* repressor and two regions suspected to be DNA-binding domains in 13 other DNA-regulatory proteins. In this first region (residues 10–25), 15 out of 16 residues correspond to the "consensus sequence" of Gicquel-Sanzey and Cossart (1982); in the second region (residues 50–65) the fit is only four out of 16. Hence the *rbt* repressor seems to be a typical repressor.

### 7.4. The *dal* Repressor Protein

Figure 14 shows the DNA and protein sequence of this repressor. The translation start is not clear, since there are two ATGs in the $NH_2$-terminal region. In the absence of $NH_2$-terminal sequences one cannot be sure, but for reasons discussed below we believe that the second Met is the $NH_2$ terminus.

This protein also has sequences that match the assumed DNA-binding regions of a "consensus" repressor, residues 13–28 (fit = $\frac{11}{16}$) and residues 42–57 (fit = $\frac{6}{15}$), but the homology between the *rbt* repressor and the *dal* repressor is not remarkable. It remains possible that tertiary structure comparisons would reveal some similarity, but the evidence so far cannot support a recent evolutionary origin for these two control proteins.

One other feature of these two repressor genes is of interest. About 0.5 kb of noncoding DNA separates them, and this contains the secondary λ attachment site sequenced by Loviny *et al.* (1981). It is not impossible that such a site might have been involved in the genesis of the two operons.

### 7.5. The *dal* promoter

This is very efficient in transcription and translation and was shown to be sensitive to catabolite repression, as for the *rbt* promoter (Knott, 1982). Its sequence shows several interesting features. Figure 15 shows

```
353
CTG GCG TGG AAA GGA TGC TCA ATC GAT GAA GCT ATG TTG CTC TCG CTT CAT CGA TTG AGC
                                                              +1

413
AAG TGG GGA AAT TAA CAG CGA CAT GAT TTT GTG ACG CAA GTT GGA AAA ATG AAG AAG ATA
                                                              M   K   K   I

473
ACG ATT TAC GAC CTG GCG GAA CTT TCC GGC GTT TCG GCG AGC GCG GTG AGC GCT ATC CTC
 T   I   Y   D   L   A   E   L   S   G   V   S   A   S   A   V   S   A   I   L

533
AAC GGT AAC TGG AAA AAG CGC CGC ATC AGC GCC AAG CTT GCG GAA AAG GTG ACG CGG ATT
 N   G   N   W   K   K   R   R   I   S   A   K   L   A   E   K   V   T   R   I

593
GCC GAG GAG CAG GGC TAT GCG ATT AAC CGT CAG GCC AGC ATG CTG CGC AGT AAA AAA TCA
 A   E   E   Q   G   Y   A   I   N   R   Q   A   S   M   L   R   S   K   K   S

653
CAC GTC ATC GGT ATG ATT ATC CCC AAA TAT GAT AAC CGC TAC TTC GGT TCC ATC GCC GAA
 H   V   I   G   M   I   I   P   K   Y   D   N   R   Y   F   G   S   I   A   E

713
CGC TTT GAG GAG ATG GCC CGC GAG CGC GGC CTG CTG CCG ATT ATC ACC TGT ACG CGC CGG
 R   F   E   E   M   A   R   E   R   G   L   L   P   I   I   T   C   T   R   R

773
CGT CCG GAA CTG GAG ATC GAA GCG GTA AAG GCG ATG CTC TCC TGG CAG GTT GAC AGG GTG
 R   P   E   L   E   I   E   A   V   K   A   M   L   S   W   Q   V   D   R   V

833
GTC GCG ACC GGG GCG ACC AAT CCG GAT AAA ATT TCC GCG CTA TGC CAG CAG GCG GGC GTG
 V   A   T   G   A   T   N   P   D   K   I   S   A   L   C   Q   Q   A   G   V

893
CCA ACG GTC AAT CTC GAC CTG CCG GGC TCT CTC TCG CCG TCG GTG ATT TCG GAT AAC TAC
 P   T   V   N   L   D   L   P   G   S   L   S   P   S   V   I   S   D   N   Y

953
GGC GGC GCG AAA GCG TTG ACT CAT AAG ATC CTG GCT AAT AGC GCC CGC CGC CGC GGA GAG
 G   G   A   K   A   L   T   H   K   I   L   A   N   S   A   R   R   R   G   E

1013
CTG GCG CCG CTG ACC TTT ATC GGC GGG CGC AGA GCG ACC ATA ACA CCA GCG AGC GTT TAC
 L   A   P   L   T   F   I   G   G   R   R   A   T   I   T   P   A   S   V   Y

1073
GCG GCT TCC ACG ATG CGC ATC GCG AGC TGG GGC TTA GCG TGC CGC AGG CGA ATA TTC TGG
 A   A   S   T   M   R   I   A   S   W   G   L   A   C   R   R   R   I   F   W

1133
CTC CCG GCT ATT CGA AAG GCC ACG TTG AGG ACT GCC TGC AGG AGC GGT TTG GCA GCG AGA
 L   P   A   I   R   K   A   T   L   R   T   A   C   R   S   G   L   A   A   R

1193
AGA CGC TGT TGC AGG GGA TAT TTG TTA ACT CGA CGA TAT CCC TGG AAG GGG TTG TGC GCT
 R   R   C   C   R   G   Y   L   L   T   R   R   Y   P   W   K   G   L   C   A

1253
GGC TGT CGC AGG TGG GTC TGA CCG GCA GCG AGC AGC CGC CGA TGG GCT GCT TCG ACT GGG
 G   C   R   R   W   V   *

1313
ATC CTT TTG TCT
```

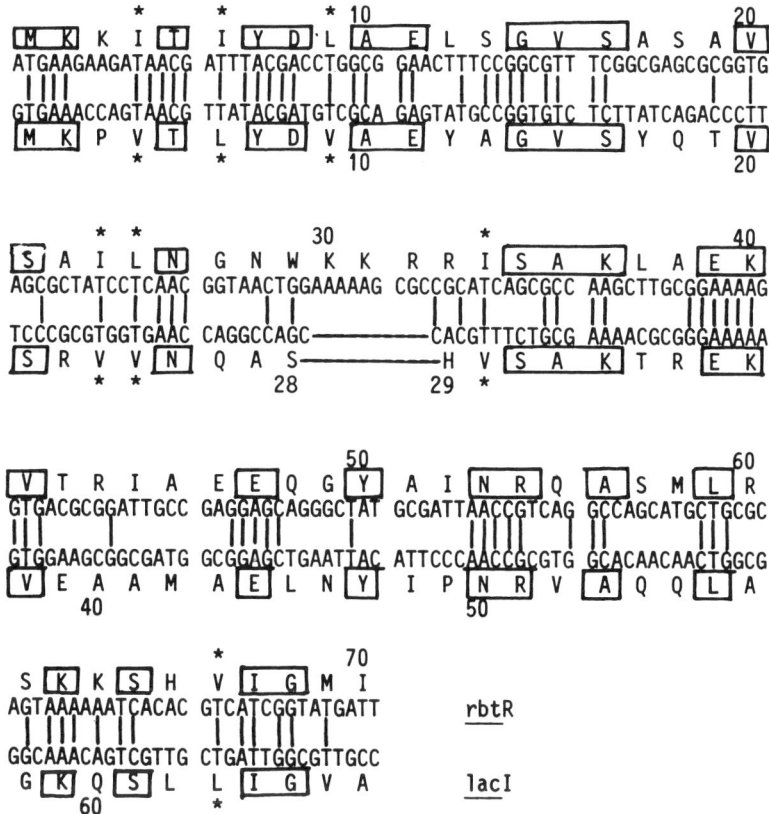

**Figure 13.** A comparison of the NH$_2$-terminal sequences of *rbtR* and *lacI*. Identical amino acids are boxed. Similar amino acids are marked by asterisks. Identical nucleotides are joined by vertical lines.

the origin of mRNA transcription and a good Shine–Dalgarno sequence within a weak hairpin, consistent with efficient translation (Shine and Dalgarno, 1974), which is separated by 7 bp from the ATG of the NH$_2$-terminal methionine. The "10 region" shows an excellent fit to the consensus sequence TATPuPuPTPu of Pribnow (1975), but the "−35 region," also assumed to be involved in binding of RNA polymerase, is less close

**Figure 12.** Nucleotide sequence of the *rbtR* gene and the predicted amino acid sequence. Translation of *rbtR* gene begins at codon ATG (nucleotides 459–461) and stops at codon TGA (nucleotides 1261–1271), which codes for 270 amino acid residues. The connected boxes indicate the 35 and 10 regions; "+1" indicates the mRNA transcription start site. The dyad symmetry is indicated by the underlines, and a string of Ts is dotted; both of them are termination signals for translation.

```
                                                                              3
GCT TGC CAG TTA TGC CAA CCA CGG CGA GAA TGC GGG TAT TCG CCG GGC CGG TTG CGT GCC
   ─────────              ─────────
     -35                    -10       +1

 63
                                            MET SER LYS GLU ASP ASP ILE ARG LEU ASP GLN
CTG GCC AGT AAC TTA GGG CGG GAA ACC ATG AGT AAA GAA GAC GAT ATC CGG TTG GAT CAG
123
LYS VAL ARG ALA ALA TRP MET TYR TYR ILE ALA GLY GLN ASN GLN SER GLU ILE ALA SER
AAG GTG CGT GCC GCA TGG ATG TAC TAC ATC GCC GGC CAG AAT CAG AGC GAG ATT GCC AGC
183
GLN LEU GLY THR SER ARG PRO VAL VAL GLN ARG LEU ILE ALA ALA ALA LYS GLU GLU GLY
CAG CTG GGC ACC TCC AGA CCG GTG GTG CAA CGG CTG ATC GCC GCC GCG AAA GAA GAA GGC
243
ILE VAL SER ILE ASN LEU HIS HIS PRO VAL ALA ASN CYS LEU ASP TYR ALA GLN LEU LEU
ATT GTG TCG ATT AAT CTG CAC CAT CCG GTA GCG AAC TGC CTC GAT TAT GCG CAG TTG CTG
303
GLN GLU LYS TYR GLY LEU ILE GLU CYS ASN VAL VAL PRO ALA PHE SER GLU GLU SER THR
CAG GAA AAA TAC GGC TTG ATC GAG TGC AAT GTG GTC CCC GCC TTT AGC GAA GAA AGC ACC
363
LEU ASP SER VAL SER PHE GLY CYS TYR GLN LEU MET ALA ARG TYR LEU GLN ASP ASP LYS
CTC GAC AGT GTG TCA TTT GGC TGC TAT CAG CTG ATG GCT CGC TAT CTG CAG GAT GAT AAA
423
GLU LYS ILE ILE CYS LEU GLY SER GLY LEU THR LEU LYS LYS ALA LEU GLN ARG ILE ASP
GAG AAA ATC ATC TGT CTG GGC TCG GGC CTG ACC CTG AAA AAA GCG CTG CAG CGC ATC GAT
483
PHE ASP SER LEU ASN THR ARG CYS VAL ALA LEU ILE SER ALA MET ASN ALA ASP GLY GLN
TTT GAC AGC CTG AAT ACC CGC TGC GTG GCG TTG ATC AGC GCC ATG AAC GCC GAC GGG CAG
543
CYS ASN TYR TYR ASP ASP VAL PRO LEU LEU LEU THR ARG LYS ILE LYS ALA LYS TYR TYR
TGC AAT TAT TAC GAT GAC GTG CCC CTG CTG CTG ACC CGC AAA ATT AAG GCC AAG TAC TAT
603
GLN TRP PRO ALA PRO ARG TYR ALA GLN SER ALA ASP GLU TYR GLU MET TRP CYS THR ASN
CAG TGG CCG GCG CCG CGC TAC GCC CAA AGC GCG GAT GAG TAT GAA ATG TGG TGC ACC AAT
663
ARG LEU PHE ARG SER VAL SER GLY VAL ALA ALA ARG ARG THR ***
CGT TTA TTC CGC AGC GTC TCC GGC GTC GCC GCG AGG CGG ACG TGA TTT TTG TCG GCA TTG
```

**Figure 14.** The *dalR* structural gene and promoter region.

to the consensus sequence of Siebenlist *et al.* (1980). However, a good "consensus site" TGTG($N_8$)CTCT for binding catabolite repressor protein is present at 56–71. The rest of the sequence is unusual in that there are three closely repeating sequences in the $-20$ to $-100$ region and a total of six regions that resemble putative promoters (Fig. 15). The concept of RNA polymerase "stacking" during rapid transcription is suggested, as for the *gal* operon (Willmund and Kneser, 1973).

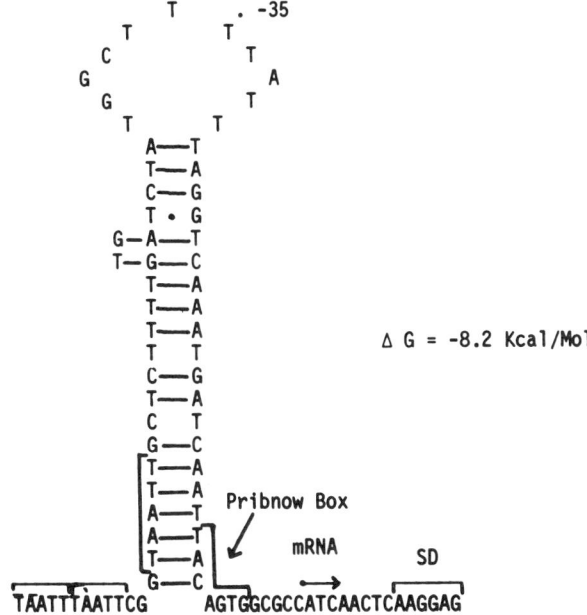

**Figure 15.** A possible secondary structure for the *dalDK* promoter region.

### 7.6. The *rbt* Promoter

There is no sequence similarity to the *dal* promoter and some of the structural features are different (Fig. 16). A good SD sequence 7 bp from the $NH_2$-terminal ATG explains efficient translation, but a very good fit to the consensus "10" and "35" promoter regions lies between $-77$ and $-103$ from the $NH_2$-terminal Met codon. A large element of dyad symmetry between $-33$ and $-57$, which could form a very stable loop, separates the promoter from the SD site. A potential cAMP–CRP binding site $TGAG(N_8)CGCG$ is present at $-49$ to $-64$ and another $TGAG(N_8)CACG$ is present at $-77$ to $-93$. As in the *dal* promoter, there are two other putative promoters, at $-28$ to $-58$ and $-65$ to $-89$, so the possibility of "polymerase stacking" is again apparent.

### 7.7. The *rbt* Repressor Promoter

Can we learn anything about the efficiency of protein expression by comparing the efficient *rbt* and *dal* operon promoters with the inefficient *rbtR* and *dalR* promoters? The sequence preceding the $NH_2$-terminal co-

**Figure 16.** A possible secondary structure for the *rbtDK* promoter region.

don is ACGC*AA*GT*TGG*AAAA, offering a reasonable Shine–Dalgarno sequence. However, the best "−35 to −10" region for polymerase binding lies at residues −70 to −103 and this is contained within an extremely stable hairpin (−47 to −94), as shown in Fig. 17. Presumably this structure must "melt" to allow efficient transcription.

There are only 254 nucleotides between the ATG of *rbtD* on one strand and that of *rbtB* on the other. Since the effective promoter regions seem to extend about 100 bp into this region, it is likely that transcription of one gene would interfere with that of the other. This makes good biological sense.

### 7.8. The *dal* Repressor Promoter

The sequence preceding the coding sequence for the *dal* repressor protein contains signals similar to those recognized as important in transcription and translation of other proteins (Fig. 18). There is a good "consensus sequence" for the "−35" and "−10" regions involved in RNA polymerase binding and a respectable Shine–Dalgarno sequence for translation of the mRNA. From the DNA sequence alone it is difficult to see why this protein could not be expressed at high levels. The *trp* repressor is regulated autogenously by binding of the repressor protein itself to an operator region within its own promoter region (Gunsalus and Yanofsky, 1980). No such operator region can be recognized within the *dalR* promoter, so it is unlikely that there is analogous self-regulation.

However, it is possible that the levels of this repressor protein are controlled by the level of D-arabitol, which is the inducer of the *dalD–dalK* operon. As in the ribitol operon, there is only a short length of DNA separating the *dalR* and *dalD* genes, which are transcribed from opposite strands; in fact, position −35 of the *dalR* promoter is equivalent to residue −100 of the *dalDK* promoter. If there is "polymerase stacking" for the latter, as postulated, repressor synthesis would be switched off when the enzymes of the *dalDK* operon are being actively synthesized.

## 8. Translation of the Two Kinases

The intercistronic region between *rbtD* and *rbtK* is shown in Fig. 10 and that between *dalD* and *dalK* in Fig. 19. In both cases there are three stop codons terminating translation in each reading frame that precedes the $NH_2$-terminal ATG, which is very close to the ribosome-binding sequence. Although there are no precise figures, it is clear that the kinase genes are translated less well than their respective dehydrogenases. The

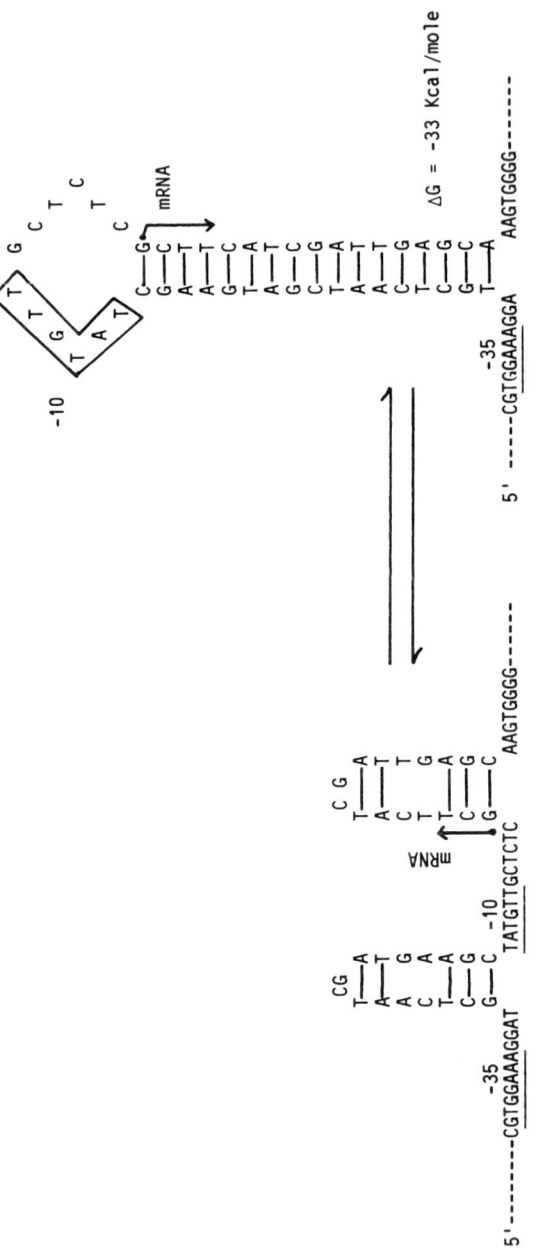

**Figure 17.** A possible secondary structure for the *rbtR* promoter region.

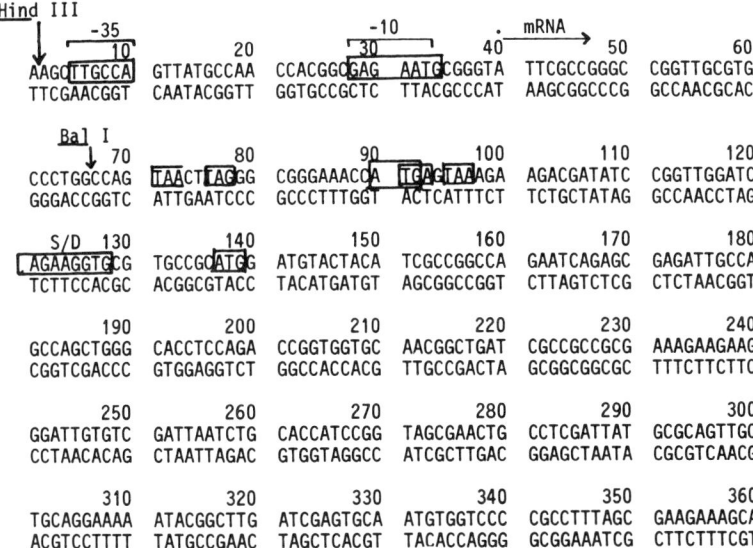

**Figure 18.** The DNA sequence of the *dalR* promoter region. Potential −35 and −10 sequences are indicated. Translational initiation and termination codons are boxed, together with the most obvious Shine–Dalgarno (S/D) sequence.

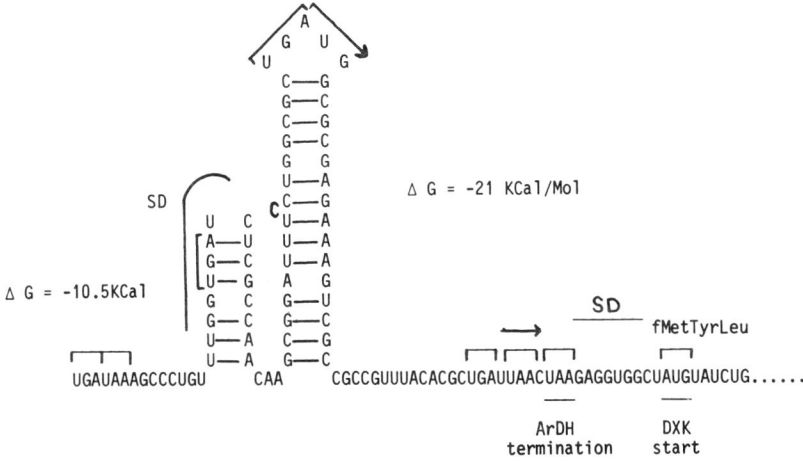

**Figure 19.** Secondary structures for the mRNA around the *dalD–dalK* intercistronic region. All translational termination signals are enclosed in brackets. Arrows indicate the beginning and end of a 12-amino acid peptide that could be synthesized following an initiation event at the first Shine–Dalgarno sequence.

ribosomes engaged in dehydrogenase translation may pause at the stop signals, and the reinitiation sequence for the kinases is closer to the initial ATG than is considered optimal for good translation.

This fact alone may account for weaker expression, but there is some evidence (discussed above) for premature transcription termination following the ribitol dehydrogenase gene. Moreover, Fig. 19 shows potential base-paired loops within the *dal* mRNA that might relate to transcription termination. The first loop has a potential Shine–Dalgarno sequence, which could lead to a 12-residue peptide, if translated. The second loop bears a UGA stop codon overlapping an AUG triplet. Considering their position, the similarity of these structures to known transcription terminators and RNA processing sites may be significant.

## 9. Invert Repeat Sequences Enclose the Two Operons

Since the DNA sequences of the two kinase genes are not yet complete, it is too early to say whether the enzymes arose by invert gene duplication followed by divergence to different substrate specificities. Their subunit size and structure allows this possibility, unlike the case of the two dehydrogenases. However, there is no $NH_2$-terminal homology, and Neuberger and Hartley (1979) were unable to detect hybridization even under relatively nonstringent conditions between the *BstC–BstD* fragment, which contains the D-xylulokinase gene, and the *SalB–SalC* fragment, which contains the D-ribulokinase gene (see Fig. 9). What did surprisingly emerge from these hybridization studies was that there was strong homology between sequences in the *SalB–SalC* fragment and in the *BstD–BstE* fragment, which follows the D-xylulokinase gene. This suggested, but did not prove, that the two operons were sandwiched between invert repeats of at least 100 bp with strong DNA homology.

Hence the 0.99-kb *BstD–BstE* fragment was isolated from plasmid pRD251 (*BstC–BstE* insert) and its DNA sequence was determined. Figure 20 shows a portion of this sequence extending in the 3′–5′ direction from the *Bst*E site compared with that of another fragment 3′ to the *Kpn*B site at the opposite end of the *rbt* operon. There is 85% identity between the 364 residues that can be thus compared. It is obvious that the homology probably extends both beyond the *Bst*E site and toward the D-xylulokinase gene, but these sequences are not yet available. This is particularly frustrating, since there are large, open reading frames within each of these fragments that could extend in either direction. It is possible that the repeats might encode other proteins, possibly pentitol permeases. If so, there would be obvious functional significance for the strong conservation

## THE STRUCTURE AND CONTROL OF THE PENTITOL OPERONS

```
     0         10         20         30         40         50
     GGATCCCCC ACAGCACGTT AGAGAAAATG GTGGTGAAGA AGAAGACTGC CCATACCTGG
     ********  *  *       *****      **  **  **  **  *******  ****  *  **  **********
     ACAATCCCCC AGAAAACGTT CGAAAAGATA GTAGTGAAGA AGAATGCCGC CCATACCTGC
143            153        163        173        183        193
 60             70         80         90        100        110
     AGCCATTCTG AAGTGCTGAA GCCCAGCTCA TCGACAAACA TCATCGGCAT AATCACCGCA
     ********  *  *  ***  *  **       *********  **  *******  **********   *********
     AGCCATTCCG AGGTGGTAAA TCCCAGCTCA TCAACAAACA TCATCGGCAT GATCACCGCA
203            213        223        233        243        253
120            130        140        150        160        170
     AAGCCGAACA GCGAGAGGGT ATTGATGATC CTCACCATGC TCGACAGCAG AATATTGCGG
     **  **  ****  ****  *****       *****  **  *  ***  ****  ****  *  ***  ***       *****
     AAACCAAACA GCGACAGGGT GTTGATAATG CGCACAATGC TCGAGAACAG AATGCTGCGG
263            273        283        293        303        313
180            190        200        210        220        230
     TTGGTATAGA GCAGCGTCGC GCGCGTCCCA GCTCGGAAAA CTTCTCACGG GTGGTGAGAT
     **********  **********  *****  ****  *  ****  **   **  *****   *****  **  *
     TTGGTATAGA GCAGCGTCGC GCGCGCCCCA GTTCGGCGAA TTTTTCACGA GTGGTCAGGT
323            333        343        353        363        373
240            250        260        270        280        290
     TCTGCATATG CTGCGGCGTT TGAATATGGC GCAGGGAAAC CAGGGCAATC ACGCCCCCGG
     **********   ********  *       ****  *  ***   **  **  ***  *****     *  ***  ****
     TCTGCATATG GCGCGGCGTT TCGGTATGAC GCATCGACAC CAGCGCAATA ATGCCGCCGG
383            393        403        413        423        433
300            310        320        330        340        350
     TAAGGCAGAA GGCCAGCGCC AGCCACAGGG TGCCCATTTC GCCAATGTGA GGAATGGTAA
     **   ***  **  *    *******   **********  **********  ********   *********   **********
     TA--GCAAAA GAGCAGCGCC AGCCACAGGG TGCCCATTTC GCCAATGTGG GGAATGGTAA
443            453        463        473        483        493
360            370        380        390        400        410
     AGCT     ---- extends for 506 towards Bst D
     ****
     AGCT
     503
```

Figure 20. The inverted repeat sequence. The top strand is a part of the *BstE–BstD* fragment; the bottom strand is a part of the *SalB–SalC* fragment.

of DNA sequence within these invert repeats. In any case, it is clear that the two pentitol operons form a structure resembling the remnants of a "metabolic transposon," though there is no evidence that the genes do actually jump around today.

## 10. Structure of an Experimentally Evolved Gene Duplication

With this detailed knowledge of the two pentitol operons we can now return to one of the questions raised in the previous chapter: what are the molecular events leading to duplication of the ribitol dehydrogenase gene during continuous culture on xylitol?

Neuberger and Hartley (1981) approached this problem by isolating chromosomal DNA from several *K. aerogenes* strains that superproduce this enzyme, digesting them with different restriction endonucleases, and probing the resulting agarose gels with radioactive nick-translated λp *rbt* DNA (Southern, 1975). If the whole of the pentitol operon region were duplicated, this would be revealed only by enzymes that cut outside this region (e.g., *Eco*RI) and the size of the resulting fragments might be too large to be adequately resolved on gels. However, a partial duplication of the *rbt* operon would be revealed by the appearance of an additional hybridizable band containing the novel duplication joint, as shown in Fig. 21. Moreover, the theory predicts that this fragment would have the same size for any enzyme that cuts only once within the region duplicated.

Among seven RDH-superproducing strains screened in this way by Neuberger and Hartley (1979), one, strain A3, showed the predicted pattern, and Fig. 21 shows the restriction endonuclease mapping, which locates the novel joint within a region inside each of the kinase genes. An *Eco*RI digest of DNA from strain A3 gave a single hybridizable fragment of about 45 kb, compared with a 20-kb band with wild-type DNA from strain A'. This suggests that there are about five copies of the region shown in Fig. 21 in tandem array.

The structure of A3 predicts that it should contain elevated levels of constitutive ribitol dehydrogenase and inducible D-arabitol dehydrogenase, but wild-type levels of D-ribulokinase and D-xylulokinase. This was observed with strain A3 (RDH level is 6.1 times wild type; ArDH is 6.9 times; DRK is <1.05 times), whereas with another RDH-superproducing strain, A211, enzyme levels of both RDH and DRK were elevated. Moreover, strain A3 readily segregated revertants with wild-type levels of all three enzymes, showing that the duplication is indeed reversible, as would be predicted.

The DNA sequence of the duplication joint in strain A3 was determined by cloning a *Hin*dIII digest of its DNA into pBR322 and transforming into *E. coli* strain SK1592 ($F^-$ *gal thi hsdR endA sbc B15 tonA*) (David *et al.*, 1984). The *Hin*B–*Hin*A "joint" fragment should express D-arabitol dehydrogenase, which would allow growth on D-arabitol, since the D-xylulose so produced will be converted to D-xylose by constitutive levels of D-xylose isomerase, which will then fully induce the D-xylulokinase of the *xyl* operon. Five colonies grew on D-arabitol, but four proved to be *K. aerogenes* contaminants, since they also grew on inositol (which is not a substrate for *E. coli* K12). The remaining transformant proved to harbor a plasmid pJD24 containing the desired 3.9-kb *Hin*B–*Hin*A "joint" fragment together with an adventitiously cloned 3.8-kb fragment from elsewhere in the *K. aerogenes* chromosome. Subcloning a *Bst*I/*Hin*dIII digest of this plasmid into the *tet* gene in pAT153 yielded pJD243 containing the 2.6-kb *Bst*C–*Hin*A fragment with the phenotype Rbt$^+$ Xyl$^+$ Dal$^-$Ap$^R$ Tc$^S$ (Fig. 22). It is clear that *E. coli* K12 requires only the dehydrogenase genes for growth on ribitol, xylitol, or D-arabitol and that host permeases and kinases can be recruited to establish a new metabolic pathway.

Figure 23 compares the DNA sequence in the "joint" region from strain A3 with the sequences in comparable regions of the wild-type *dalK* and *rbtK* genes. Sequences to the right and left of this region were identical, as expected. It is clear that the gene duplication in strain A3 arises from a 7-bp region of precise identity within the *dalK* and *rbtK* coding regions, which are separated by 5.8 kb within the *K. aerogenes* genome.

A repeat of this size within such a distance is not statistically unusual, so the intervening 5.8-kb sequence (which is almost entirely known) was screened for other identical GCCTGCC sequences. None were found, but it is obvious that homologies of this sort must be common within bacterial genomes. Indeed, a comparable short repeat of a CACCAC sequence was found at a duplication joint for the *ampC* gene (Edlund and Normark, 1981), which contains a 3-bp repeat like that in the above *rbtD* duplication, and in 7 out of 12 spontaneous deletions in the *lacI* gene of *E. coli* similar 3-bp repeats were found (Farabaugh *et al.*, 1978). It is too early to speculate about the mechanisms involved, but evidence is abundant that gene duplications of this sort could arise and become established if the selective advantage were sufficiently high.

## 11. Evolutionary Lessons from the Pentitol Operons

If we now return to some of the questions that were posed at the beginning of the previous chapter, we see that the research described has answered some of these and left others open.

There is no doubt that gene duplications can arise spontaneously and quite frequently in microbial systems and that these are quite readily selected by selective pressures akin to those within a natural ecological niche. Provided that the protein so superproduced remains inducible, the duplications could survive for a reasonable time. Stable fixation into the chromosome would demand translocation to another site or to another strand to create an invert duplication such as that observed flanking the pentitol operons. Mechanisms for translocating these genes can be readily inferred from the transposonlike nature of the invert repeat sequences,

CHAPTER 3

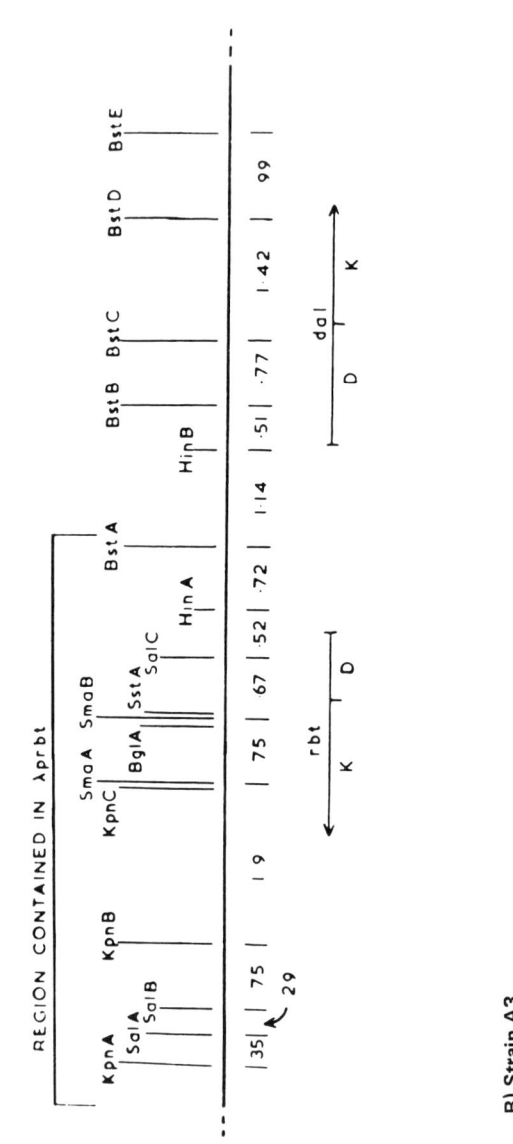

A) Strain A′

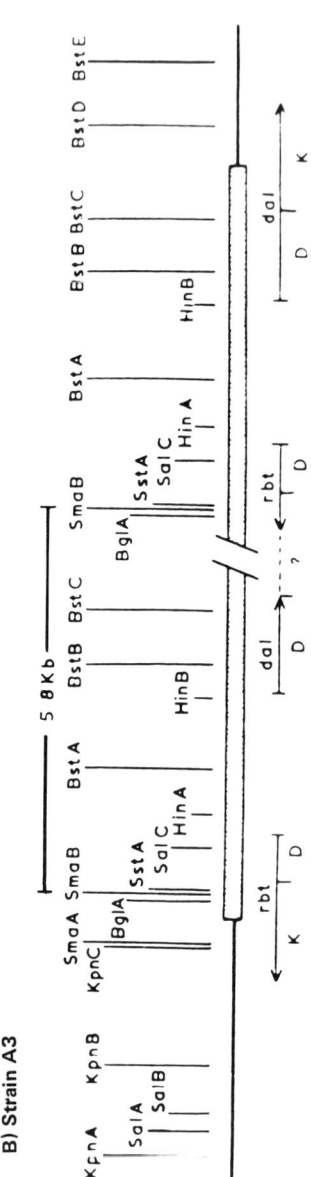

B) Strain A3

**Figure 21.** Physical map of the pentitol operons on the chromosomes of strains A' and A3. (A) Strain A' DNA. The locations of the target sites for endonucleases *Bst*I, *Hin*dIII, *Sal*I, and *Sma*I were determined by restriction mapping of phages λp *rbt* and λp *rbt dal* as described in Neuberger and Hartley (1979). The positions of *Bgl*II, *Kpn*I, and *Sst*I sites were elucidated by comparison of the fragmentation patterns of λp *rbt*, λp *rbt dal*, and λ627 (the progenitor of λp *rbt*) DNAs by agarose gel electrophoresis of appropriate single and double digests. The sizes of the relevant fragments were determined using an *Eco*RI + *Bst*I digest of λc I857 S7 DNA or a *Hin*fI or *Hae*III digest of pBR322 to generate marker fragments of known sizes (Haggerty and Schleif, 1976; Sutcliffe, 1978). Distances between sites are given in kb. The whole region contains no target sites for *Eco*RI. The locations of the genes encoding the pentitol dehydrogenase (*D*) and the pentulokinase (*K*) of each operon are indicated, as is the region of DNA included in the genome of phage λp *rbt*. (B) Strain A3 DNA. The structure of the primary duplication is presented, with the boxed region designating the duplicated DNA and the slanted double lines indicating the novel DNA joint. Although strain A3 contains several copies of the primary duplication, only the basic duplication structure is indicated, for reasons of clarity.

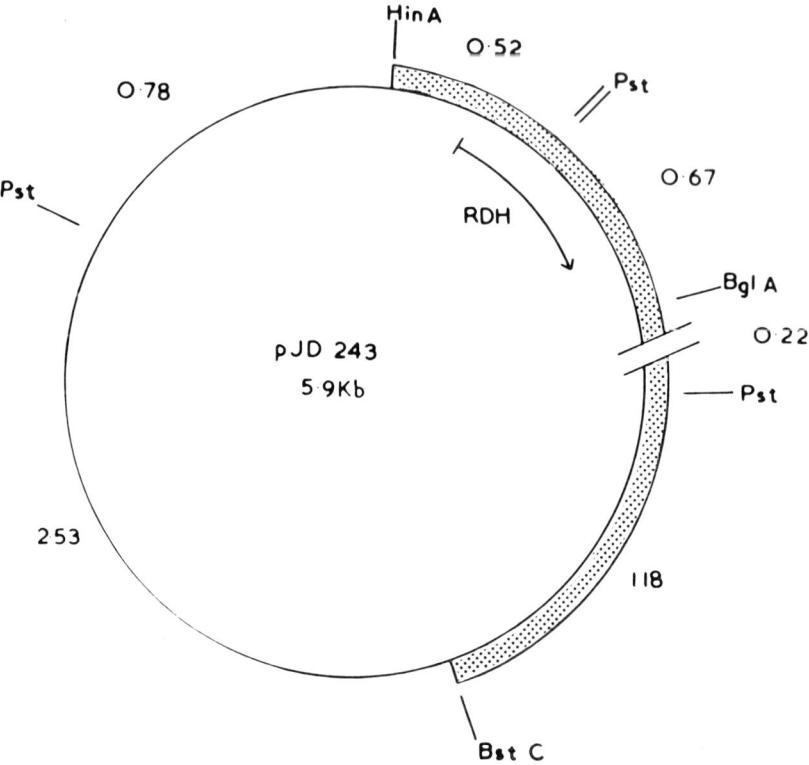

**Figure 22.** Restriction map of plasmid pJD243. The plasmid contains the *Hin*A to *Bst*C restriction fragment from pJD24 (double line) cloned between the *Hin*dIII and *Bst* sites of pAT153 (single line). The sizes shown are in kb pairs. The start point for *rbt*D transcription is indicated.

```
        140        150        160        170        180        190
a    GCGGCGCGCT GGGCGCGGCG CGGCTGGCCT GCCTTGCGGC GGGCAAACCC ATTGCCGCGG
     ********** ********** ******ˌ**** ***  *    *    *    *  *  *
A3   GCGGCGCGCT GGGCGCGGCG CGGCTGGCCT GCCAGATCTC ACGGCTCGAC TGCTCGACCT
     ** **   *    *   * *      ****  ***ˌ******** ********** **********
b    CCGCCGTTTT AATGCAGTAG CACACCGCCT GCCAGATCTC ACGGCTCGAC TGCTCGACCT
         10         20         30         40         50         60
```

**Figure 23.** Structure of the novel joint of the duplication in the chromosome of strain A3. The sequences are (a) the parental sequence in the DXK gene of strain A'; (A3) the sequence of the novel joint in the duplicated region of strain A3; (b) the parental sequence in the DRK gene of strain A'. The direction of the three sequences is *Bst*C → *Bgl*A (Fig. 21). Asterisks indicate homologous nucleotides and the box contains the 7-bp direct repeat present in the three sequences and used for alignment of the three sequences.

or from the phage-mediated transduction of the genes from *K. aerogenes* into *E. coli* and thereafter into phage λ by secondary site attachment.

It is also demonstrated that mutations altering enzyme specificity can arise in response to simple selective pressures. Indeed, the surprise is that so many different mutations can make an improved xylitol dehydrogenase from ribitol dehydrogenase, and that these do not impair the original enzyme activity. There is little fuel for the view that "silent genes" are a prerequisite for evolution of new enzyme specificities.

There is one case—the change of Ala–Val to Ser–Ser at positions 146 and 147 in the ribitol dehydrogenase gene—that could be seized on by advocates of neutral mutations, since it does not change observable properties of the enzyme, or by selectionists, since it occurred following a transfer of the gene to *E. coli* rather than in extensive chemostat growth of *K. aerogenes* populations.

However, the detailed structure of the pentitol operons speaks more for selection for function than to a history of evolutionary accidents. The repressor and dehydrogenase genes appear to have been recruited to their present positions rather than to have diverged from a primeval invert gene duplication. The intriguing pattern of transcription from opposite strands has implications for control of the operons that would offer selective advantage. The only area of extensive homology recognized so far is within the invert repeat sequences, and here the surprise is that the homology is so extensive and so close. Admittedly, we have not yet recognized any function for these regions that would explain this conservation of sequence, but I believe that they will prove to have an important biological role.

In summary, therefore, this study advocates humility in interpreting the structures of proteins and genes in terms of evolutionary history. We still understand little of the mechanisms controlling gene expression and protein function. The more we learn, the more exquisite these appear to be. Darwin would not have been surprised.

ACKNOWLEDGMENTS. This chapter is dedicated to my students, who pushed me from protein chemistry into genetic engineering: Michael S. Neuberger, who made and exploited λp *rbt* and λp *rbt dal*; Rosemary A. Patterson, who helped him to isolate the proteins; Mohammad (Bahman) Bahramian, who purified the mRNA; Thérèse Loviny and Timothy J. Knott, who developed the DNA sequencing and determined most of the structure; Peter J. David, who elucidated the gene duplication; and Jing-Cai Wu, who solved the *rbt* repressor. They appeared to enjoy this investigation of obscure phenomena.

# References

Anderson, W. F., Ohlendorf, D. H., Takeda, Y., and Matthews, B. W., 1981, Structure of the *cro* repressor from bacteriophage λ and its interaction with DNA, *Nature* **290**:754–758.
Bahramian, M. B., and Hartley, B. S., 1980, A switch from translational control to transcriptional control of protein synthesis in mid-exponential growth phase of bacterial cultures, *Eur. J. Biochem.* **110**:507–519.
Bahramian, M. B., and Hartley, B. S., 1982, Ribitol dehydrogenase mRNA from an enzyme superproducer strain of *K. aerogenes*, *Eur. J. Biochem.* **122**:271–279.
Bahramian, M. B., Loviny, T., and Hartley, B. S., 1982, Chemical and biological stability *in vivo* of ribitol dehydrogenase mRNA from an enzyme superproducer strain of *K. aerogenes*, *Eur. J. Biochem.* **122**:279–282.
Bailey, J. M., and Davidson, N., 1976, Methylmercury as a reversible denaturing agent for agarose gel electrophoresis, *Anal. Biochem.* **70**:75–85.
Bidwell, K., and Landy, A., 1979, Structural features of λ site-specific recombination at a secondary *att* site in gal/T, *Cell* **16**:397–406.
Bolivar, F., Rodriguez, R. L., Betlach, M. C., and Boyer, H. W., 1977, Construction and characterization of new cloning vehicles I. Ampicillin-resistant derivatives of the plasmid pMB9, *Gene* **2**:75–93.
Burleigh, B. D., Rigby, P. W. J., and Hartley, B. S., 1974, A comparison of wild-type and mutant ribitol dehydrogenases from *K. aerogenes*, *Biochem. J.* **143**:341–352.
Campbell, A. M., 1962, Episomes, *Adv. Genet.* **11**:101–145.
Catterall, J. F., and Welker, N. E., 1977, Isolation and properties of a thermostable restriction endonuclease (Endo R. Bst 1503), *J. Bacteriol.* **129**:1110–1120.
Charnetzky, W. T., and Mortlock, R. P., 1974, Close genetic linkage of the determinants of the ribitol and D-arabitol catabolic pathways in *K. aerogenes*, *J. Bacteriol.* **119**:176–182.
Chou, P. Y., and Fasman, G. D., 1974, Prediction of protein conformation, *Biochemistry* **13**:222–245.
Christie, G. E., and Platt, T., 1979, A secondary attachment site for bacteriophage λ in *trp* C of *E. coli*, *Cell* **16**:407–413.
Cleland, W. W., 1963, The kinetics of enzyme-catalysed reactions with two or more substrates or products: I. Nomenclature and rate equations, *Biochim. Biophys. Acta* **67**:104–137.
Cowie, D. B., Spiegelman, S., Roberts, R. B., and Duerksen, J. D., 1961, Ribosome-bound β-galactosidase, *Proc. Natl. Acad. Sci. USA* **47**:114–122.
Csordas-Toth, E., Boros, I., and Venetianer, P., 1979, Nucleotide sequence of a secondary attachment site for bacteriophage lambda on the *E. coli* chromosome, *Nucleic Acids Res.* **7**:1335–1341.
David, P. J., Loviny, T., and Hartley, B. S., 1984, Molecular structure of a gene duplication in *K. aerogenes*, (in preparation).
Dothie, J. M., Giglio, J. R., Moore, C. H., Taylor, S. S., and Hartley, B. S., 1984, Ribitol dehydrogenase from *K. aerogenese*: Sequences and properties of wild-type and mutant strains, *Biochem. J.* (submitted).
Edlund, T., and Normark, S., 1981, Recombination between short DNA homologies causes tandem duplication, *Nature* **292**:269–271.
Farabaugh, P. J., Schmeissner, U., Hofer, M., and Miller, J. H., 1978, Genetic studies of the *lac* repressor. VII: On the molecular nature of spontaneous hotspots in the *lacI* gene of *Escherichia coli*, *J. Mol. Biol.* **126**:847–863.

Gicquel-Sanzey, B., and Cossart, P., 1982, Homologies between different procaryotic DNA-binding regulatory proteins and between their sites of action, *EMBO J.* **1**:591–595.

Godson, G. N., 1976, A technique of rapid lysis for the preparation of *E. coli* polyribosomes, *Meth. Enzymol.* **12A**:503–516.

Godson, G. N., and Sinsheimer, R. L., 1967, Use of brij lysis as a general method to prepare polyribosomes from *Escherichia coli, Biochim. Biophys. Acta* **149**:489–495.

Guerrini, F., 1969, On the asymmetry of λ integration sites, *J. Mol. Biol.* **46**:523–542.

Gunsalus, R. P., and Yanofsky, C., 1980, Nucleotide sequence and expression of *Escherichia coli trp R*, the structural gene for the *trp* aporepressor, *Proc. Natl. Acad. Sci. USA* **77**:7117–7121.

Haggerty, D. M., and Schleif, R. F., 1976, Location in bacteriophage lambda DNA of cleavage sites of the site-specific endonuclease from *Bacillus amyloliquefaciens H, J. Virol.* **18**:659–663.

Hamlin, J., and Zabin, I., 1972, β-Galactosidase: Immunological activity of ribosome-bound, growing, polypeptide chains, *Proc. Natl. Acad. Sci. USA* **69**:412–416.

Kamp, D., Kahmann, R., Zipser, D., and Roberts, R. J., 1977, Mapping of restriction sites in the attachment site region of bacteriophage lambda, *Mol. Gen. Genet.* **154**:231–248.

Knott, T. J., 1982, The D-Arabitol Operon of *Klebsiella aerogenes*, Ph.D. Thesis, University of London.

Koch, A. L., 1971, The adaptive responses of *Escherichia coli* to a feast and famine existence, *Adv. Microb. Physiol.* **6**:147–217.

Landy, A., and Ross, W., 1977, Viral integration and excision: Structure of the lambda *att* sites, *Science* **197**:1147–1160.

Lin, E. C. C., 1961, An inducible D-arabitol dehydrogenase from *Aerobacter aerogenes, J. Biol. Chem.* **236**:31–36.

Lindahl, L., Yamamoto, M., Nomura, M., Kirschbaum, J. B., Allet, B., and Rochaix, J.-D., 1977, Mapping of a cluster of genes for components of the transcriptional and translational machineries of *Escherichia coli, J. Mol. Biol.* **109**:23–47.

Loviny, T., Neuberger, M. S., and Hartley, B. S., 1981, Sequence of a secondary phage λ attachment site located between the pentitol operons of *K. aerogenes, Biochem. J.* **193**:631–637.

Loviny, T., Norton, P. M., and Hartley, B. S., 1984, Ribitol dehydrogenase of *Klebsiella aerogenes*: Sequence of the structural gene, *Biochem. J.* (submitted).

McKay, D. B., and Steitz, T. A., 1981, Structure of catabolite gene activator protein at 2.9 Å resolution suggests binding to left-handed B-DNA, *Nature* **290**:744–749.

Maxam, A., and Gilbert, W., 1977, A new method for sequencing DNA, *Proc. Natl. Acad. Sci. USA* **74**:560–564.

Messing, J., and Vieira, J., 1982, A new pair of M13 vectors for selecting either DNA strand of double digest restriction fragments, *Gene* **19**:269–276.

Messing, J., Crea, R., and Seeburg, P. H., 1981, A system for shotgun DNA sequencing, *Nucleic Acids Res.* **9**:309–321.

Miller, H. I., and Friedman, D. I., 1967, in: *DNA Insertion Elements, Plasmids and Episomes* (A. I. Bukhari, J. A. Shapiro, and S. L. Adhya, eds.), Cold Spring Harbor Laboratory, New York, pp. 349–356.

Morris, H. R., Williams, D. H., Midwinter, G. G., and Hartley, B. S., 1974, A mass-spectrometric sequence study of the enzyme ribitol dehydrogenase from *Klebsiella aerogenes, Biochem. J.* **141**:701–713.

Muller-Hill, B., 1975, *Lac* repressor and *lac* operator, *Prog. Biophys. Mol. Biol.* **30**:227–252.

Neuberger, M. S., and Hartley, B. S., 1979, Investigations into the *K. aerogenes* pentitol operons using specialised transducing phages $\lambda_{p\ rbi}$ and $\lambda_{p\ rbi\ dal}$, *J. Mol. Biol.* **132**:435–470.
Neuberger, M. S., and Hartley, B. S., 1981, Structure of an experimentally evolved gene duplication in a mutant of *Klebsiella aerogenes*, *J. Gen. Microbiol.* **122**:181–191.
Neuberger, M. S., Patterson, R. A., and Hartley, B. S., 1979, Purification and properties of *K. aerogenes* D-arabitol dehydrogenase, *Biochem. J.* **183**:31–42.
Neuberger, M. S., Hartley, B. S., and Walker, J. E., 1981, Purification and properties of D-ribulokinase and D-xylulokinase from *K. aerogenes*, *Biochem. J.* **193**:513–524.
Pabo, C. O., and Lewis, M., 1982, The operator-binding domain of λ repressor: Structure and DNA recognition, *Nature* **298**:443–447.
Parkinson, J. S., and Huskey, R. J., 1971, Deletion mutants of bacteriophage lambda I: Isolation and initial characterization, *J. Mol. Biol.* **56**:369–384.
Pribnow, D., 1975, Nucleotide sequence of an RNA polymerase binding site at an early T7 promoter, *Proc. Natl. Acad. Sci. USA* **72**:784–788.
Reiner, A. M., 1975, Genes for ribitol and D-arabitol metabolism in *E. coli*: Their loci in C strains and absence in K-12 and B strains, *J. Bacteriol.* **123**:530–536.
Robinson, L. H., and Landy, A., 1977, *Hin*dII, *Hin*dIII, and *Hpa*I restriction fragment maps of bacteriophage λ DNA, *Gene* **2**:1–31.
Sanger, F., Nicklen, S., and Coulson, A. R., 1977, DNA sequencing with chain terminating inhibitors, *Proc. Natl. Acad. Sci. USA* **74**:5463–5467.
Scangos, G. A., and Reiner, A. M., 1978a, Ribitol and D-arabitol metabolism in *E. coli*, *J. Bacteriol.* **134**:492–500.
Scangos, G. A., and Reiner, A. M., 1978b, Acquisition of ability to utilise xylitol: Disadvantages of a constitutive catabolic pathway in *E. coli*, *J. Bacteriol.* **134**:501–505.
Schechter, I., 1974, Use of antibodies for the isolation of biologically pure messenger ribonucleic acid from fully functional eucaryotic cells, *Biochemistry* **13**:1875–1885.
Shimada, K., Weisberg, R. A., and Gottesman, M. E., 1972, Prophage lambda at unusual chromosomal locations 1: Location of the secondary attachment sites and the properties of the lysogens, *J. Mol. Biol.* **63**:483–503.
Shimada, K., Weisberg, R. A., and Gottesman, M. E., 1973, Prophage lambda at unusual chromosomal locations II. Mutations induced by bacteriophage lambda in *Escherichia coli* K 12, *J. Mol. Biol.* **80**:297–314.
Shine, J., and Dalgarno, L., 1974, The 3'-terminal sequence of *E. coli* 16S ribosomal RNA: Complementarity to nonsense triplets and ribosome binding sites, *Proc. Natl. Acad. Sci. USA* **71**:1342–1346.
Siebenlist, U., Simpson, R. B., and Gilbert, W., 1980, *E. coli* RNA polymerase interacts homologously with two different promoters, *Cell* **20**:269–281.
Southern, E. M., 1975, Detection of specific sequences among DNA fragments separated by gel electrophoresis, *J. Mol. Biol.* **98**:503–517.
Sutcliffe, J. G., 1978, pBR322 restriction map derived from the DNA sequence: Accurate DNA size markers up to 4361 nucleotide pairs long, *Nucleic Acids Res.* **5**:2721–2728.
Thomas, M., and Davis, R. W., 1975, Studies on the cleavage of bacteriophage lambda DNA with *Eco*RI restriction endonuclease, *J. Mol. Biol.* **91**:315–328.
Weber, I. T., McKay, D. B., and Steitz, T. A., 1982, Two helix DNA binding motif of CAP found in *lac* repressor and *gal* repressor, *Nucleic Acids Res.* **10**:5085–5102.
Williams, J. G. K., Wulff, D. L., and Nash, H. R., 1977, in: *DNA Insertion Elements, Plasmids and Episomes* (R. I. Bukhari, J. A. Shapiro, and S. L. Adhya, eds.), Cold Spring Harbor Laboratory, New York, pp. 357–361.
Williamson, A. R., and Askonas, B. A., 1967, Biosynthesis of immunoglobulins: The sep-

arate classes of polyribosomes synthesizing heavy and light chains, *J. Mol. Biol.* **23**:201–216.

Willmund, R., and Kneser, H., 1973, Different binding of RNA polymerase to individual promoters, *Mol. Gen. Genet.* **126** 165–175.

Wilson, B. L., and Mortlock, R. P., 1973, Regulation of D-xylose and D-arabitol catabolism by *Aerobacter aerogenes, J. Bacterol.* **113**:1404–1411.

Wu, T. T., Lin, E. C. C., and Tanaka, S., 1968, utants of *Aerobacter aerogenes* capable of using xylitol as a novel carbon source, *J. Bacteriol.* **96**:447–456.

CHAPTER 4

# The Development of Catabolic Pathways for the Uncommon Aldopentoses

ROBERT P. MORTLOCK

## 1. The Structure of the Aldopentoses and Their Occurrence in Nature

### 1.1. The Structure of the Aldopentoses

Those aldehyde sugars that are five carbons in length are known as aldopentoses. Alternate positions of the hydroxyl groups can give rise to eight possible epimeric structures: D- and L-ribose, D- and L-arabinose, D- and L-lyxose, and D- and L-xylose. The different hydroxyl configurations of these eight sugars can be seen in Fig. 1.

### 1.2. D-Ribose and L-Ribose

D-Ribose is common in nature and is the sugar in ribonucleic acid (RNA), whereas deoxyribose is the normal sugar in deoxyribonucleic acid (DNA). It also has been reported present as a component of polysaccharides from certain bacteria (Schaffer, 1972) and in the cell walls of a few fungi (Laskin and Lechevalier, 1973). L-Ribose does not appear to occur naturally.

---

ROBERT P. MORTLOCK • Department of Microbiology, New York State College of Agriculture and Life Sciences, Cornell University, Ithaca, New York 14853.

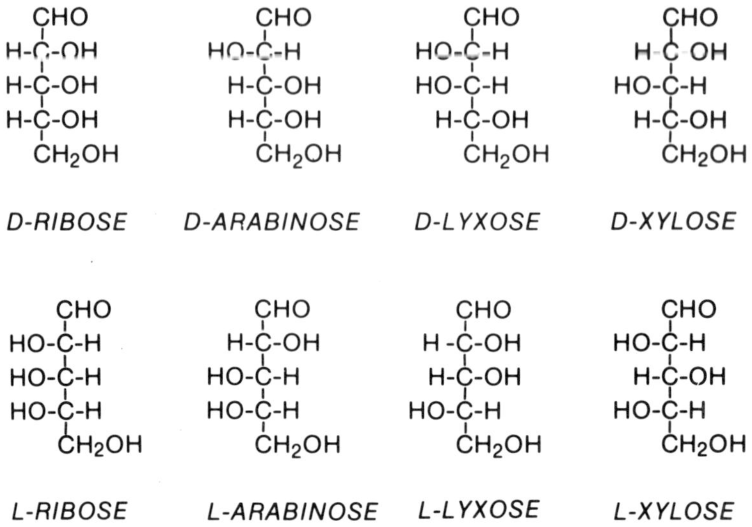

```
        CHO              CHO              CHO              CHO
     H-C-OH           HO-C-H            HO-C-H           H-C-OH
     H-C-OH            H-C-OH           HO-C-H           HO-C-H
     H-C-OH            H-C-OH            H-C-OH           H-C-OH
      CH₂OH             CH₂OH             CH₂OH            CH₂OH

     D-RIBOSE         D-ARABINOSE        D-LYXOSE         D-XYLOSE


        CHO              CHO              CHO              CHO
     HO-C-H            H-C-OH            H-C-OH           HO-C-H
     HO-C-H            HO-C-H            H-C-OH            H-C-OH
     HO-C-H            HO-C-H            HO-C-H           HO-C-H
      CH₂OH             CH₂OH             CH₂OH            CH₂OH

     L-RIBOSE         L-ARABINOSE        L-LYXOSE         L-XYLOSE
```

Figure 1. The structures of the eight aldopentoses.

### 1.3. D-Xylose and L-Xylose

D-Xylose is widely distributed in nature and has been commonly referred to as wood sugar. Several large-scale industrial procedures are available for its preparation from such raw materials as corncobs, cottonseed hulls, and straw (Schaffer, 1972). L-Xylose is not considered to occur naturally in any significant quantity, although the 2-keto form, L-xylulose, is a metabolic intermediate in certain pathways and may be excreted in the urine of persons suffering from the congenital disease pentosuria (Horecker, 1962; Schaffer, 1972).

### 1.4. D-Arabinose and L-Arabinose

D-Arabinose is present in nature but is rarely encountered. It has been found as a minor component of the cell-wall polysaccharide material of some microorganisms. On the other hand, L-arabinose is widely distributed in plant material and can be considered very common in nature. It has been purified from many materials, including mesquite gum, cherry gum, rye and wheat bran, and spent beet pulp (Schaffer, 1972).

### 1.5. D-Lyxose and L-Lyxose

The occurrence of D-lyxose is very unusual in nature. L-Lyxose must also be considered very rare, although it has been reported as a product of hydrolysis of the antibiotic curamycin (Schaffer, 1972).

## 2. The Pathways of Degradation of Aldopentoses by Coliform Bacteria

### 2.1. Pathways for the Degradation of Those Sugars Commonly Found in Nature

Many bacteria have been found to possess inducible enzyme pathways for the degradation of the most commonly occurring aldopentoses: D-ribose, D-xylose, and L-arabinose. The normal pathways of catabolism of these sugars by coliform bacteria are shown in Fig. 2. The strategy found in each case was the conversion of the aldopentose into a 2-ketopentose 5-phosphate, which was then further degraded to fructose 6-phosphate and glyceraldehyde 3-phosphate by the transketolase and transaldolase rearrangement reactions.

D-Ribose was phosphorylated at the C-5 position and then isomerized to yield D-ribulose 5-phosphate. The degradation of D-xylose and L-arabinose, however, followed a different strategy. For each of these aldopentoses an existing active transport system brought the sugar through the cell membrane. An isomerization reaction then formed the 2-ketopentose or pentulose and a kinase catalyzed a phosphorylation to form the ketopentose 5-phosphate, as shown in Fig. 2. This same reaction sequence, isomerization followed by phosphorylation, was also found to exist for the degradation of the less common aldopentoses by *Klebsiella* strains that had gained the ability to utilize such sugars (Mortlock, 1976).

The regulation of the L-arabinose pathway has been carefully studied in *E. coli* and, to a lesser extent, in *Klebsiella* (Englesberg, 1971; Leblanc and Mortlock, 1973). L-Arabinose was found to function as the inducer of the three enzymes of the L-arabinose operon: L-arabinose isomerase,

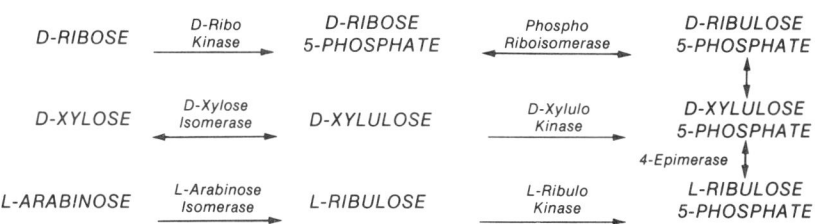

**Figure 2.** The inducible pathways of degradation of the more commonly occurring aldopentoses. The pathways illustrated are those found in *K. pneumoniae* PRL-R3 and are the normal catabolic pathways found in coliform bacteria. For D-Xylose and L-arabinose the first reaction is an isomerization reaction converting the aldehyde sugar to a 2-ketopentose (pentulose). The second reaction is catalyzed by a kinase with the formation of the pentulose 5-phosphate. D-Xylulose 5-phosphate is the common intermediate and is further degraded by the transketolase and transaldolase rearrangement reactions.

L-ribulokinase, and L-ribulose 5-phosphate 4-epimerase. The regulation of the D-xylose catabolic pathway has not been studied as carefully in *Klebsiella*, but data indicate that D-xylose is the inducer of D-xylose isomerase and D-xylulokinase (Wilson and Mortlock, 1972). For all three common pentoses, D-ribose, L-arabinose, and D-xylose, the incubation of *K. pneumoniae* cells with the pentose led to the induction of the appropriate enzyme activities, and growth of the cells was observed after only a few hours (Mortlock and Wood, 1964a,b).

## 2.2. Pathways for the Degradation of Those Sugars Not Commonly Found in Nature

In contrast to the observations for the common sugars, the incubation of cells of *K. pneumoniae* strain PRL-R3 with one of the less common pentoses as a potential growth substrate resulted in growth taking place only following a delay ranging from about 1 day for such sugars as D-arabinose and L-arabitol to 3 or more weeks for L-xylose (Mortlock and Wood, 1964a). The time required to complete growth on a minimal-salts medium supplemented with one of these sugars is shown in Table I. Unusual delays in growth were observed for all of the less common sugars. Unfortunately, L-ribose was not available in sufficient quantity to test in such an experiment.

Since the carbohydrate added to the medium was the only potential source of carbon and energy for cell growth, it was apparent that in order to complete growth in that medium the organism must have acquired the ability to use that carbohydrate as the carbon and energy source to support

Table I
The Growth of *Klebsiella pneumoniae* Strain PRL-R3 on Aldopentoses[a]

| Growth substrate | Time required to complete growth |
| --- | --- |
| D-Xylose | ≤1 day |
| L-Arabinose | |
| D-Ribose | |
| D-Arabinose | 2–3 days |
| D-Lyxose | 4 days |
| L-Lyxose | 2–4 days |
| L-Xylose | 3–4 weeks |

[a]The culture of 5.5 ml of minimal salts medium was supplemented with 0.5% of the indicated carbohydrate and inoculated to contain $2 \times 10^6$ glucose-grown cells per ml. Tubes containing the cultures were shaken at 26°C until growth was complete. [Data are from Mortlock and Wood (1964a) and unpublished observations of the author.]

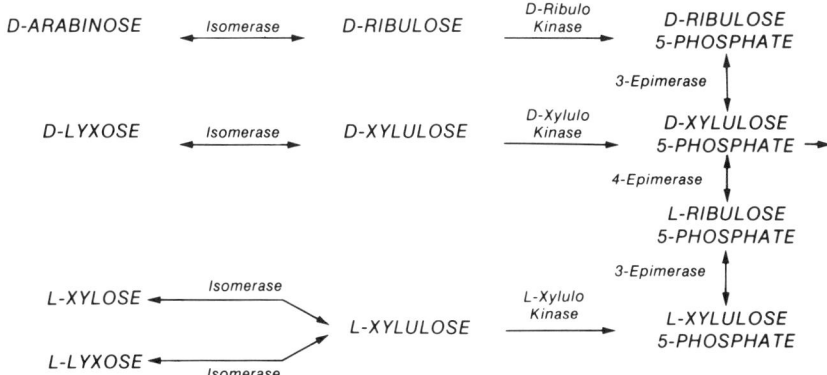

Figure 3. The pathways of degradation of the less common pentoses by mutants of *K. pneumoniae* PRL-R3. The first reaction in each pathway is an isomerization converting the aldopentose to a 2-ketopentose, while the second reaction is catalyzed by a kinase with the formation of the pentulose 5-phosphate. D-Xylulose 5-phosphate is also the common intermediate in these reactions.

its growth. Once growth had been completed on one of these less common sugars, the bacterial cells in that particular culture could be examined for the presence of enzyme activities capable of degrading the new substrate. In such a manner, one by one, the enzymatic pathways for the degradation of D-arabinose, D-lyxose, L-lyxose, and L-xylose were established in strains that had grown on these sugars, as shown in Fig. 3.

In each case the metabolic route used for degradation of the sugar involved an isomerization to the 2-ketopentose (or pentulose) followed by phosphorylation to the 5-phosphate. Thus, cultures grown on D-arabinose contained one enzyme to catalyze the isomerization of D-arabinose to D-ribulose and another enzyme to catalyze the phosphorylation of D-ribulose to D-ribulose 5-phosphate, as shown in Fig. 3. Cells grown on D-lyxose possessed an isomerase to convert D-lyxose to D-xylulose and a kinase to make D-xylulose 5-phosphate. Cells grown on L-xylose possessed isomerase activity to convert L-xylose to L-xylulose, and cells grown on L-lyxose appeared to have an isomerase to make L-xylulose from that rare aldopentose. In each of the latter two cases the cells also possessed an L-xylulokinase activity to form L-xylulose 5-phosphate. The cells also had the required epimerase activities to convert the respective ketopentose 5-phosphates to D-xylulose 5-phosphate, so the transaldolase and transketolase rearrangements could result in the formation of fructose 6-phosphate and glyceraldehyde 3-phosphate. These latter two compounds could be further degraded by other common cellular enzymes.

Eventually it was established that mutations were necessary to establish growth on the less common aldopentoses, such as D-arabinose, D-lyxose, L-lyxose, and L-xylose, but this was not at all obvious in early studies. If a mutation was truly required to permit growth on a sugar such as D-arabinose, it would be expected that the frequency of such mutants in the parent strain would be low. Only a small proportion of the cells, perhaps $10^{-6}$–$10^{-8}$, would actually consist of such mutants and be capable of growth with D-arabinose as the substrate. Yet, in initial experiments using *K. pneumoniae* PRL-R3 and examining the organism's growth capabilities, it was observed that when a small number of cells was plated onto the surface of a petri dish containing minimal-salts agar supplemented with one sugar such as D-arabinose, after 2 weeks of incubation each cell formed a visible colony. It appeared from such experiments that each cell was capable of utilizing the sugar as a growth substrate. However, control experiments employing salts–agar containing no added carbohydrate also gave rise to one colony per cell in the same time period, showing the organism's capability to grow utilizing agar alone as a carbon and energy source. Once growth on salts–agar had resulted in the formation of a small colony with enough cells, then the occurrence of a single mutant within each of these micro-colonies could result in the apparent formation of a carbohydrate, positive colony from each original cell placed on the agar surface. With the most purified agar then available, growth of the cells on the surface of salts–agar plates could not be prevented.

When liquid medium was used for growth it sometimes was especially difficult to tell the difference between the alternate possibilities of a very slow induction and the selection of carbohydrate-positive mutants (Table I). With D-lyxose or L-xylose as potential substrates, the very long periods required before growth occurred strongly suggested the requirement for the selection of at least one mutational event to permit growth. For D-arabinose, however, the time of about 24 hr required before growth was observed made it difficult to distinguish between the selection of a few D-arabinose-positive mutants from the population of cells in the inoculum and the slow or delayed induction of all of the cells in the inoculum.

Once a culture was grown on an uncommon pentose, the transfer of cells to a new medium containing the same sugar resulted in growth initiated after a greatly shortened lag time. If cells were transferred back to a glucose medium for about 30 generations, however, and then placed back on the uncommon sugar, the original long lag time was usually observed before growth was apparent.

Further studies eventually showed that growth on all four of these uncommon aldopentoses (D-arabinose, D-lyxose, L-lyxose, and L-xylose) required at least one mutation. When media containing one of these four sugars as the carbon source was inoculated with cells of the parent strain, the time required before growth was observed was actually the time for the selection and growth of the mutants. After growth was finally completed, the majority of the cells in the culture consisted of mutants and if that culture was then used as inoculum for a medium containing the same sugar, growth was initiated much more rapidly. However, even though the parent strain could not utilize the carbohydrate as a growth substrate, some of those parent cells still remained. If the culture was used to inoculate a nutrient broth or peptone broth medium, there would be selection against the mutants, with the parent strain reestablishing itself in the majority of the population. Thus, after 30 or more generations in the absence of the selective carbohydrate, a long lag period might once again be required to establish growth on that uncommon sugar.

## 2.3. Enzyme Activities Establishing Growth on the New Aldopentose Substrates

### 2.3.1. Establishment of Isomerase Activity

Investigation showed that each pathway for the degradation of the new aldopentose substrates followed the same strategy as previously observed for the catabolism of D-xylose and L-arabinose. After transport of the aldopentose through the cell membrane, an enzyme catalyzed an isomerization reaction to form the 2-ketopentose. D-Arabinose-positive mutants grown on D-arabinose possessed an isomerization activity of unknown origin that catalyzed the isomerization of D-arabinose to the 2-ketopentose, D-ribulose, as shown in Fig. 3 (Mortlock and Wood, 1964a). Mutants that had gained the ability to grow on D-lyxose and were grown on that sugar possessed isomerase activity to convert D-lyxose to D-xylulose (Anderson and Allison, 1965). In a similar manner, L-lyxose-positive mutants grown on L-lyxose possessed an enzyme activity converting L-lyxose to L-xylulose, and L-xylose-positive mutants grown on L-xylose contained an isomerase to convert L-xylose to L-xylulose (Anderson and Wood, 1960; Mortlock and Wood, 1964a). These isomerase activities were not found in the parent strain even when cells were incubated for several hours with the uncommon sugar.

### 2.3.2. Availability of Kinase Activity

The parent strain already possessed the ability to synthesize kinase activity to complete each of the new pathways. The ketopentose D-ribulose formed by the isomerization of D-arabinose was known to be an intermediate in the ribitol catabolic pathway and was the actual inducer of the enzymes of that pathway, including D-ribulokinase (see Chapter 1). The D-xylulose formed from D-lyxose was an intermediate in the catabolic pathways for the degradation of both D-arabitol (Chapter 1) and D-xylose (Fig. 2), and thus the organism possessed the ability to make two different D-xylulokinases, either one of which could catalyze the phosphorylation of D-xylulose during growth on D-lyxose. The cells were also able to synthesize an L-xylulokinase that was used during growth on L-arabitol but that also could be used for growth on L-xylose or L-lyxose. Thus, this *Klebsiella* strain already had the information to synthesize kinase activities to complete each of these new pathways, providing the regulatory systems of the cell permitted expression of that information during growth on the new substrate.

## 3. The Biochemical and Genetic Bases for the Establishment of New Enzymatic Pathways for the Degradation of Aldopentoses

### 3.1. The Utilization of D-Lyxose

#### 3.1.1. A New Pathway for D-Lyxose Catabolism in *Klebsiella*

3.1.1a. Selection of Mutants Capable of Growth on D-Lyxose. When glucose-grown cells of *K. pneumoniae* PRL-R3 were used for the inoculation of a medium containing D-lyxose as the only carbon source, about 4 days was required before growth was completed, as shown in Table I. Table II gives a more detailed examination of the events taking place after *K. pneumoniae* PRL-R3 was inoculated into a medium containing D-lyxose. Initially there was a slight increase in the direct cell count and a decrease in the cell mass, normal results for a culture starved for carbon and energy. The first detectable increase in cell number and mass came after about 60 hr of incubation. At the same time there occurred a measurable disappearance of D-lyxose from the medium and the culture gained the ability to respire (utilize oxygen) with D-lyxose as the carbon source (Mortlock and Wood, 1964a,b). These effects resulted from the growth of D-lyxose-positive mutants, which were present in the original inoculum in small numbers and which eventually (after about 60 hr) reached a large enough population to influence the apparent behavior of the entire culture. Once a pure culture of these D-lyxose-positive mutants was isolated, it was able to grow on D-lyxose without such a long lag time and retained that ability, even after cultivation on D-glucose in the absence of D-lyxose (Anderson and Allison, 1965).

Table II
Adaptation of *Klebsiella pneumoniae* PRL-R3 to Growth on D-Lyxose[a]

| Time (hr) | Direct cell count ($\times 10^{-9}$) | Cell dry weight (mg/ml) | D-Lyxose (mg/ml) | Oxygen uptake ($\mu$l/hr) |
|---|---|---|---|---|
| 0  | 2.5 | 0.75 | 4.6 | 0 |
| 12 | 3.8 | 0.60 | 4.7 | 0 |
| 17 | 3.0 | 0.58 | 4.8 | 0 |
| 35 | 2.6 | 0.58 | 4.5 | 0 |
| 41 | 2.5 | 0.50 | 4.7 | 0 |
| 60 | 3.4 | 0.51 | 4.6 | 0 |
| 83 | 3.6 | 0.80 | 2.5 | 55 |
| 89 | 7.0 | 1.08 | 0   | 122 |

[a]The inoculum was grown on glucose, washed, and added to a mineral salts medium supplemented with D-lyxose as the sole carbon and energy source. Oxygen uptake was expressed in $\mu$l $O_2$ used/hr per mg dry weight, for cells removed and incubated separately with 0.5% D-lyxose. [Data are from Mortlock and Wood (1964b).]

Cells grown on D-lyxose possessed the enzyme activities required to degrade that sugar (Fig. 3), as shown in Table III. In addition to D-lyxose isomerase and D-xylulokinase activities, however, the cell extracts also possessed very low levels of activity for ribitol dehydrogenase and D-ribulokinase, the enzymes of the ribitol operon, and low but significant activity for both D-xylose isomerase and D-arabitol dehydrogenase (Mortlock and Wood, 1964a; Anderson and Allison, 1965).

3.1.1b. *Origin of the Kinase of the D-Lyxose Pathway.* The parent *Klebsiella* strain was known to have inducible enzyme pathways for the

Table III
Enzyme Activities Found in Cells of *Klebsiella pneumoniae* PRL-R3 Grown on D-Lyxose[a]

| Enzyme | Activity ($\mu$mole/min per mg protein) |
|---|---|
| D-Lyxose isomerase | 0.033–0.05 |
| D-Xylose isomerase | 0.016–0.067 |
| D-Xylulokinase | 0.03–0.08 |
| D-Ribulokinase | <0.02–0.02 |
| D-Arabitol dehydrogenase | 0.07–0.41 |
| Ribitol dehydrogenase | 0.02–0.12 |

[a]Cell-free extracts were prepared from D-lyxose-grown cells and assayed for enzyme activity. The assay for isomerase activity was less sensitive than the other enzyme assays, resulting in lower numbers. [Data are from Mortlock and Wood (1964a).]

degradation of both D-arabitol and D-xylose. The structural genes coding for the D-arabitol dehydrogenase and D-xylulokinase of the D-arabitol catabolic pathways are believed to be located on a single operon with D-arabitol as the apparent inducer of the operon (see Chapter 1). The two enzymes involved in the D-xylose catabolic pathway, D-xylose isomerase and a separate D-xylulokinase, were coordinately controlled, probably on a single operon, with D-xylose as the apparent inducer. It is probable that some of the D-xylulose formed by the isomerization of D-lyxose was converted to D-arabitol, causing the partial induction of the D-arabitol pathway. Some of the D-xylulose was also converted to D-xylose, resulting in the partial induction of the D-xylose pathway. In any event, the presence of D-arabitol dehydrogenase indicated that the D-xylulokinase of the D-arabitol pathway must have been present in equivalent amounts, while the presence of D-xylose isomerase indicated that the D-xylulokinase associated with the D-xylose catabolic pathway must have been induced to the same extent. Thus, the cells had been partially induced for two separate D-xylulokinase activities, either serving to phosphorylate the D-xylulose formed from the isomerization of D-lyxose.

3.1.1c. Origin of the Isomerase of the D-Lyxose Pathway. With the origin of the D-xylulokinase activity (or activities) of the new D-lyxose pathway explained, the mutation leading to the ability to isomerize D-lyxose also required an explanation. In an attempt to find the origin of this enzyme, a search was made for more naturally occurring sugars of similar structure. Of the naturally occurring carbohydrates, D-mannose possessed a structure most similar to that of D-lyxose, but cells grown on either D-mannose, D-fructose, or D-glucose possessed no detectable D-lyxose isomerase activity (Anderson and Allison, 1965).

Anderson and Allison (1965) purified D-lyxose isomerase 130-fold from a crude extract of a mutant of *K. pneumoniae* PRL-R3 grown on that pentose. The purified enzyme had an apparent molecular weight of 40,000 and required a metal ion for activity, manganese being best. The purified enzyme also possessed activity for the isomerization of D-mannose, suggesting that D-mannose might be the natural evolved substrate for the enzyme, but D-mannose was actually a poorer substrate for the enzyme than D-lyxose. The $K_m$ for D-lyxose was $3.6 \times 10^{-3}$, while the $K_m$ for D-mannose was $10^{-2}$, showing that the enzyme had a superior binding affinity for D-lyxose. The isomerase also achieved a higher maximum velocity with D-lyxose as the substrate.

Palleroni and Doudoroff (1956) described a D-mannose isomerase, isolated from a mutant of *Pseudomonas saccharophila,* that was able to isomerize D-lyxose at 11% the rate at which it isomerized D-mannose.

With the enzyme from the *Klebsiella* mutant, however, no substrate was found for the enzyme with greater activity than D-lyxose. Also, no other inducer of the enzyme other than D-lyxose could be found (Anderson and Allison, 1965). Thus, the origin of the D-lyxose isomerase remained obscure, as did the nature of the mutation or mutational events permitting the mutant strain to synthesize the enzyme and grow on D-lyxose.

### 3.1.2. A New Pathway for D-Lyxose Catabolism in *Escherichia*

Stevens and Wu (1976) reported a mutant of *E. coli* K12 that had gained the ability to grow on D-lyxose. The parent strain was unable to utilize D-lyxose as a growth substrate, but the mutant grew on a concentration of 0.2% D-lyxose with a cell doubling time of 3.1 hr. The route of D-lyxose catabolism in the *E. coli* mutant was the same as reported for *Klebsiella*, with D-lyxose isomerized to D-xylulose and D-xylulokinase catalyzing the phosphorylation to form D-xylulose 5-phosphate. Although *E. coli* K12 differed from the *Klebsiella* in that it did not have a D-arabitol pathway, it did possess a pathway for catabolism of D-xylose. Stevens and Wu found that the D-xylulokinase of this D-xylose pathway was used for the phosphorylation of D-xylulose during growth on D-lyxose. They also found that the permease of the D-xylose pathway, normally employed to bring D-xylose through the cell membrane, was used to transport D-lyxose through the cell membrane.

The same investigators found that the D-lyxose-positive mutants had acquired an isomerase to convert D-lyxose to D-xylulose. This enzyme was constitutive in the D-lyxose-positive mutants but was not constitutive in the D-lyxose-negative parent strain and could not be induced in the parent strain by any substrate tested, including D-mannose.

Using this D-lyxose-positive mutant as the parent strain, Stevens *et al.* (1981) obtained a second mutation permitting a faster growth rate when cells grew on D-lyxose. The new doubling time was decreased to about 2 hr and, upon examination, this new strain was found to produce twice as much isomerase as the original D-lyxose-positive strain. The enzyme was purified and found to have four subunits, each with a molecular weight of about 40,000. The $K_m$ of the tetramer was 80 mM for D-mannose, while it was 300 mM for D-lyxose. Since mannose appeared to be the better substrate, these authors termed the enzyme a D-mannose isomerase.

Neither the origin of this enzyme nor the nature of the mutational event leading to its constitutive synthesis could be identified. Although many bacteria metabolize D-mannose by isomerizing it to D-fructose, coliform bacteria such as *Klebsiella* and *Escherichia* normally transport D-mannose through the cell membrane and into the cell by an active transport system known as the phosphoenolpyruvate:sugar phosphotransferase system. This transport system phosphorylates D-mannose as it passes through the cell membrane, resulting in the entrance of the carbohydrate into the cell as mannose phosphate (Dills *et al.*, 1980). D-Mannose phosphate is then isomerized to fructose phosphate by an enzyme termed mannose-phosphate isomerase. Thus, *E. coli* K12 was not known to possess the ability to synthesize a true D-mannose isomerase enzyme and seemed to have no need for such a catalytic activity.

Stevens *et al.* suggested that this D-mannose isomerase with the ability to isomerize D-lyxose might be an evolutionary remnant, produced by a structural gene that was once utilized by an ancestral organism that metabolized D-mannose by isomerizing the unphosphorylated form to D-fructose. They postulated that random mutations accumulating in this gene since it was last required for growth had decreased the affinity of the enzyme for D-mannose. Such mutations, accumulating over time, might also have eliminated the normal regulation of the enzyme so that incubation of cells with D-mannose no longer resulted in the induction of the enzyme. A mutation permitting the constitutive synthesis of the enzyme allowed it to be utilized for the isomerization of D-lyxose.

If the above speculations were true and the gene coding for the structure of this isomerase was an evolutionary remnant from a previous metabolic pathway, the genetic information for the synthesis of the enzyme had survived in the absence of selective pressure with only minor alterations in structure. Zipkas and Riley have suggested that *E. coli* K12 might have increased its DNA content in the past by duplicating its entire chromosome, thus doubling the amount of genetic material and providing duplicate genes that could then diverge in function (Zipkas and Riley, 1975; Riley and Anilionis, 1978; also see Chapter 10, this volume). According to this speculation, this past duplication of the chromosome resulted in the tendency for biochemically related genes with a common ancestry to lie 180° apart on the circular *E. coli* chromosome. Stevens *et al.* (1981) mapped their new D-mannose (D-lyxose) isomerase structural gene at 85 min on the *E. coli* K12 chromosome. With all these speculations, it is most interesting to note that the structural gene for mannosephosphate isomerase in the same organism has been mapped at 36 min on the cell chromosome, almost a perfect 180° away from the newly discovered isomerase for D-lyxose and D-mannose (Bachmann and Low, 1975).

### 3.2. The Utilization of D-Arabinose

#### 3.2.1. A New Pathway for D-Arabinose Catabolism in *Klebsiella*

3.2.1a. A Regulatory Mutation Permitting Constitutive Synthesis of an Isomerase. The pathway for D-arabinose catabolism by coliform bacteria such as *Klebsiella* and *Escherichia* was reported to involve isomerization to D-ribulose followed by phosphorylation to yield D-ribulose 5-phosphate, as shown in Fig. 3 (Cohen, 1953). In their studies of the enzyme activities of cells grown on various five-carbon sugars, Mortlock and Wood (1964a) reported that cells of *K. pneumoniae* PRL-R3 grown on D-arabinose also possessed both of these enzyme activities, as shown in Table IV. D-Arabinose isomerase activity varied from 0.28 to 0.65 U/mg protein, while D-ribulokinase activity varied from 0.24 to 0.46 U/mg protein. In addition to these required enzyme activities, however, large amounts of ribitol dehydrogenase and small amounts of L-xylose isomerase activity were also present.

Bisson *et al.* (1968) showed that the D-ribulose produced from the isomerization of D-arabinose caused the induction of the enzymes of the ribitol operon, resulting in the synthesis of both ribitol dehydrogenase and D-ribulokinase (Fig. 2; see Chapter 1). This observation explained the presence of both enzymes in cells grown on D-arabinose as well as the origin of the kinase activity required for catabolism of D-arabinose. The reason for the presence of the L-xylose isomerase activity, however, could not be explained at that time.

When the parent strain of *K. pneumoniae* PRL-R3 was used to inoculate a medium of mineral salts with D-arabinose as the only potential carbon and energy source to support growth, no significant increase in cell mass could be observed until after 12 hr of incubation, as shown in Table V. A significant increase in cell number could not be detected until after 30 hr of incubation, but with the unusually large inoculum used in

Table IV
Enzyme Activities Found in Cells of *Klebsiella pneumoniae* PRL-R3 Grown on L-Xylose and D-Arabinose[a]

| Enzyme | Activity ($\mu$mole/min per mg protein) | |
|---|---|---|
| | L-Xylose | D-Arabinose |
| L-Xylose isomerase | 0.13–0.25 | 0.008–0.016 |
| D-Arabinose isomerase | 3.98–9.73 | 0.28–0.65 |
| L-Xylulokinase | 0.02–0.19 | <0.02 |
| D-Ribulokinase | 0.02–0.12 | 0.24–0.46 |
| D-Xylulokinase | <0.02 | <0.02–0.062 |
| Ribitol dehydrogenase | 0.02–0.10 | 0.91–3.49 |
| D-Arabitol dehydrogenase | <0.02 | <0.02–0.58 |

[a] The enzyme activities given represent the high and low values for duplicate experiments. If no activity was detected, it is indicated as less than 0.02, which was the lower limit of detection of the assay. [Data are from Mortlock and Wood (1964a).]

## Table V
### Adaptation of *Klebsiella pneumoniae* PRL-R2 to Growth on D-Arabinose[a]

| Time (hr) | Direct cell count ($\times 10^{-9}$) | Cell dry weight (mg/ml) | D-Arabinose (mg/ml) | Oxygen uptake ($\mu$l/hr) |
|---|---|---|---|---|
| 0 | 2.2 | 0.56 | 5.2 | 0 |
| 2 | 2.2 | 0.57 | 5.2 | 0 |
| 4 | 2.2 | 0.56 | 5.1 | 0 |
| 6 | 2.2 | 0.60 | — | 32 |
| 12 | 2.2 | 0.73 | 5.0 | 91 |
| 17 | 2.2 | 1.01 | — | 73 |
| 24 | 2.2 | 1.16 | 3.5 | — |
| 35 | 6.9 | 1.70 | 3.2 | 89 |
| 41 | 6.4 | 2.00 | 0.7 | 169 |
| 48 | 7.6 | 2.12 | — | — |

[a] The inoculum was grown on glucose, washed, and added to a mineral salts medium supplemented with D-arabinose as the sole carbon and energy source. Oxygen uptake was expressed in $\mu$l $O_2$ used/hr per mg dry weight for cells removed and incubated separately with 0.5% D-arabinose. [Data are from Mortlock and Wood (1964b).]

this experiment the culture was able to show some detectable respiration with D-arabinose as the carbon source after only 6 hr. When a smaller inoculum was used, growth on D-arabinose was usually complete in 2–3 days, as shown in Table I. If a culture grown on D-arabinose was inoculated into a new D-arabinose medium, growth was then complete in about 1 day.

It would have been easy to accept the above data as resulting from the slow induction of a previously existing pathway for the catabolism of D-arabinose, especially since after a D-arabinose-grown culture was transferred back to a glucose medium for 20–30 generations it once again would require the longer lag period to initiate growth. However, careful examination showed that growth on D-arabinose actually resulted from the selection of a mutant that occurred spontaneously and was present in the inoculum at a relatively high frequency. Incubation with D-arabinose as the substrate selected for these mutant cells, but after transfer to a medium without D-arabinose, selection could occur against the mutant with the reestablishment of the parent strain as the majority of the population. Therefore, the culture would lose the ability to complete growth on D-arabinose in the shorter incubation time.

Camyre and Mortlock (1965), studying the growth of *Klebsiella pneumoniae* PRL-R3 on D-arabinose, found that the enzyme activity that isomerized D-arabinose to D-ribulose was synthesized constitutively in the D-arabinose-positive mutants. Suspecting that the isomerase might be one that evolved for the isomerization of a more common sugar and was obtained for use for the D-arabinose pathway by a mutation permitting its constitutive synthesis, they examined the literature for naturally occurring carbohydrates of similar structure. The structure of the methyl pentose, L-fucose, was found to be very similar to the structure of D-arabinose, as shown in Fig. 4. Furthermore, L-fucose was widely distributed in nature. It could be found in certain seaweeds (Briggs et al., 1982), in bacterial lipopolysaccharide material, associated with the cell membrane of some eukaryotic cells, and in the jelly coat of sea urchin eggs (Hotta and Kurokawa, 1973). It was also reported as the primary milk carbohydrate of the monotremes (egg-laying mammals), such as the echidna and the platypus (Messer and Kerry, 1973). In addition, many bacterial strains, including strains of *Klebsiella* and *Escherichia,* were known to be able to metabolize L-fucose by means of an inducible enzyme pathway that included an isomerization reaction as shown in Fig. 5A (Green and Cohen, 1956).

Examination of extracts of cells grown on either D-arabinose or L-fucose showed that they possessed activity for the isomerization of both L-fucose and D-arabinose, as shown in Table VI, and the ratio of D-arabinose to L-fucose isomerization activity was similar in both extracts. The common identity of the two isomerase activities was confirmed by Oliver and Mortlock (1971b), who partially purified the isomerase and showed both isomerase activities to be due to a single enzyme, the L-fucose isomerase of the L-fucose pathway. In addition to the data indicating that the isomerase catalyzed the isomerization of D-arabinose, there was evidence that the permease of the L-fucose pathway brought D-arabinose through the cell membrane into the cell.

The relationship between the pathways of catabolism of L-fucose, D-arabinose, and ribitol by this *Klebsiella* strain are shown in Fig. 5. L-Fucose was catabolized by an inducible pathway consisting of an isomerase to convert L-fucose to the 2-keto sugar L-fuculose, a kinase to form L-fuculose 1-phosphate, and an aldolase to cleave the latter com-

$$
\begin{array}{cc}
\text{H-C=O} & \text{H-C=O} \\
\text{HO-C-H} & \text{HO-C-H} \\
\text{H-C-OH} & \text{H-C-OH} \\
\text{H-C-OH} & \text{H-C-OH} \\
\text{HO-C-H} & \text{H}_2\text{-C-OH} \\
\text{CH}_3 & \\
\textit{L-FUCOSE} & \textit{D-ARABINOSE}
\end{array}
$$

**Figure 4.** A comparison of the structures of L-fucose and D-arabinose.

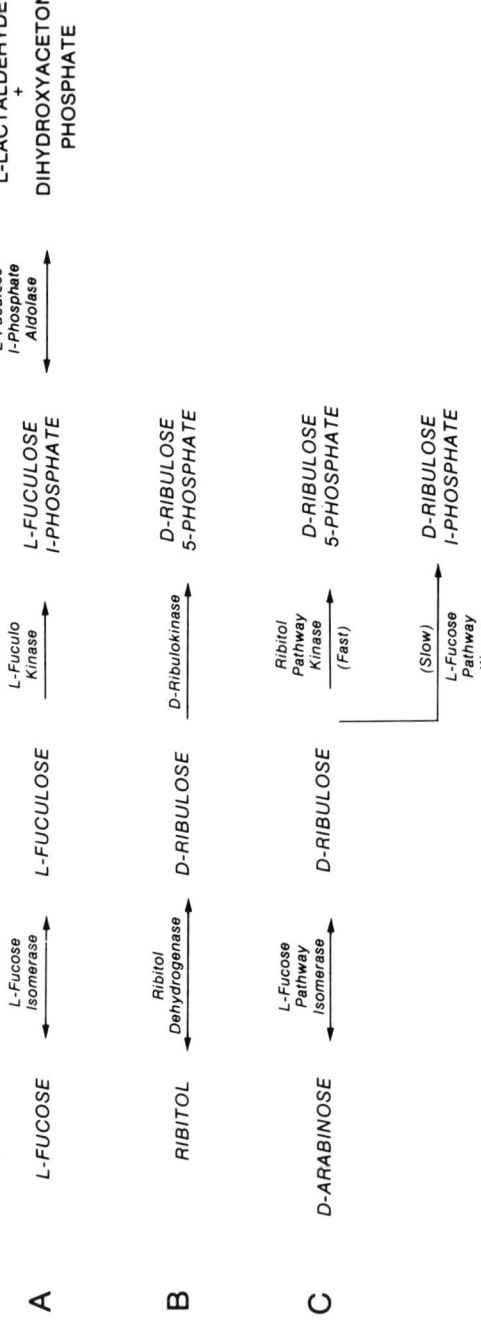

**Figure 5.** Pathways for the degradation of L-fucose, ribitol, and D-arabinose. (A) Pathway found in *Klebsiella* and *Escherichia* strains for the degradation of L-fucose. (B) Pathway found in *Klebsiella* strains for the degradation of ribitol. (C) Alternate pathways for D-arabinose degradation in D-arabinose-positive mutants. For strains possessing the ribitol pathway, D-ribulokinase can catalyze the rapid phosphorylation of its natural substrate, D-ribulose. Because of structural similarities between L-fuculose and D-ribulose, L-fuculokinase can catalyze a slow phosphorylation of D-ribulose to form the 1-phosphate.

## Table VI
### Isomerase Activity found in Cell-Free Extracts of *Klebsiella pneumoniae* PRL-R3[a]

| Strain | Growth substrate | Activity (U/mg protein) | | |
| --- | --- | --- | --- | --- |
| | | L-Fucose isomerase | D-Arabinose isomerase | L-Xylose isomerase |
| Parent | Casein hydrolysate | <0.16 | <0.016 | <0.016 |
| | L-Fucose | 4.48 | 2.25 | 0.073 |
| D-Arabinose positive mutant | D-Arabinose | 1.77 | 0.87 | 0.043 |
| L-Xylose positive mutant | L-Xylose | 5.03 | 2.26 | 0.072 |

[a] Data are from Oliver and Mortlock (1971a).

pound to yield dihydroxyacetone phosphate and lactaldehyde (Fig. 5A). These last two compounds were further metabolized by other enzymes of the cell. The first two enzymes in the L-fucose catabolic pathway, L-fucose isomerase and L-fuculokinase, have been shown to be coordinately controlled by a common regulator gene, and the inducer was the intermediate in the pathway, L-fuculose 1-phosphate (St. Martin and Mortlock, 1976). Even when the cells were uninduced, small or basal levels of the enzymes were still present. When L-fucose was available as a potential substrate, these basal levels of enzyme converted a small amount of the L-fucose to the inducer, L-fuculose 1-phosphate, resulting in the induction of higher levels of both enzymes. When cells were given D-arabinose as a potential growth substrate, the inducer could not be formed and the L-fucose pathway enzymes could not be induced. However, any mutants present in the inoculum that were constitutive for the synthesis of the isomerase could utilize the enzyme to isomerize D-arabinose to D-ribulose. Since D-ribulose was the inducer of the ribitol operon, the formation of D-ribulose within these spontaneous mutant cells caused the induction of the ribitol operon (Fig. 5B). This resulted in the synthesis of D-ribulokinase, thereby providing the second enzyme activity of the new pathway. Thus, the pathway for D-arabinose catabolism consisted of an isomerase from the L-fucose catabolic pathway, obtained by means of a regulatory mutation that permitted the constitutive synthesis of the enzyme, and a ribulokinase borrowed from the ribitol pathway.

3.2.1b. *A Regulatory Mutation Permitting Induced Synthesis of an Isomerase.* A study of D-arabinose-positive mutants of another organism,

K. aerogenes W70, revealed that not all of the isolates were constitutive for L-fucose isomerase. Some of the mutants were still found to be inducible for isomerase activity despite the fact they were able to use the enzyme to grow on D-arabinose. The isomerase activities of four of the D-arabinose-positive mutants from K. aerogenes W70, strains C110, C111, C104, and C107, are shown in Table VII.

All of the D-arabinose-positive mutants isolated possessed a mutation in the regulation of the synthesis of L-fucose isomerase. In some cases this mutation resulted in the constitutive synthesis of the enzyme, as shown for strains C104 and C107, where isomerase activity was present when cells were grown on casein hydrolysate without carbohydrate. For the other two strains shown, C110 and C111, the regulatory mutation did not result in the constitutive synthesis of the enzyme, but permitted the isomerase to be induced when the cells were incubated with D-arabinose. These inducible mutants now recognized D-arabinose, or some product derived from D-arabinose, as a second, alternate inducer of the L-fucose isomerase (St. Martin and Mortlock, 1977).

St. Martin and Mortlock produced evidence that the normal inducer of the L-fucose isomerase and L-fuculokinase enzymes was the intermediate in the pathway, L-fuculose 1-phosphate (Fig. 5A). When cells were presented with L-fucose as a potential growth substrate, the low endog-

Table VII
Isomerase Activity Found in D-Arabinose-Positive Isolates of *Klebsiella aerogenes* W70[a]

| Strain | Growth medium | Isomerase activity (U/mg protein) |
|---|---|---|
| Parent strain (araK4) | CH | <0.07 |
| | CH + L-Fucose | 1.19 |
| | CH + D-Arabinose | <0.07 |
| D-Arabinose-positive isolates | | |
| C110 | CH | <0.003 |
| | D-Arabinose | 0.483 |
| C111 | CH | <0.003 |
| | D-Arabinose | 0.595 |
| C104 | CH | 0.677 |
| | D-Arabinose | 0.391 |
| C107 | CH | 0.569 |
| | D-Arabinose | 0.719 |

[a]Cells were grown at 37°C on a mimimal salts medium supplemented with the indicated carbon and energy sources. CH, Casein hydrolysate. [Data are from St. Martin and Mortlock (1977).]

enous levels of isomerase and kinase that could be found even in uninduced cells converted some of the L-fucose to fuculose 1-phosphate. Once this latter compound was formed, induction resulted and much higher levels of the enzymes could be synthesized.

The D-arabinose-positive strains, such as C110 and C111, were still inducible for isomerase activity and had acquired a mutation in a regulatory gene so that D-ribulose 1-phosphate was also recognized as an inducer. When these mutant cells of the D-arabinose inducible phenotype were presented with D-arabinose as a growth substrate, endogenous levels of L-fucose isomerase and fuculokinase were able to convert some of the D-arabinose to D-ribulose and then ribulose 1-phosphate, as shown in Fig. 5C. The formation of the latter caused the induction of higher amounts of both the isomerase and kinase, resulting in the more rapid isomerization of D-arabinose to D-ribulose. D-Ribulose, however, was the inducer of the ribitol operon and the formation of D-ribulose from D-arabinose resulted in the synthesis of the enzymes of that operon, including D-ribulokinase. That kinase from the ribitol pathway phosphorylated most of the ribulose to D-ribulose 5-phosphate, which could be converted to D-xylulose 5-phosphate for further degradation. Some of the 1-phosphate had to continue to be made, however, in order to maintain the induction of L-fucose isomerase (St. Martin and Mortlock, 1976). Thus, a delicate balance was required between the amount of D-ribulose that was phosphorylated to the 5-phosphate and the amount that was converted to the 1-phosphate (Fig. 5C).

### 3.2.2. A New Pathway for D-Arabinose Catabolism in Escherichia coli

Mutants of *E. coli* K12 that had gained the ability to grow on D-arabinose could also be isolated, but since this strain of *E. coli* did not possess a pathway for the degradation of ribitol, the origin of the kinase activity to complete the new pathway in these mutants was a mystery. Investigation showed that this organism used all three of the enzymes from the L-fucose catabolic pathway (L-fucose isomerase, L-fuculokinase, and L-fuculose 1-phosphate aldolase) to establish the new D-arabinose pathway. In those mutants that had gained the ability to grow on D-arabinose, the isomerase catalyzed the conversion of D-arabinose to D-ribulose, L-fuculokinase phosphorylated the ribulose at the C-1 position, and the aldolase cleaved the ribulose 1-phosphate to form dehydroxyacetone phosphate and glycolaldehyde, compounds that were further degraded by other cellular enzymes. The isomerase activity of these enzymes is shown in Table VIII.

All of the D-arabinose-positive mutants of *E. coli* K12 isolated were

**Table VIII**
Isomerase Activity Found in D-Arabinose-Positive Isolates of *Escherichia coli* K12[a]

|  |  | Enzyme activity (U/mg protein) | | |
| --- | --- | --- | --- | --- |
| Strain | Growth substrate | Isomerase | Kinase | Aldolase |
| Parent strain (1000) | Casein hydrolysate | 1–7 | 1–3 | 6 |
|  | L-Fucose | 62–249 | 40–45 | 222 |
|  | D-Arabinose | 3–7 | 4 | 1–3 |
| D-Arabinose-positive isolate (1102) | Casein hydrolysate | 1–2 | 4 | 15 |
|  | L-Fucose | 110–214 | 25–31 | 206 |
|  | D-Arabinose | 165–449 | 28–30 | 340 |

[a]All specific activities are expressed as nanomoles per minute per milligram protein. Cells were grown at 37°C on a mimimal salts medium supplemented with the indicated carbon and energy sources. [Data are from LeBlanc and Mortlock (1971b).]

of the inducible phenotype, similar to the C110 and C111 mutants of *K. aerogenes* W70, where incubation of the mutant cells with D-arabinose resulted in the induction of the L-fucose pathway enzymes (LeBlanc and Mortlock, 1971a,b, 1972). Recent evidence has indicated that the natural inducer of the L-fucose pathway enzymes in this strain of *E. coli* is also the intermediate L-fuculose 1-phosphate and that as a result of the mutation in a regulator gene, these D-arabinose-positive mutants also recognize D-ribulose 1-phosphate as an alternate inducer (Bartkus and Mortlock, 1983). Skjold and Ezekiel (1982) have reported that the L-fucose pathway in *E. coli* K12 is regulated under the positive control of an activator protein rather than simply under the negative control of a repressor.

### 3.2.3. A New Pathway for D-Arabinose Catabolism in *Salmonella typhimurium*

A study by Gutnick *et al.* (1969) on the ability of a strain of *Salmonella typhimurium* to utilize various compounds as carbon or nitrogen sources for growth reported the use of L-fucose but not D-arabinose. Old and Mortlock (1977) were able to obtain mutants from several different biotypes of *S. typhimurium*, all of which gained the ability to utilize D-arabinose. The pathway for D-arabinose degradation in the mutants was similar to that described for *Escherichia coli*, with the D-ribulose formed by the isomerization of D-arabinose being converted to the 1-phosphate. The mutation establishing growth on D-arabinose also resembled the mutation found in the D-arabinose-positive mutants of *E. coli*, in that the incubation of the mutant cells with D-arabinose resulted in the induction of the enzymes of the L-fucose catabolic pathway.

## 3.3. The L-Lyxose and L-Xylose Pathways in *Klebsiella pneumoniae*

Anderson and Wood (1962) reported the growth of *K. pneumoniae* PRL-R3 on L-xylose. When cells grown on glucose were used to inoculate a medium containing only L-xylose as the carbon and energy source, a very long lag time was required before growth was observed. These authors did not determine whether growth on L-xylose resulted from a very slow adaptation of cells for the utilization of L-xylose or from the selection of L-xylose-positive mutants.

Cell-free extracts prepared from cells grown on L-xylose contained an isomerase activity capable of converting L-xylose to L-xylulose, as shown in Fig. 3. The extracts also contained an L-xylulokinase to catalyze the phosphorylation to L-xylulose 5-phosphate. The latter compound was converted to L-ribulose 5-phosphate by a 3-epimerase activity and then to D-xylulose 5-phosphate by an L-ribulose 5-phosphate 4-epimerase (Anderson and Wood, 1960, 1962a,b).

Sufficient L-lyxose was not available to grow cells for enzyme assay, but cells grown on L-xylose possessed an isomerase activity that would convert L-lyxose to L-xylulose. Since the same extract also had enzyme activities to convert the L-xylulose to D-xylulose 5-phosphate, a possible pathway for the degradation of L-lyxose was identified in the L-xylose-grown cells (Anderson and Wood, 1962a).

Later Mortlock and Wood (1964a) confirmed the ability of *K. pneumoniae* PRL-R3 to grow on L-xylose, but found that 5–6 weeks normally was required after inoculation of an L-xylose medium before any growth could be observed (Table I). Once growth had occurred, however, the transfer of cells to a new L-xylose medium resulted in growth being complete in several days. A pure culture, obtained after growth on L-xylose was completed, retained the ability to grow on L-xylose with the shorter lag time even after it had been cultured on glucose, suggesting that at least one mutation had taken place to permit the utilization of L-xylose as a growth substrate. A much shorter lag time appeared necessary to select mutants capable of growing on L-lyxose, although the lack of availability of sufficient quantities of L-lyxose hampered growth experiments with that substrate.

Mortlock and Wood also tested for isomerase, dehydrogenase, and kinase activity in extracts of cells grown on a variety of five-carbon sugars and detected both L-xylose isomerase and L-xylulokinase activity in cells grown on L-xylose, as shown in Table IV. In addition, the L-xylose-grown cells contained low but detectable levels of both enzymes of the inducible ribitol operon, ribitol dehydrogenase and D-ribulokinase (see Chapter 1), and very high amounts of activity for the isomerization of D-arabinose. Later it was shown that the isomerization of both L-xylose and D-arabinose was catalyzed by the same enzyme, the L-fucose isomerase of the catabolic pathway for the utilization of L-fucose (Camyre and Mortlock, 1965; Oliver and Mortlock, 1971a,b). Although L-fucose isomerase was normally an inducible enzyme, a mutation permitting constitutive synthesis allowed the use of the enzyme for the isomerization of L-xylose as well as D-arabinose. All of the L-xylose-positive mutants tested were found to be constitutive for the synthesis of the enzyme.

Separate experiments showed that the formation of L-xylulokinase and all other required enzymes for the degradation of L-xylulose were induced when L-xylulose was formed within the cells (LeBlanc and Mortlock, 1973). Yet, when L-fucose isomerase-constitutive mutants were obtained by other methods and given L-xylose as a potential growth substrate, a lag period still was required before growth could occur, although the lag period was greatly decreased to about 1 week. It has been postulated, but not experimentally confirmed, that two separate mutations were required to yield the L-xylose-positive mutant. One of the mutations provided the mutant with constitutive isomerase activity, while the second mutation provided some transport system to bring the pentose through the cell membrane. According to this suggestion, when an L-xylose medium was inoculated with the parent strain, those isomerase-constitutive mutants present in the cell population could only grow very slowly since they were dependent upon the simple diffusion of the pentose through the cell membrane. When the population of these constitutive mutants eventually reached a sufficient number, a second mutation occurred to permit the active transport of L-xylose through the membrane. This latter mutation greatly increased the growth rate on L-xylose, and this new strain with two mutations was able to complete growth on L-xylose in only a few days.

Anderson and Wood (1962a) had shown that extracts prepared from L-xylose-grown cells contained all the enzyme activities required for the degradation of L-lyxose, including L-lyxose isomerase activity. Mortlock and Oliver found that L-lyxose-grown cells were also constitutive for L-fucose isomerase activity, but they were unable to show any activity of the purified isomerase for L-lyxose as a substrate. The nature of the mutation or mutations permitting growth on L-lyxose is still not known.

## 4. Summary

Mutants of *K. pneumoniae* PRL-R3, *K. aerogenes* W70, and *E. coli* K12 have been isolated that were able to grow using D-arabinose as the only carbon and energy source. In all cases the initial pathway for D-arabinose catabolism was found to be catalyzed by unaltered enzymes, borrowed by means of a mutation in regulation from previously existing pathways that had evolved for the degradation of more common pentoses.

For *K. pneumoniae* PRL-R3 and some of the *K. aerogenes* W70 mutants, the regulatory mutation resulted in the constitutive synthesis of the isomerase activity required for the initial reaction in the newly evolved pathway for D-arabinose. For *E. coli* K12, *S. typhimurium,* and some of the *K. aerogenes* W70 mutants, the mutation in a regulator gene allowed a product of D-arabinose degradation to induce the required enzyme activity. Although mutations in the isomerase structural gene can alter the enzyme and permit more efficient isomerization of D-arabinose (Oliver and Mortlock, 1971b), such mutations are not required to establish growth on D-arabinose. In all cases the mutation required to initiate growth on D-arabinose was a mutation in the regulation of a previously existing pathway.

Mutants of *K. pneumoniae* PRL-R3 have also been isolated that grow on L-xylose, and in those cases as well a regulatory mutation established the constitutive isomerase activity in the mutant cells. A second mutation for the transport of L-xylose through the cell membrane may also be required for growth on L-xylose, although this has not been confirmed. Mutants of the same organism capable of growth on L-lyxose have also been studied, but even though such mutants have been found to be constitutive for certain enzymes of the L-fucose catabolic pathway, the role of those enzymes in the metabolism of L-lyxose has not been established.

Mutants capable of growth on D-lyxose have been isolated from both *K. pneumoniae* PRL-R3 and *E. coli* K12. In both cases the mutants acquired an enzyme that catalyzed the isomerization of D-lyxose to D-xylulose, an enzyme activity the parent strains were not known to possess. Since this enzyme activity was constitutive in the *E. coli* mutants but not in the parent strain, it would appear that a regulatory mutation had also occurred in this situation, a mutation that permitted the structural gene for this enzyme to be expressed. This newly observed isomerase also possessed activity for the isomerization of the more commonly occurring sugar D-mannose. However, D-mannose is normally transported into *E. coli* cells by a transport system that catalyzes its phosphorylation as it passes through the cell membrane. Once inside the cells, the mannose phosphate is isomerized by a mannosephosphate isomerase enzyme. It has been suggested that the isomerase utilized to establish the new D-lyxose pathway may have been produced by a silent gene, a gene that was not expressed or used but was still retained on the chromosome of the cell. It also has been suggested that this gene and the structural gene for mannosephosphate isomerase may share a common evolutionary history, a concept that is strengthened by the observation that the structural genes for these two enzymes are located almost exactly 180° apart on the *E. coli* K12 chromosome.

In each of the above cases it would appear that the critical mutation to establish growth was a mutation in regulation. While in some cases the structural gene for the needed catalytic activity originated from a catabolic pathway normally employed for a known pathway, in one case, that of D-lyxose, the data suggest that the new isomerase activity may have been obtained by activating a "silent gene," a gene not normally utilized for any current metabolic pathway.

ACKNOWLEDGMENT. The author's research on this topic has been supported by Public Health Services Grant AI 15328 from the National Institute of Allergy and Infectious Diseases and by Hatch Project 144-436 from the U.S. Department of Agriculture.

## References

Anderson, R. L., and Allison, D. P., 1965, Purification and characterization of D-lyxose isomerase, *J. Biochem.* **240**:2367–2372.

Anderson, R. L., and Wood, W. A., 1960, L-Xylulokinase and L-xylulose 5-phosphate-L-ribulose 5-phosphate 3-epimerase in *Aerobacter aerogenes, Biochim. Biophys. Acta* **42**:374–376.

Anderson, R. L., and Wood, W. A., 1962a, Pathway of L-xylose and L-lyxose degradation in *Aerobacter aerogenes, J. Biol. Chem.* **237**:296–303.

Anderson, R. L., and Wood, W. A., 1962b, Purification and properties of L-xylulokinase, *J. Biol. Chem.* **237**:1029–1033.

Bachmann, B. J., and Low, K. B., 1975, Linkage map of *Escherichia coli* K-12, edition 6, *Microbiol. Rev.* **44**:1–56.

Bartkus, J. M., and Mortlock, R. P., 1983, Induction of L-fucose isomerase in wild-type and D-arabinose utilizing strains of *Escherichia coli* K-12, *Abstr. Annu. Rev. Microbiol.* K 238, p. 216.

Bisson, T. M., Oliver, E. J., and Mortlock, R. P., 1968, Regulation of pentitol metabolism by *Aerobacter aerogenes*. II. Induction of the ribitol pathway, *J. Bacteriol.* **95**:932–936.

Briggs, J., Finch, P., Percival, E., and Weigel, H., 1982, Assignment of the L configuration to the fucose elaborated by brown seaweeds, *Carbohydrate Res.* **103**:186–189.

Camyre, K. P., and Mortlock, R. P., 1965, Growth of *Aerobacter aerogenes* on D-arabinose and L-xylose, *J. Bacteriol.* **90**:1157–1158.

Cohen, S. S., 1953, Studies on D-ribulose and its enzymatic conversion to D-arabinose, *J. Biol. Chem.* **201**:71–83.

Dills, S. S., Apperson, A., Schmidt, M. R., and Saier, Jr., M. H., 1980, Carbohydrate transport in bacteria. *Microbiol. Rev.* **44**:385–418.

Englesberg, E., 1971, Regulation in the L-arabinose system, in: *Metabolic Pathways*, Vol. 1 (J. Vogel, ed.), Academic Press, New York, pp. 257–294.

Green, M., and Cohen, S. S., 1956, Enzymatic conversion of L-fucose to L-fuculose, *J. Biol. Chem.* **19**:557–568.

Gutnick, D., Calvo, J. M., Klopotowski, T., and Ames, B. N., 1969, Compounds which serve as the sole source of carbon and nitrogen for *Salmonella typhimurium* LT2, *J. Bacteriol.* **100**:215–219.

Horecker, B. L., 1962, Oxidative pathways, in: *Pentose Metabolism in Bacteria. CIBA Lectures in Microbial Biochemistry*, Wiley, New York.

Hotta, K., and Kurokawa, M., 1973, A novel sialic acid and fucose-containing disaccharide isolated from the jelly coat of sea urchin eggs, *J. Biol. Chem.* **248**:629–631.

Laskin, A. I., and Lechevalier, H. A., 1973, in: *Handbook of Microbiology*, Vol. II: *Microbial Composition* (A. I. Laskin, and H. A. Lechevalier, eds.), CRC Press, Cleveland, Ohio.

LeBlanc, D. J., and Mortlock, R. P., 1971a, Metabolism of D-arabinose: Origin of a D-ribulokinase activity in *Escherichia coli*, *J. Bacteriol.* **106**:82–89.

LeBlanc, D. J., and Mortlock, R. P., 1971b, Metabolism of D-arabinose: A new pathway in *Escherichia coli*, *J. Bacteriol.* **106**:90–96.

LeBlanc, D. J., and Mortlock, R. P., 1972, The metabolism of D-arabinose: Alternate kinases for the phosphorylation of D-ribulose in *Escherichia coli* and *Aerobacter aerogenes*, *Arch. Biochem. Biophys.* **150**:774–781.

LeBlanc, D. J., and Mortlock, R. P., 1973, Regulation of the L-arabinose catabolic pathway in *Aerobacter aerogenes*, *Arch. Biochem. Biophys.* **156**:390–396.

Messer, M., and Kerry, K. R., 1973, Milk carbohydrates of the echidna and the platypus, *Science* **180**:201–203.

Mortlock, R. P., 1976, Catabolism of unnatural carbohydrates by microorganisms, *Adv. Microb. Phys.* **13**:1–55.

Mortlock, R. P., and Wood, W. A., 1964a, Metabolism of pentoses and pentitols by *Aerobacter aerogenes* I. Demonstration of pentose isomerase, pentulokinase, and pentitol dehydrogenase enzyme families, *J. Bacteriol.* **88**:838–844.

Mortlock, R. P., and Wood, W. A., 1964b, Metabolism of pentoses and pentitols by *Aerobacter aerogenes*. II. Mechanism of acquisition of kinase, isomerase, and dehydrogenase activity, *J. Bacteriol.* **88**:845–849.

Old, D. C., and Mortlock, R. P., 1977, The metabolism of D-arabinose by *Salmonella typhimurium*, *J. Gen. Microbiol.* **101**:341–344.

Oliver, E. J., and Mortlock, R. P., 1971a, Growth on *Aerobacter aerogenes* on D-arabinose: Origin of the enzyme activities, *J. Bacteriol.* **108**:287–292.

Oliver, E. J., and Mortlock, R. P., 1971b, Metabolism of D-arabinose by *Aerobacter aerogenes:* Purification of the isomerase, *J. Bacteriol.* **108**:293–299.

Palleroni, N. J., and Doudoroff, M., 1956, Mannose isomerase of *Pseudomonas saccharophila*, *J. Biol. Chem.* **218**:535–548.

Riley, M., and Anilionis, A., 1978, Evolution of the bacterial genome, *Annu. Rev. Microbiol.* **32**:519–560.

St. Martin, E. J., and Mortlock, R. P., 1976, Natural and altered induction of the L-fucose catabolic enzymes in *Klebsiella aerogenes*, *J. Bacteriol.* **127**:91–97.

St. Martin, E. J., and Mortlock, R. P., 1977, A comparison of alternate metabolic strategies for the utilization of D-arabinose, *J. Mol. Evol.* **10**:111–122.

Schaffer, R., 1972, Naturally occuring monosaccharides, in: *The Carbohydrates*, Vol. IA (W. Pigman and D. Horton, eds.), Academic Press, New York, pp. 69–111.

Skjold, A. C., and Ezekiel, D. H., 1982, Regulation of D-arabinose utilization in *Escherichia coli* K-12, *J. Bacteriol.* **152**:521–523.

Stevens, F. J., and Wu, T. T., 1976, Growth on D-lyxose of mutant strain of *Escherichia coli* K12 using a novel isomerase and enzymes related to D-xylose metabolism, *J. Gen. Microbiol* **97**:257–265.

Stevens, F. J., Stevens, P. W., Hovis, J. G., and Wu, T. T., 1981, Some properties of D-mannose isomerase from *Escherichia coli* K12, *J. Gen. Microbiol.* **124**:219–223.

Wilson, B. L., and Mortlock, R. P., 1972, Regulation of D-xylose and D-arabitol catabolism in *Aerobacter aerogenes*, *J. Bacteriol.* **113**:1404–1411.

Zipkas, D., and Riley, M., 1975, Proposal concerning mechanism of evolution of the genome of *Escherichia coli*, *Proc. Natl. Acad. Sci. USA* **72**:1354–1358.

CHAPTER 5

# Functional Divergence of the L-Fucose System in Mutants of Escherichi coli

## E. C. C. LIN AND T. T. WU

## 1. Introduction

The catabolic system for L-fucose of *Escherichia coli* has given rise to a number of novel metabolic functions. Studies on this system have yielded illustrations of three concepts for biochemical evolution: (1) a pyridine nucleotide-linked oxidoreductase can be elected by the cell to serve as either a dehydrogenase or a reductase, depending primarily on the mode of regulating the expression of the structural gene and on the nature of the preceding and following reactions in the pathway; (2) genetic mobilization of components of an established metabolic systems for a novel function can lead to the extinction of the remaining genes that become superfluous; and (3) once the expression of a structural gene is liberated from its normal regulatory constraint to provide a new service, the gene product can act as steppingstone for the elaboration of other novel metabolic pathways—the principle of preadaptation (Lin *et al.*, 1976; Wu, 1978; Lin, 1979, 1981). The evolutionary studies of the fucose system, as well as metabolic systems derived from or related to it, will be presented partly in a historical framework, since the ways in which knowledge unfolds and converges are often of heuristic value themselves.

---

*E. C. C. LIN* • Departments of Microbiology and Molecular Genetics, Harvard Medical School, Boston, Massachusetts 02115.  *T. T. WU* • Departments of Biochemistry and Molecular Biology, Northwestern University, Evanston, Illinois 60201.

## 2. Reversibility of NAD-Linked Reactions

Interconversions of alcohols and their corresponding aldehydes and ketones at neutral pH are overwhelmingly in favor of alcohol formation when coupled with pyridine nucleotides [typically by a factor of $10^4$ (Krebs and Kornberg, 1957)]. Because of this thermodynamic bias, it is a simple matter for enzymes catalyzing this class of reactions to serve physiologically as carbonyl group reductases. Classical cases are the reduction of acetaldehyde to ethanol and of pyruvate to lactate during anaerobiosis. (In this context, alcohol dehydrogenase and lactate dehydrogenase, the enzymes that catalyze these respective reactions, are functionally misnomers.) On the other hand, many pyridine nucleotide-linked enzymes do serve a reverse physiological role as hydydrogenases of alcohols. The conversion of D-arabitol or xylitol to D-xylulose and of ribitol to D-ribulose by such enzymes are good examples (see Chapter 1). Cellular NAD/NADH ratios are not radically different in aerobically or anaerobically growing cells (Wimpenny and Firth, 1972), which is not surprising in view of the fact that under any growth conditions there are numerous metabolic reactions that require either one or the other form of the coenzyme. Therefore, drastic changes in the relative concentrations of the oxidized and reduced forms of this coenzyme are not available as a general mechanism for determining whether an enzyme is to act as a dehydrogenase or a reductase. In the utilization of the pentitols, the cell partially surmounts the energetic obstacle of the dehydrogenation step by active transport of the substrate to an elevated intracellular concentration, thus providing a chemical "push." More importantly, a "pull" for the reaction is provided by using an ATP-dependent kinase in the subsequent step to trap the pentose as a phosphorylated product. The equilibrium constant of the pulling reaction is about $10^3$ in favor of pentose phosphate formation. Because pyridine nucleotide-linked enzymes can function in opposite manners, they are often referred to as oxidoreductases.

## 3. A Mutant That Uses an NAD-Linked Dehydrogenase to Grow on L-1,2-Propanediol

Wild-type *E. coli* strains K12, B, ML, and W do not utilize L-1,2-propanediol (propylene glycol), but mutants that do can be readily isolated. In a pilot experiment, about $10^{10}$ *E. coli* K12 cells derived from a previous population treated with a chemical mutagen (ethyl methanesulfonate) were inoculated into 1 liter of mineral medium containing 0.2% of DL-1,2-propanediol (13 mM of the L isomer) as the sole potential source of carbon and energy and incubated aerobically in a 2-liter flask at 37°C on a rotary shaker. At the end of 4 days, the culture was visibly more turbid. A portion was withdrawn and subcultured repeatedly in a propanediol medium to allow approximately 100 mass doublings. A sample of the final culture was diluted and plated on agar containing 0.2% of propanediol. After 2 days of growth at 37°C, a large colony was picked and after purification by recloning was designated as strain 3. This mutant does not grow at all when tested with D-1,2-propanediol, but grows efficiently on the L isomer, giving about the same cell yield as glycerol or glucose on the basis of per milligram of carbon.

Mutants of *E. coli* B, ML, and W similar to strain 3 were also isolated (D. Brandon, unpublished observation). But so far, no mutants capable of growth on D-1,2-propanediol have been isolated.

### 3.1. Characterization of the novel Biochemical Pathway in the Mutant

#### 3.1.1. The First Enzyme

When extracts of strain 3 grown on different carbon and energy sources were examined, there appeared a *constitutively* synthesized NAD-linked enzyme that catalyzes the dehydrogenation of L-1,2-propanediol but not the D isomer. The purified enzyme acts also on 1,3-propanediol, which suggests that the dehydrogenation involves the primary alcohol group of the substrate. This is consistent with the ability of the protein to catalyze the reoxidation of NADH with L-lactaldehyde (the expected product of L-1,2-propanediol dehydrogenation) but not with hydroxyacetone. The oxidoreductase has a $K_m$ of about 2–3 mM for L-1,2-propanediol (varies with the condition of assay), a $K_m$ of 0.07 mM for L-lactaldehyde, and a $K_m$ of close to 0.01 mM for both NAD and NADH. Despite the stereospecificity for the L isomer, the enzyme is not highly specific in other respects. It dehydrogenates a number of alcoholic compounds, including glycerol, ethylene glycol, and ethanol. Unlike many NAD-linked dehydrogenases that use $Zn^{2+}$ at the catalytic center, this enzyme apparently uses $Fe^{2+}$ or $Mn^{2+}$, and the values of $K_m$ for different substrates are significantly dependent on the metal ion added. The enzyme has a molecular weight of 76,000 and is composed of two identical subunits (Sridhara *et al.*, 1969; Boronat and Aguilar, 1979, 1981a).

The responsibility of the enzyme for the new growth ability was confirmed by examining the enzymic profile of mutants of strain 3 that lost the acquired function. Some lack the activity of the oxidoreductase; others have defects further on in the pathway (Sridhara *et al.*, 1969).

### 3.1.2. The Second Enzyme

The enzyme that utilizes lactaldehyde also turned out to be NAD-linked; it catalyzes the formation of L-lactate as the product. Unlike the pyridine nucleotide-coupled interconversion of alcohols and their carbonyl counterparts, the NAD-linked oxidation of an aldehyde to a carboxylic acid is highly favored [by a factor of $10^4$ in the case of conversion of acetaldehyde to acetate at pH 7 (Krebs and Kornberg, 1957)]. Indeed, when tested in an enzyme assay mixture, no reverse activity was demonstrated with L-lactate and NADH. The $K_m$ for L-lactaldehyde is 10 mM and that for NAD is 0.01 mM (Sridhara and Wu, 1969). Thus the second enzyme in the novel metabolic pathway, even though NAD-linked, plays a role analogous to that of the ATP-dependent kinase in the pentitol pathways already discussed: dehydrogenation catalyzed by the first enzyme is promoted by effective depletion of the product.

### 3.1.3. Permeability to Propanediol

Is there an active transport system in the cell membrane that raises the internal concentration of propanediol to favor its dehydrogenation? The answer is probably no. Studies on permeability of *E. coli* showed that the cytoplasmic membrane cannot effectively retain a small uncharged compound against a high concentration gradient. Their active transport would therefore only dissipate metabolic energy. On the other hand, the acquisition of a membrane protein that catalyzes entry of the substrate by facilitated diffusion at no metabolic cost should be advantageous; by this system, net inflow occurs only to replenish the compound being metabolically depleted (Sanno *et al.*, 1968; Richey and Lin, 1972; Heller *et al.*, 1980). Indeed, there is indirect evidence for a facilitator protein that mediates the equilibration of propanediol across the cytoplasmic membrane (Hacking *et al.*, 1978).

### 3.2. Identifying the Original Role of a Recruited Enzyme

Genes recruited for novel functions by mutations might be inexpressible (Wu, 1978; also see Chapter 6, this volume), or they might have specialized functions and be expressed only under the appropriate conditions. In bacteria useless genes are not likely to be long spared from pruning by deletion; the intense selection for efficient growth also would impose economical use of DNA. Even if a useless gene should survive, its partners might not. In such a case, its original function might elude identification. On the other hand, if a gene in the parental or progenitor strain is functional, how can the customary role of the gene be determined? In principle there are three ways to search for an answer: (1) by characterizing the substrate specificity of the enzyme; (2) by exploring the expression pattern of the gene; and (3) by mapping the gene in the chromosome. The first approach is based on the premise that the wild-type enzyme has come close to or attained "optimization" by natural selection. Thus the physiological substrate might be recognized by its having the highest ratio of $V_{max}/K_m$ among the various compounds attacked by the enzyme. It was on this basis that the enzyme active on xylitol as a novel carbon and energy source was identified as ribitol dehydrogenase (see Chapters 1 and 2). The second approach is based on the rationale that the discovery of growth conditions specifically increasing the synthesis of the enzyme might reveal its true function. In the xylitol case, the recognition of ribitol as the normal substrate of the dehydrogenase was confirmed by the ability of ribitol to induce the synthesis of the protein in the wild-type cells. The third approach banks on the chance that bacterial genes specifying a set of products with a common function are clustered together—sometimes organized into a single operon—and that if the gene of unknown function is found within a region encoding a known function, the answer might come from "guilt by association."

### 3.3. Connection of the Propanediol Oxidoreductase with the Fucose System

The broad substrate specificity of propanediol oxidoreductase rendered the first approach unfeasible. It was the last approach that yielded the clue for the role of the protein in the wild-type strain. When the genetic site that conferred the growth ability on propanediol was mapped, the position coincided with the locus for L-fucose utilization (Falkow *et al.*, 1963; Taylor and Trotter, 1967; Cocks *et al.*, 1974; Bachmann and Low, 1980).

Both L-fucose and its structural relative, L-rhamnose, were reported to be metabolized by pathways in which the phosphorylated six-carbon intermediates are cleaved to L-lactaldehyde and dihydroxyacetone phosphate (Ghalambor and Heath, 1962; Sawada and Takagi, 1964). Moreover, it was known that in some bacteria half of the rhamnose carbon fermented is excreted as propanediol (Kluyver and Schnellen, 1937). Considering these findings and the similarity between the rhamnose and fucose pathways, it appeared that the lactaldehyde produced from the two sugars is used anaerobically as a hydrogen sink and discarded as propanediol. For this an NAD-linked enzyme would be required. Indeed, enzyme assays showed that a propanediol oxidoreductase activity in wild-type cells is inducible by fucose, and only anaerobically. It then became evident that mutations in strain 3 allow this enzyme to be synthesized *constitutively* and *aerobically* (Cocks *et al.*, 1974).

## 4. Biochemistry of the Fucose System

A better understanding of the mutational steps leading to the recruitment of a lactaldehyde reductase for the reverse role of propanediol utilization thus requires further knowledge of the fucose pathway and its regulation in wild-type cells. Figure 1 shows this biochemical pathway converging with a parallel pathway for rhamnose utilization. The latter is discussed in Section 8. Fucose is metabolized by the sequential action of fucose permease (Hacking and Lin, 1976), fucose isomerase (Green and Cohen, 1956), fuculose kinase (Heath and Ghalambor, 1962), and fuculose 1-phosphate aldolase (Ghalambor and Heath, 1962). The aldolase cleaves the six-carbon substrate into dihydroxyacetone phosphate and lactaldehyde. Anaerobically, lactaldehyde is completely reduced to propanediol by an NAD-linked oxidoreductase. For each mole of fucose fermented, 1 mole of propanediol is excreted via a facilitator protein (Hacking et al., 1978) into the medium and is irretrievably lost (Cocks et al., 1974). The sacrifice of one-half of the carbon skeleton of fucose in this way permits more dihydroxyacetone phosphate to be used as a carbon source. Aerobically, lactaldehyde is completely oxidized by an NAD-linked dehydrogenase to lactate, which is converted to pyruvate by a dehydrogenase of the flavoprotein class (Cocks et al., 1974). As a bonus, dehydrogenation by such a flavoprotein results in the generation of metabolic energy by electron transport.

Synthesis of the enzymes in the trunk pathway is inducible by fucose in the growth medium both aerobically and anaerobically. But the two enzymes in the branches of the system are also controlled by respiratory conditions: whereas induction of lactaldehyde dehydrogenase activity requires aerobiosis, the induction of propanediol oxidoreductase activity requires anaerobiosis (Table I). This explains why excreted or exogenous propanediol cannot be utilized under aerobic conditions.

## 5. Enzymic Changes in the Fucose System in Mutants and Revertants

### 5.1. Propanediol-Positive Mutants Exploit Both Branches of the Fucose System

Once the pivotal genetic change of converting the oxidoreductase from a lactaldehyde-reducing enzyme to a propanediol-oxidizing enzyme is brought about, the rest of the metabolic reactions in the novel pathway can follow the course established in the wild-type strain, because the first

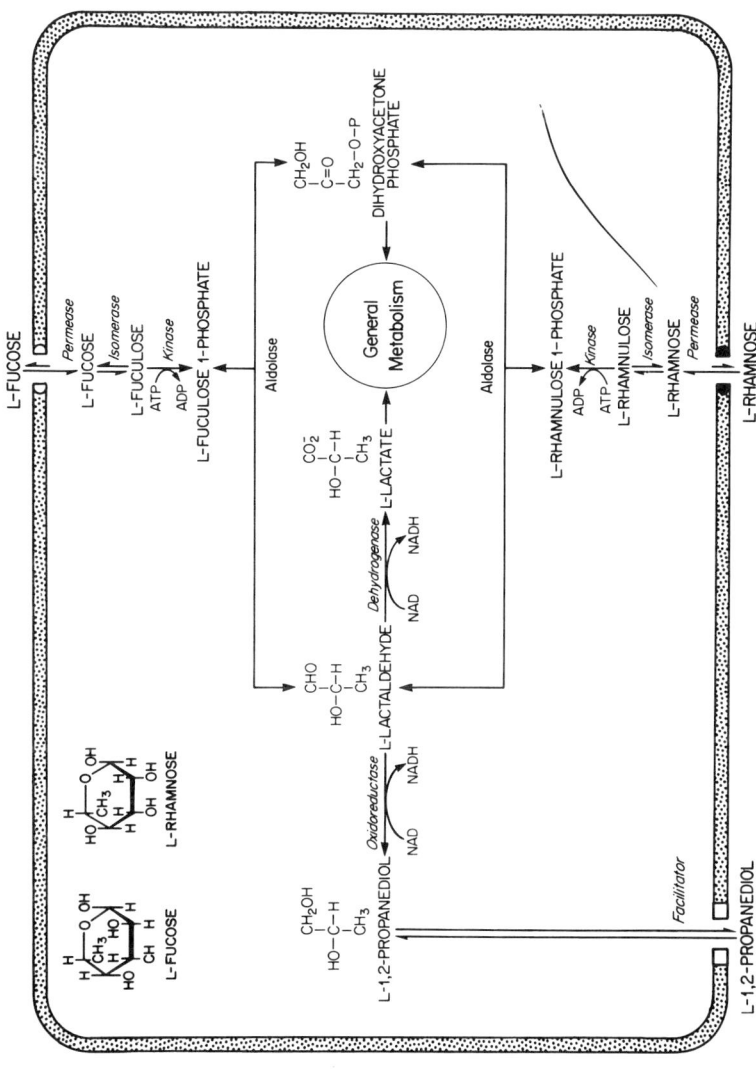

**Figure 1.** Pathways of L-fucose and L-rhamnose dissimilation. The enzyme catalyzing the interconversion of propanediol and lactaldehyde is referred to as propanediol oxidoreductase since the actual role of this protein depends upon the strain in which it is found. FAD and $FADH_2$ are the oxidized and reduced forms of flavin adenine dinucleotide; NAD and NADH are the oxidized and reduced forms of nicotinamide adenine dinucleotide.

## Table I
### Activity of Fucose Pathway Enzymes in Crude Extracts of Wild-Type E. coli as a Function of Growth Conditions[a]

| | | | | | Specific activity[c] | | |
|---|---|---|---|---|---|---|---|
| Carbon[b] | $O_2$ | Fucose permease | Fucose isomerase | Fuculose kinase | Fuculose 1-phosphate aldolase | Propanediol oxidoreductase | Lactaldehyde dehydrogenase |
| CAA | + | 0 | 36 | 27 | 0 | 20 | 93 |
| CAA + fucose | + | 55 | 500 | 440 | 240 | 80 | 100 |
| Fucose | + | 58 | 520 | 420 | 130 | 42 | 280 |
| Glucose | + | | | | | 6 | 15 |
| CAA + pyruvate | − | | | | | 89 | 0 |
| Fucose | − | 64 | 750 | 310 | 270 | 240 | 32 |
| Glucose | − | | | | | 0 | 0 |

[a] Adapted from Hacking and Lin (1976).
[b] CAA, Casein hydrolysate.
[c] Fucose permease activity is expressed in nanomoles per minute per milligram of dry weight at 37°C; all other activities are expressed in nanomoles per minute per milligram of protein at 25°C.

metabolite, lactaldehyde, induces the same NAD-linked dehydrogenase used aerobically for fucose dissimilation (Sridhara and Wu, 1969; Hacking and Lin, 1976). The second metabolite, lactate, in turn induces its dehydrogenase and is converted to pyruvate. The trunk pathway for fucose dissimilation is superfluous under this condition of growth.

Before discussing the kinds of obstacles that must be overcome by mutations in order to have active propanediol oxidoreductase synthesized constitutively and aerobically, it would be useful to describe the successive phenotypic changes associated with improved aerobic growth on propanediol and to consider what is known about the genetic organization of the fucose system.

## 5.2. A Primary Stage Mutant

In the original pilot experiment, no attempt was made to determine the number of mutations that led to the appearance of strain 3. Therefore, it is not known at which stage the ability to grow on fucose was lost; nor is it known whether this loss is an inevitable consequence of repeated selection on propanediol. To gain more precise information, the selection experiment was repeated with nine independent wild-type clones (this time without the use of mutagenesis), and the succession of mutants that arose from one clone was closely monitored and collected for study.

The basal aerobic activity of the oxidoreductase is very low in wild-type strain 1, but the presence of L-fucose during growth increases the activity fourfold. Strain 413, a primary propanediol-positive mutant from one of the nine clones, shows a tenfold elevation in the basal activity and the presence of L-fucose during growth raises this activity level another four times. The inducibility of the trunk enzymes (permease, isomease, kinase, and aldolase) in the fucose pathway remains unperturbed (Table II).

## 5.3. A Secondary Mutant

Strain 421, a derivative of strain 413 with improved growth rate, shows a further increase in the basal activity, which cannot be further increased by induction. It is the constitutivity of this activity that is associated with the fucose-negative phenotype. (Strain 418 was another derivative of strain 413 which still remained fucose-positive.) In all nine independent selections, populations of the initial mutants were eventually replaced by a mutant that lost the ability to grow on fucose. The same defects in enzyme synthesis are responsible for the phenotype: fucose permease, fucose isomerase, and fuculose kinase become noninducible,

## Table II
## Activity of Fucose Pathway Enzymes in Extracts of Various Mutants Grown Aerobically on Different Media[a]

| Strain | Carbon source[b] | Fucose permease | Fucose isomerase | Fuculose kinase | Fuculose 1-phosphate aldolase | Propanediol oxidoreductase | Lactaldehyde dehydrogenase |
|---|---|---|---|---|---|---|---|
| 1 | CAA | 0 | 36 | 27 | 0 | 20 | 93 |
|  | CAA + fucose | 55 | 500 | 440 | 240 | 80 | 100 |
| 413 | CAA | 0 | 20 | 40 | 10 | 210 | 160 |
|  | CAA + fucose | 50 | 460 | 420 | 190 | 990 | 160 |
|  | Propanediol | 0 | 10 | 0 | 0 | 260 | 330 |
| 421 | CAA | 0 | 15 | 20 | 390 | 1,300 | 120 |
|  | CAA + fucose | 0 | 15 | 20 | 380 | 1,200 | 130 |
|  | Propanediol | 0 | 10 | 0 | 180 | 1,300 | 300 |
| 3 | CAA | 0 | 43 | 80 | 490 | 410 | 130 |
|  | CAA + fucose | 0.08 | 30 | 46 | 490 | 390 | 110 |
|  | Propanediol | — | 43 | 57 | 490 | 420 | 260 |
| 430 | CAA | — | — | — | 0 | 750 | 380 |
|  | CAA + fucose | 0 | 20 | 48 | 0 | 680 | 420 |
|  | Propanediol | — | — | — | 0 | 720 | 410 |

[a] Adapted from Hacking and Lin (1977) and Hacking et al. (1978).
[b] CAA, Casein hydrolysate.
[c] Fucose permease activity is expressed in nanomoles per minute per milligram of dry weight at 37°C; all other activities are expressed in nanomoles per minute per milligram of protein at 25°C.

but the aldolase becomes constitutive (Hacking and Lin, 1977). These typical changes in the regulatory pattern of gene expression are exemplified by strain 421 and by the prototype mutant, strain 3, studied earlier (Table II).

### 5.4. Pseudorevertants That Regained the Ability to Grow on Fucose

The lost ability to grow on fucose by mutants such as strain 3 can be recovered by back selection on fucose as carbon and energy source. The mechanism of restoration of this lost ability, however, is not by exact reversion. For instance, strains 54, 55 and 56 selected in this way from strain 3 show constitutive rather than inducible synthesis in the permease, isomerase, and kinase, as well as the aldolase and the oxidoreductase (Table III). The permease activity, however, is not fully restored in all these cases. These pseudorevertants can grow aerobically on both fucose and propanediol.

### 5.5. A Mutant with Superior Scavenger Power for Propanediol

The mutant strains 3 and 421 were selected by serial transfer of cells into media containing initially 12–20 mM L-1,2-propanediol. To see what kind of mutation would be favored under more stringent nutritional conditions, cells of strains 3 were cultured in media in which the propanediol concentration was not allowed to exceed 0.5 mM. A mutant, strain 430, that emerged from this selection was identified as a large colony on agar containing 0.25 mM L-1,2-propanediol. Strain 430 shows a growth $K_m$ of 1 mM on the compound, in contrast to strain 3, which shows a growth $K_m$ of 7 mM (Boronat and Aguilar, 1981a). At the enzymatic level, strain 430 shows four changes: (1) the aldolase gene is inactive or deleted; (2) the constitutive oxidoreductase activity level is increased; (3) lactaldehyde dehydrogenase is synthesized constitutively and at an elevated level; and (4) an additional permeation system is mobilized for the entry of propanediol [data not presented in Table II; see Hacking et al. (1978)]. The last three changes should act in concert to hasten the entry of propanediol and its conversion to lactate, which is more retainable because of its negative charge [the phospholipid cell membrane is less permeable to compounds bearing electrical charges (Davis, 1958)]. The significance of the loss of the aldolase activity is not yet known. It is possible that increasing the expression of the gene for the oxidoreductase yet another notch, rendered the aldolase gene inexpressible. Alternatively, an independent mutation inactivated the aldolase gene, thereby preventing gratuitous synthesis of the protein. Evidently because of this additional mutation, strain 430, in contrast to strain 3, can no longer be reverted to

## Table III
### Activity of Fucose Pathway Enzymes in Crude Extracts of Revertants of Strain 3 Grown Aerobically[a]

| Strain number | Carbon source[b] | Specific activity[c] | | | |
|---|---|---|---|---|---|
| | | Fucose permease | Fucose isomerase | Fuculose kinase | Fuculose 1-phosphate aldolase |
| 54 | CAA | 0.3 | 110 | 110 | 470 |
| | CAA + fucose | 0.3 | 100 | 120 | 500 |
| 55 | CAA | 8.0 | 230 | 160 | 500 |
| | CAA + fucose | 7.0 | 240 | 160 | 440 |
| 56 | CAA | 15.0 | 600 | 550 | 370 |
| | CAA + fucose | 15.0 | 630 | 400 | 420 |

[a] Adapted from Hacking and Lin (1976).
[b] CAA, Casein hydrolysate.
[c] Fucose permease activity is expressed in nanomoles per minute per milligram of dry weight; all other activities are expressed in nanomoles per minute per milligram of protein.

grow on fucose (Hacking et al., 1978); it is not known how many genetic events separated strain 430 from its progenitor. In any case, an evolutionary point of no return is apparently reached, setting the stage for the deletion of functionless genes in the trunk pathway.

## 5.6. Changes in the Property of the Oxidoreductase

Selection for rapid growth on propanediol changed not only the expression of the oxidoreductase gene, but also the structure of the protein. As shown in Table IV, the $V_{max}$ of the mutant enzyme of strain 3 for propanediol dehydrogenation is 1.3 times higher than that of the wild-type enzyme of strain 1; the $K_m$ of the mutant enzyme is twice the wild-type value. There is no change in the $K_m$ values for NAD and NADH (Boronat and Aguilar, 1981a). An improvement of the catalytic rate accompanied by reduction of apparent affinity of the substrate is not out of the ordinary, in view of the Haldane equation (1930, pp. 80–83) which takes the following form in this particular case:

$$K_{eq} = \frac{[L]}{[P]} = \frac{V^P K_m^L K_m^{NADH}}{V^L K_m^P K_m^{NAD}}$$

where $K_{eq}$ is the equilibrium constant of the interconversion of P (propanediol) and L (lactaldehyde); $V^P$ is the maximal velocity of the forward reaction; $V^L$ is the maximal velocity of the reverse reaction; and $K_m^P$, $K_m^L$, $K_m^{NAD}$, and $K_m^{NADH}$ are the apparent dissociation constants of the enzyme from the substrate, the product, NAD, and NADH, respectively. In general terms, the Haldane relationship predicts that for a given chemical reaction, an enzyme cannot improve its forward rate of catalysis without (1) sacrificing some apparent affinity for one or both of the substrates, (2) increasing the rate of the reverse reaction (which becomes self-defeating unless at least one of the products is rapidly removed), or (3) increasing the apparent affinity for one or both of the products (which

Table IV
Kinetic Parameters of Mutant and Wild-Type Propanediol Oxidoreductases[a]

| Source of enzyme | L-1,2-Propanediol | | L-Lactaldehyde | |
|---|---|---|---|---|
| | $V_{max}$ | $K_m$ | $V_{max}$ | $K_m$ |
| Strain 1 | 3.9 | 1.3 | 12 | 0.030 |
| Strain 3 | 5.1 | 2.7 | 24 | 0.071 |

[a]Units of $V_{max}$ are μmole/min per mg protein at 25°C, and $K_m$ is in mM. [From Boronat and Aguilar (1981a).]

can eventually become obstructive by making the catalytic site unavailable to the substrates). Unfortunately, the kinetic data taken from the mutant and wild-type enzymes cannot be strictly applied to the equation, because the forward and reverse conditions were not measured under the same conditions (for example, pH 9.5 for assaying the forward reaction and pH 7.0 for assaying the reverse reaction in different chemical solutions). [For a more rigorous discussion of the evolution of enzymes towards perfection, see Albery and Knowles (1976).]

A structural alteration of the strain 3 enzyme is further indicated by moderate changes in substrate specificity, decreased thermal stability, and optimal activity at a higher than normal pH. The simplest inference to be drawn is that in a growth medium not limited in substrate concentration (about ten times the $K_m$ value), the selective pressure placed a premium on increasing the $V^P$ (Boronat and Aguilar, 1981b). An alternative explanation would be that the changes of kinetic characteristics are incidental to the structural modification enabling the protein to be synthesized in an active form aerobically (see Section 6.5 on posttranscriptional control).

## 6. Genetic Organization and Regulation of the Fucose System

### 6.1. A Regulon Comprised of Closely Linked Operons

The changes in enzymic patterns in various mutants and pseudo-revertants revealed unusual complexity of the fucose system. The manner in which the enzymic profiles varied in the different strains suggests a regulon with five operons encoding (1) the permease, (2) the isomerase and the kinase, (3) the aldolase, (4) the oxidoreductase (and likely also the propanediol facilitator), and (5) the regulator protein. A set of deletion-mutations generated from the direction of the neighboring *argA* locus indicated the counterclockwise gene order in the circular *E. coli* chromosome: the regulator gene (*fucC*; assignment tentative), the kinase gene (*fucK*), the isomerase gene (*fucI*), the permease gene (*fucP*), the aldolase gene (*fucA*), and the oxidoreductase gene (*fucO*), as shown in Fig. 2 (Hacking and Lin, 1976, 1977; Skjold and Ezekiel, 1982a; Y. Zhu and T. Chakrabarti, personal communication). It is of interest to note here that in the related bacterium *Klebsiella pneumoniae* [previously *K. aerogenes* (Ørskov, 1974)] W70, the control of the isomerase and the kinase can also be dissociated from that of the aldolase (St. Martin and Mortlock, 1976). [The DNA of *K. pneumoniae* and that of *E. coli* show 25–50% homology (McCarthy and Bolton, 1963).]

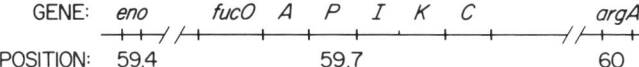

**Figure 2.** Tentative map of the fucose region. Proteins encoded by the genetic markers: *eno*, enolase (grows on glycerol plus succinate, but not on glucose); *fucO*, propanediol oxidoreductase; *fucA*, fuculose 1-phosphate aldolase; *fucP*, fucose permease; *fucI*, fucose isomerase; *fucK*, fuculose kinase; *fucC*, positive regulator protein; *argA*, N-acetylglutamate synthetase (requires arginine for growth).

## 6.2. Positive Control

Although direct evidence is not yet available, arguments can be presented for the joint control of the genes specifying the fucose trunk pathway and the oxidoreductase by an activator rather than a repressor protein. First, fucose-constitutive mutants are rare (LeBlanc and Mortlock, 1971b; T. Chakrabarti, unpublished observation). Repressor-controlled systems readily give rise to constitutive mutants (frequencies of roughly $10^{-6}$), because many if not most random changes demolish gene function. This is not true for activator-controlled systems, because few mutations (roughly $10^{-9}$) should be able to remodel the protein in such a way that it can function with the target promoter without the inducer. (For the same reason, mutations that change inducer specificity should be very rare for both repressor and activator proteins.) Second, the homologous L-rhamnose system is under positive control (see discussion of this system in Section 8).

## 6.3. The Inducer

As already mentioned, synthesis of the enzymes in the trunk pathway for fucose can be induced either aerobically or anaerobically. The actual inducer of this pathway in both *E. coli* (Bartkus and Mortlock, 1983) *K. pneumoniae* (St. Martin and Mortlock, 1976) is not likely to be fucose, but fuculose 1-phosphate. *Klebsiella pneumoniae* mutants lacking only the isomerase or the kinase have not been isolated but mutants blocked only in the aldolase have been obtained. This means that for the fucose system to be induced, the substrate and its metabolites have to satisfy the stringent composite specificities of the permease, the isomerase, the kinase, and the activator protein. Induction by the substrate itself or the first product in a catabolic pathway is common in bacteria, but induction by the second product is less frequent. Multiple screening might be especially important to ensure that a substrate can be utilized; partial metabolism of structural analogs of the substrate might lead to toxic accumulation of dead-end products. However, other factors being equal, multiple-step screening delays the response to the substrate. [It takes *E. coli* cells about 3.5hr to resume growth when they are transferred from a glucose to a fucose medium (LeBlanc and Mortlock, 1971b). Growth lags on other sugars metabolized by inducible systems usually are less than 1 hr.] An alternative explanation for the unusual induction by a remote product based on mechanistic fixation during the course of evolution in unappealing in light of the discovery that induction specificity can be altered by a single mutation (see Section 10.1 on D-arabinose-positive mutants).

### 6.4. Lactaldehyde Dehydrogenase under Separate Control

The structural gene for lactaldehyde dehydrogenase is in all likelihood under the control of a separate inducer (lactaldehyde itself) and a regulatory protein, since the enzyme is inducible in strain 413 without concomitant increases in the levels of enzymes of the trunk pathway. Furthermore, the dehydrogenase remains inducible by propanediol in strains 3 and 421, in which all the other enzymes specifically related to fucose utilization are either noninducible or constitutive (Table II). This separate control is also consistent with the ability of wild-type cells to grow on lactaldehyde as the source of carbon and energy (Sridhara and Wu, 1969). The position of this gene is close to the genes of the fucose system (Chakrabarti; *et al.*, 1984).

### 6.5. Posttranscriptional Control of the Oxidoreductase Activity

An unexpected finding is that the regulation of the oxidoreductase activity occurs at two different levels. Expression of the gene is actually inducible (presumably by fuculose 1-phosphate) irrespective of the presence or absence of molecular oxygen, but only under anaerobic conditions is the enzymatically active protein formed. The induced aerobic synthesis of an enzymically inactive protein was demonstrated with the aid of specific antibodies (Boronat and Aguilar, 1981b). Aerobic induction of gene expression was confirmed by the behavior of three different hybrid operons with the promoter of the oxidoreductase gene joined to the structural genes for lactose utilization in a bacterial host strain with its normal lactose operon deleted. In these cells, fucose induces the synthesis of $\beta$-galactosidase both aerobically and anaerobically (Chen *et al.*, 1983; Chen and Lin, 1984a).

In wild-type cells, the oxidoreductase induced anaerobically remains enzymically active during aerobic growth in the absence of fucose. On the other hand, cells fully induced aerobically by fucose and then allowed limited anaerobic growth on glucose do not gain any oxidoreductase activity. This would indicate that once the protein is completely formed, no further modification takes place (Chen *et al.*, 1983). How the enzymic activity of the protein is determined prior to completion remains to be discovered.

## 7. Sequential Mutations Changing Propanediol and Fucose Utilization

Until fine structures of the fucose genetic system and its regulatory elements are characterized, only conjectures can be offered to explain the changes in the various mutants. Constraints on the activator protein to interact with several nonidentical promoters of the fucose regulon might impose an unusually severe limit on tolerable amino acid substitutions. As a consequence, very few missense mutations could allow the dispensation of the inducer without undermining the activator functions (at least in *E. coli*). This makes an "up-promoter" mutation in the oxidoreductase operon gene a more frequent event that permits growth on propanediol. Such a solution might have been used by strain 413, which has undisturbed inducibility of the permease, isomerase, kinase, aldolase, as well as the retention of a fourfold inducibility of the oxidoreductase gene. Alternatively, the first mutation permitting growth on propanediol might be a change in the oxidoreductase structural gene that results in the synthesis of an active enzyme under aerobic conditions without changing the basal level.

Once a genetic change is introduced in the promoter of the oxidoreductase gene, the next likely change that can further increase its expression might then be a mutation in the activator gene that directs the synthesis of an altered regulator protein capable of activating the promoter of the oxidoreductase gene, as well as the promoter of the aldolase gene, without the inducer. However, as an unavoidable consequence, this altered protein fails to recognize the promoters of the remaining operons, rendering the permease, isomerase, and kinase noninducible. It is impressive that in a total of ten independent series of selection, improvement of growth rate on propanediol inevitably led to the loss of the ability to grow on fucose. Thus the dice of this genetic game seem heavily loaded (Hacking and Lin, 1977).

In strains 54, 55, and 56, selected from three separate clones of strain 3 as fucose-positive revertants, the synthesis of the permease, the isomerase, and the kinase all became constitutive, although the activity of the permease is not restored to the normal induced level. The suppressor mutation responsible for the expression of the genes specifying the three proteins is outside of the *eno* and *argA* boundary shown in Fig. 2. This suppressor gene probably codes for a substitute activator protein for the *fucP* and *fucIK* operons, but the normal function of the protein is unknown (Chen, Y.-M., *et al.*, 1984).

## 8. Relationship of the Fucose and the Rhamnose Systems

The genes of the fucose and the rhamnose systems (Fig. 1) might have evolved by duplication of segments of the *E. coli* chromosome (Riley and Anilionis, 1978; see also Chapter 10, this volume). In parallel with the pathway for fucose, the one for rhamnose is also mediated sequentially by a permease (yet to be characterized), an isomerase (Wilson and Ajl, 1957a; Takagi and Sawada, 1964a), a kinase (Wilson and Ajl, 1957b; Chiu and Feingold, 1964; Takagi and Sawada, 1964b), and an aldolase (Sawada and Takagi, 1964; Chiu and Feingold, 1965, 1969; Schwartz and Feingold, 1972). There is one interesting difference in the trunk pathway: rhamnose, but not fucose, is an attractant in chemotaxis. This would suggest that the transport of fucose does not require a protein component in the periplasmic compartment between the inner and outer membranes, whereas the transport of rhamnose does. If so, the fucose transport system would probably have a higher $K_m$ and a broader specificity in comparison with the rhamnose transport system (Adler *et al.*, 1973; Andrews and Lin, 1976). The rhamnose genes cluster at 87.2 min on the chromosome. The gene order is *rhaC*, *rhaK*, *rhaI*, and *rhaA*, encoding the activator protein, the kinase, the isomerase, and the aldolase, respectively. Mutations in gene C prevent the expression of the structural genes, but the defect is recessive to the wild-type allele, which is diagnostic of a positive regulatory protein (Power, 1967).

If the fucose and the rhamnose systems were derived from a common ancestral system, the duplication is incomplete. This was revealed by the inability of rhamnose to induce propanediol oxidoreductase in mutants deleted in the fucose region. In contrast, the fucose system is self-sufficient; no effect on the inducibility of propanediol oxidoreductase is seen in mutants deleted in the rhamnose region. The results from this study seem to indicate that the two systems share a common structural gene of the enzyme. There appears to be a gene in the *rha* locus whose product is necessary for activating the *fucO* gene (Chen and Lin, 1984b). However, results from another study indicate that the rhamnose system possesses its own propanediol oxidoreductase structural gene (Ros and Aguilar,

1984). The exact nature of the interaction between the *fuc* and *rha* systems has yet to be clarified.

## 9. Conversion of the Fucose System for D-Arabinose Utilization

Coliform or enteric bacteria such as *E. coli*, *K. pneumoniae*, and *Salmonella typhimurium* capable of growing on L-fucose are usually not capable of growing on D-arabinose (L-fucose with a hydrogen atom instead of the methyl group). Mutants that grow on the pentose readily arise (Monod, 1947; Mortlock, 1982).

An isomerase that converts D-arabinose to D-ribulose was found in such mutants of *E. coli* (Cohen, 1953). Later enzymic studies identified this enzyme to be fucose isomerase (Green and Cohen, 1956; Boulter and Gielow, 1973). Moreover, it was shown that fuculose kinase also acts on D-ribulose, which is converted to D-ribulose 1-phosphate (Heath and Ghalambor, 1962; LeBlanc and Mortlock, 1971a), and fuculose 1-phosphate aldolase also acts on D-ribulose 1-phosphate, which is converted to dihydroxyacetone phosphate and glycoaldehyde (Ghalambor and Heath, 1962). The isomerase and the kinase were shown to have a higher apparent affinity for the physiological substrate than for the analog substrate. Genetic studies showed that fuculose kinase and fuculose 1-phosphate aldolase are indispensable for growth on the D-arabinose (LeBlanc and Mortlock, 1971b).

In the dissimilation of D-arabinose, the cleavage product glycoaldehyde is converted to glycolate (LeBlanc and Mortlock, 1971b). Glycolate is oxidized through the glyoxylate pathway (see Section 10.3 on mutants that grow on ethylene glycol).

D-Arabinose-utilizing mutants selected from *E. coli* and *S. typhimurium* respond both to L-fucose and to D-arabinose as the inducer. The lag period for growth is 12 hr from glucose to D-arabinose, in contrast to 3.5 hr for the glucose to fucose transition (LeBlanc and Mortlock, 1971b; Old and Mortlock, 1977). The mutation permitting D-arabinose to induce the fucose pathway involves a positive regulator protein (probably the product of *fucC*), which is dominant over the wild-type regulator protein in the diploid state (Skjold and Ezekiel, 1982b).

In contrast to *E. coli* K12 and *S. typhimurium*, which acquire the D-arabinose growth ability by mutation of the fucose system to dual inducibility by fucose and D-arabinose, *K. pneumoniae* PRL-R3 acquires this ability by mutation of the fucose system to constitutivity (Camyre and Mortlock, 1965; Oliver and Mortlock, 1971a). On the other hand, *K.*

*pneumoniae* W70 gives rise to D-arabinose-positive strains by mutation to either constitutivity or dual inducibility (St. Martin and Mortlock, 1977).

Continued selection of *K. pneumoniae* PRL-R3 on D-arabinose led to the isolation of mutants synthesizing an altered isomerase with increased apparent affinity for the novel substrate (Oliver and Mortlock, 1971b).

Not all strains of *E. coli* mutate from the fucose-positive to the D-arabinose-positive state. The *E. coli* strain B/r grows on D-arabinose with a doubling time of 120 min, but cannot grow on fucose. The enzymes in the pathway, however, can be induced by either compound (Boulter and Gielow, 1973). It would be important to know whether fresh isolates of *E. coli* like that can be found. If so, this would suggest that this genetic system is polymorphic in this enteric species and that D-arabinose is not as rare in nature as is generally believed.

## 10. Propanediol-Positive Mutants as Evolutionary Vanguards

### 10.1. Mutants That Grow on D-Arabitol

Even with heavy mutagenesis of wild-type *E. coli* K12 cells, attempts to select for mutants that grow on D-arabitol were unsuccessful. Such mutants, however, were obtained from strain 3 after chemical mutagenesis. Apparently in this strain, the propanediol facilitator, as well as propanediol oxidoreductase, became constitutive (consistent with the belief that the two genes belong to a single operon), and the facilitator is able to admit D-arabitol into the cell. Growth on D-arabitol would then ' require only one additional mutation that recruits an enzyme capable of converting D-arabitol to a normal metabolite. An analysis of a D-arabitol-positive mutant, strain 911, showed the constitutive synthesis of an NAD-linked dehydrogenase. this enzyme converts D-arabitol to D-xylulose, which is an intermediate in the D-xylose pathway (Fig. 3). The D-xylulose formed is probably reversibly converted to D-xylose, which then induces the pathway. [In *K. pneumoniae* PRL-R3, this pathway is inducible by the substrate D-xylose itself (Wilson and Mortlock, 1973).]

The novel dehydrogenase exhibits a $K_m$ of 480 mM for D-arabitol and a $K_m$ of 10 mM for D-galactose. A number of compounds structurally related to D-arabitol and D-galactose are inactive as substrates. Curiously, the growth $K_m$ of strain 911 on D-arabitol is two orders of magnitude lower than that of the dehydrogenase. This would suggest the presence of an active transport system, which is surprising for a membrane protein believed to catalyze facilitated diffusion of a physiological substrate. The

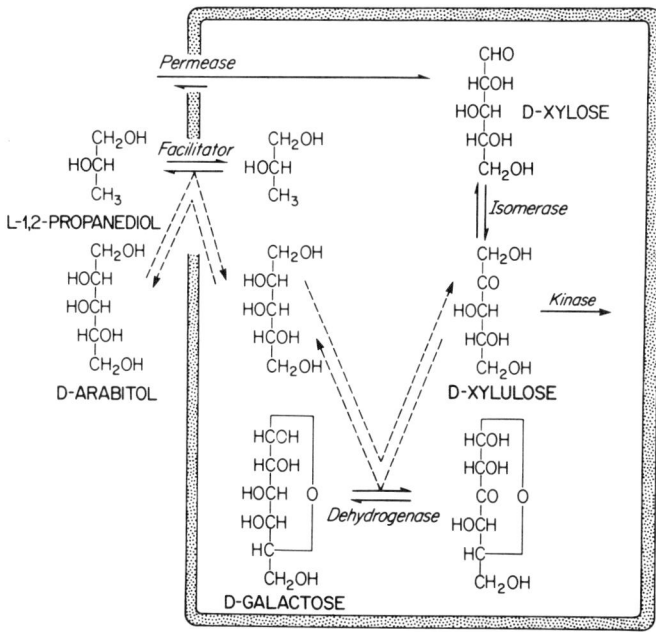

**Figure 3.** The evolved pathway for D-arabitol utilization. D-Arabitol enters through the constitutive propanediol facilitator and is acted on by a newly derepressed enzyme (most active on D-galactose) to give D-xylulose. D-Xylulose is reversibly converted by the basal isomerase activity to D-xylose, which then acts as an inducer for its pathway consisting of the permease, the isomerase, and the kinase.

sharing of a common transport by the two compounds is indicated by the ability of 10 mM DL-1,2-propanediol to inhibit about tenfold the uptake of $^{14}$C-labeled D-arabitol at low concentration. The growth yield on D-arabitol is equivalent to that on glucose (Wu, 1976a).

## 10.2. Mutants That Grow on Xylitol

The constitutive expression of propanediol oxidoreductase makes it possible for the *E. coli* mutant strains to grow on xylitol with one further mutation. This possibility was anticipated on the basis of two observations: (1) the oxidoreductase exhibited a $K_m$ of 400 mM for xylitol, converting it to the D-xylose, a utilizable pentose; and (2) the presence of catalytic amounts of D-xylose permitted strain 3 to utilize xylitol, an observation suggesting that D-xylose permease has a side specificity for xylitol and that the constitutive synthesis of this transport protein should be a sufficient condition for growth on xylitol as the *sole* source of carbon and energy (Fig. 4). (Propanediol facilitator apparently does not admit xylitol.) Indeed, single-step mutants of strain 3 that utilize D-xylose constitutively were isolated. All of them grow on xylitol without the aid of D-xylose. One such mutant, strain 821, grows in a medium initially containing 13 mM xylitol with a doubling time of 10 hr at 37°C. The growth $K_m$ on the model substrate is 48 mM and the growth yield on xylitol is equivalent to that on glucose. As expected, the mutant produces D-xylose permease, D-xylose isomerase, and D-xylulose kinase constitutively (Wu, 1976b). Other studies also indicated that genes encoding these proteins are regulated as a group (David and Wiesmeyer, 1970; Maleszka *et al.*, 1982). The selection of a secondary mutant from strain 821 with faster growth rate on xylitol should reveal the rate-limiting step for growth. It might be noted here that in the evolution of a xylitol pathway in *Klebsiella pneumoniae* strain 1033 (Lerner *et al.*, 1964), D-arabitol permease was recruited for the transport of the novel carbon source (Wu *et al.*, 1968).

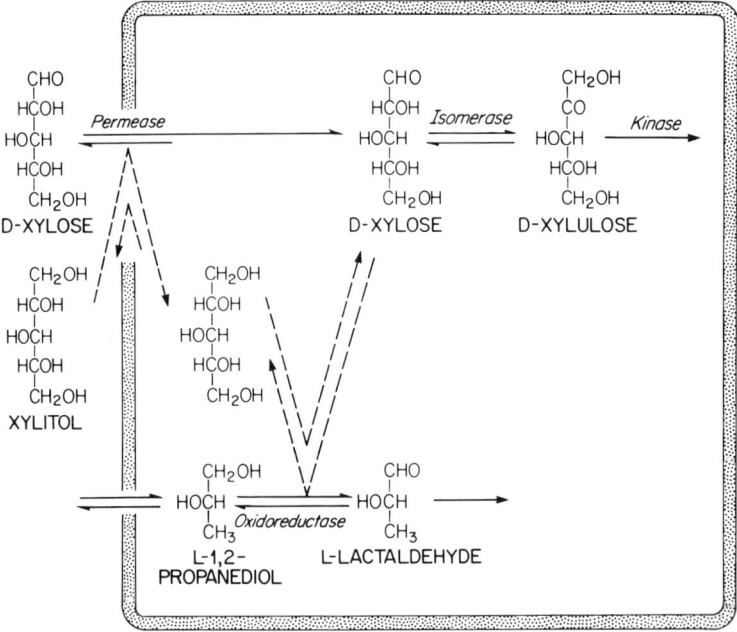

**Figure 4.** The evolved pathway for xylitol utilization. Xylitol enters through the newly derepressed D-xylose permease and is dehydrogenated by the constitutive propanediol oxidoreductase to D-xylose, which is metabolized further through its derepressed pathway.

## 10.3. Mutants That Grow on Ethylene Glycol

In contrast to mutants that exploit the constitutive propanediol facilitator to grow on D-arabitol and mutants that exploit the constitutive propanediol oxidoreductase to grow on xylitol, mutants that grow on ethylene glycol probably appropriate all three proteins involved in the consecutive utilization of propanediol: the facilitator, the oxidoreductase, and lactaldehyde dehydrogenase. An *E. coli* mutant, EG3, selected from a derivative of strain 3 for growth on ethylene glycol shows a threefold elevation in the constitutive level of propanediol oxidoreductase and constitutive expression of lactaldehyde dehydrogenase (three times the basal level). The doubling time of strain EG3 on ethylene glycol at 50 mM is 390 min, and the growth $K_m$ is 35 mM. The growth yield is similar to that obtained with equivalent carbon concentrations of glucose (Boronat *et al.*, 1983). Ethylene glycol is acted upon by propanediol oxidoreductase with a $K_m$ of about 7 mM (Sridhara and Wu, 1969). The product, glycoaldehyde, is an analogue substrate of lactaldehyde dehydrogenase (Caballero *et al.*, 1983). Glycolate, the product of glycoaldehyde, is utilizable via the glyoxalate pathway (Ornston and Ornston, 1969). The reaction sequence is diagrammed in Fig. 5.

## 11. Retrospective and Prospective Views

Studies of mutants that grow on propanediol led to the fucose system, and analysis of the fucose system revealed a link to the rhamnose system. Fucose and rhamnose are structurally similar, their internal metabolic pathways show the same sequence of reactions, and their genetic organization retains discernible homology. It is highly probable that the two systems had a common origin. Yet the duplication is not complete. Lactaldehyde dehydrogenase and propanediol oxidoreductase might be shared in common. Furthermore, the oxidoreductase is more closely associated with the fucose system in a way that has yet to be fully characterized. What has been uncovered is that the fucose locus contains either the structural gene or a common regulatory element for its expression, since rhamnose is incapable of inducing propanediol oxidoreductase activity in a mutant with a deletion in the fucose region. Thus, the fucose and the rhamnose regulons remain mechanistically connected as molecular Siamese twins.

The convergence of the two systems, by sharing the two branch enzymes, confers economy. The stereochemical differences between the metabolites of fucose and rhamnose disappear with cleavage between

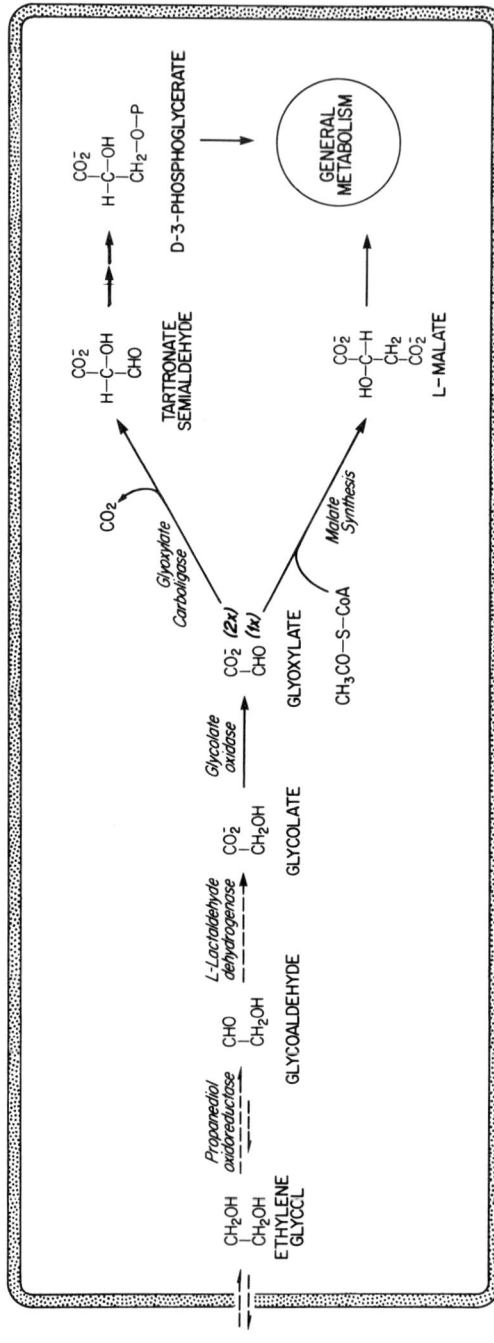

**Figure 5.** The evolved pathway for ethylene glycol utilization. Ethylene glycol is shown entering through the constitutive propanediol facilitator and is dehydrogenated by the constitutive propanediol oxidoreductase to glycoaldehyde. Glycoaldehyde is dehydrogenated by the newly derepressed lactaldehyde dehydrogenase to glycolate. Glycolate is oxidized to glyoxylate, which can be metabolized further by two different routes.

carbons 3 and 4 of the phosphorylated intermediates to yield dihydroxyacetone phosphate and lactaldehyde in both cases. It therefore would make sense that a single set of branch enzymes responding to lactaldehyde as the specific signal should take over from there on. This has definitely not happened with the oxidoreductase gene, and the reason is not clear. Perhaps the rhamnose system evolved from a partial duplication of the ancestral fucose system, followed by the translocation of the daughter genes, and for mechanistic reasons the oxidoreductase gene failed to acquire the autonomy that would allow it to be regulated directly by lactaldehyde. Alternatively, the special genetic connection of the oxidoreductase with the fucose system is maintained for a functional reason. More insight into the evolutionary history of the fucose and rhamnose systems might come with more detailed information about their genetic fine structures and organizations. In the context of duplication and divergence of genetic systems, it is of interest to note that the presumptive twinning of the ribitol and D-arabitol operons of *K. pneumoniae* W70 followed a different course. There the two sets of genes are inverted in orientation and show overlap in the regulator regions (Charnetzky and Mortlock, 1974). A similar situation is seen in *E. coli* strain C (Scangos and Reiner, 1978; Link and Reiner, 1982).

The fucose system has revealed a rich potential for giving rise to novel metabolic pathways. The depression of the anaerobic propanediol oxidoreductase branch of the pathway has already made possible the development of novel pathways for D-arabitol, xylitol, and ethylene glycol. From the trunk pathway, a route for the dissimilation of D-arabinose has been developed. Mutants that grow on two other structural analogs, L-galactose (L-fucose with a methoxy instead of the methyl group) and D-altrose (L-galactose with an inversion of the hydroxyl group at C5), have also been isolated (Y. Zhu, unpublished data). A substantial part of the requisite specificity has already been built in: both L-galactose and D-altrose are acceptable by the isomerase to form the respective 2-ketohexoses, L-tagatose and D-psicose (Green and Cohen, 1956); and L-tagatose 1-phosphate and D-psicose 1-phosphate are both acceptable to the aldolase (Ghalambor and Heath, 1962). L-Tagatose is probably acted upon by the kinase, although it has not yet been tested. D-Psicose, on the other hand, is not acted upon by the kinase (Heath and Ghalambor, 1962), but a mutation in the structural gene might confer a new specificity. Alternatively, a different kinase might be mobilized to phosphorylate D-psicose. For instance, in *E. coli* mutants missing fuculose kinase, D-arabinose utilization is made possible by the mobilization of L-ribulose kinase from the L-arabinose system (LeBlanc and Mortlock, 1972). Entry of the novel sugars would of course also be necessary. It is likely that the permease has a sufficiently wide specificity to include D-altrose snd L-galactose.

The examples we have cited should amply illustrate that all biochemical pathways have some latent metabolic potentials because enzymes cannot be absolutely perfect in their substrate specificity. The catalytic potentials of a great majority of enzymes are masked by control mechanisms that permit active enzyme synthesis only in appropriate metabolic situations. The reward for the curtailment of biochemical versatility is metabolic economy. It should therefore be small surprise that the first genetic change enabling a cell to attack a novel substrate is often a mutation that frees the expression of a structural gene encoding an enzyme with the requisite side specificity. It has been suggested that only *after* such a gene is expressed constitutively would its duplication be useful in expanding the biochemical repertoire: one copy retains its customary role, while the other copy adapts for a new function (Lin, 1970). Studies on the fucose system revealed a more attractive preparatory step preceding gene duplication: mutation to dual inducer specificity. This solution is advantageous in not requiring the destruction of the control mechanism, which eventually has to be reestablished. Specialization of the duplicated set of genes would involve parallel reimposition of nonoverlapping inducer specificity on the two regulator proteins. Indeed, it might be in this very manner that the fucose and rhamnose systems diverged originally. It should be mentioned here that the rhamnose system shows dual inducibility in wild-type *K. pneumoniae* PRL-R3. L-Mannose (L-rhamnose with the methyl group replaced by a methoxy group) also gives rise to an effector, and mutants that grow on L-mannose use rhamnose isomerase, rhamnulose kinase, and rhamnulose l-phosphate aldolase of the rhamnose system. In this case, the critical mutation probably allows the cell to accommodate one or more of the novel metabolites. In particular, the tolerance and disposition of the nonmetabolizable L-glyceraldehyde might present a special problem (Mayo and Anderson, 1968, 1969).

With what is already known about the potentials of the fucose system, it is probably not too optimistic to expect many unforeseen pathways to emerge from this metabolic system, starting from wild-type and already existing mutant strains. In this respect, experimental evolution of metabolic pathways can be expected to be as open ended as the evolution of many other processes seen in nature.

ACKNOWLEDGMENTS. We thank Sarah Monosson for editorial assistance and Yu-Mei Chen for helpful comments. This work was supported by grant PCM79-24046 from the National Science Foundation and Public Health Service grant 5-R01-GM11983 from the National Institute of General Medical Sciences.

# References

Adler, J., Hazelbauer, G. L., and Dahl, M. M., 1973, Chemotaxis toward sugars in *Escherichia coli*, *J. Bacteriol.* **115**:824–847.

Albery, W. J., and Knowles, J. R., 1976, Evolution of enzyme function and the development of catalytic efficiency, *Biochemistry* **15**:5631–5640.

Andrews, K. J., and Lin, E. C. C., 1976, Selective advantages of various bacterial carbohydrate transport mechanisms, *Fed. Proc.* **35**:2185–2189.

Bachmann, B. J., 1983, Linkage map of *Escherichia coli* K-12, Edition 7, *Microbiol. Rev.* **47**:180–230.

Bartkus, J. M., and Mortlock, R. P., 1983, Induction of L-fucose isomerase in wild-type and D-arabinose utilizing strains of *Escherichia coli* K-12, *Abst. Am. Soc. Microbiol.* 216.

Boronat, A., and Aguilar, J., 1979, Rhamnose induced propanediol oxidoreductase in *Escherichia coli*: Purification, properties and comparison with the fucose-induced enzyme, *J. Bacteriol.* **140**:320–326.

Boronat, A., and Aguilar, J., 1981a, Experimental evolution of propanediol oxidoreductase in *Escherichia coli*, *Biochim. Biophys. Acta* **672**:98–107.

Boronat, A., and Aguilar, J., 1981b, Metabolism of L-fucose and L-rhamnose in *Escherichia coli*: Differences in induction of propanediol oxidoreductase, *J. Bacteriol.* **147**:181–185.

Boronat, A., Caballero, E, and Aguilar, J., 1983, Experimental evolution of a metabolic pathway for ethylene glycol utilization by *Escherichia coli*, *J. Bacteriol.* **153**:134–139.

Boulter, J. R., and Gielow, W. O., 1973, Properties of D-arabinose isomerase purified from two strains of *Escherichia coli*, *J. Bacteriol.* **113**:687–696.

Caballero, E., Baldoma, L., Ros, J., Boronat, A., and Aguilar, J., 1983, Identification of lactaldehyde dehydrogenase and glycolaldehyde dehydrogenase as functions of the same protein in *Escherichia coli*, *J. Biol. Chem.* **258**:7788–7792.

Camyre, K. P., and Mortlock, R. P., 1965, Growth of *Aerobacter aerogenes* on D-arabinose and L-xylose, *J. Bacteriol.* **90**:1157–1158.

Chakrabarti, T., Chen, Y.-M., and Lin, E. C. C., 1984, Clustering of genes for L-fucose dissimilation by *Escherichia coli*, *J. Bacteriol.* **157**:984–986.

Charnetzky, W. T., and Mortlock, R. P., 1974, Close genetic linkage of the determinants of the ribitol and D-arabitol catabolic pathways in *Klebsiella aerogenes*, *J. Bacteriol.* **119**:176–182.

Chen, Y.-M., Lin, E. C. C., Ros, J., and Aguilar, J., 1983, Use of operon fusions to examine the regulation of the L-1,2-propanediol oxidoreductase gene of the fucose system in *Escherichia coli*, *J. Gen. Microbiol.* **129**:3355–3362.

Chen, Y.-M., and Lin, E. C. C., 1984a, Post-transcriptional control of L-1,2-propanediol oxidoreductase in the L-fucose pathway of *Escherichia coli* K-12, *J. Bacteriol.* **157**:341–344.

Chen, Y.-M., and Lin, E. C. C., 1984b, Dual control of a common L-1,2-propanediol oxidoreductase by L-fucose and L-rhamnose in *Escherichia coli*, *J. Bacteriol.* **157**:828–832.

Chen, Y.-M., Chakrabarti, T., and Lin, E. C. C., 1984, *J. Bacteriol.* (in press).

Chiu, T. H., and Feingold, D. S., 1964, The purification and properties of L-rhamnulokinase, *Biochim. Biophys. Acta* **92**:489–497.

Chiu, T. H., and Feingold, D. S., 1965, Substrate specificity of L-rhamnulose 1-phosphate aldolase, *Biochem. Biophys. Res. Commun.* **19**:511–516.

Chiu, T. II., and Feingold, D. S., 1969, L-rhamnulose 1-phosphate aldolase from *Escherichia coli*. Crystallization and properties, *Biochemistry* **8**:98–108.

Cocks, G. T., Aguilar, J., and Lin, E. C. C., 1974, Evolution of L-1,2-propanediol catabolism in *Escherchia coli* by recruitment of enzymes for L-fucose and L-lactate metabolism, *J. Bacteriol.* **118**:83–88.

Cohen, S. S., 1953, Studies on D-ribulose and its enzymatic conversion to D-arabinose, *J. Biol. Chem.* **201**:72–84.

David, J.D, and Wiesmeyer, H., 1970, Control of xylose metabolism in *Escherichia coli*, *Biochim. Biophys. Acta* **201**:497–499.

Davis, B. D., 1958, On the importance of being ionized, *Arch. Biochem. Biophys.* **87**:497–509.

Falkow S., Schneider, H., Baron, L. S., and Formal, S. B., 1963, Virulence of *Escherichia–Shigella* genetic hybrids for the guinea pig, *J. Bacteriol.* **86**:1251–1258.

Ghalambor, M. A., and Heath, E. C., 1962, The metabolism of L-fucose. II. The enzymatic cleavage of L-fuculose-1-phosphate, *J. Biol. Chem.* **237**:2427–2433.

Green, M., and Cohen, S. S., 1956, The enzymatic conversion of L-fucose to L-fuculose, *J. Biol. Chem.* **219**:557–568.

Hacking, A. J., and Lin, E. C. C., 1976, Disruption of the fucose pathway as a consequence of genetic adaptation to propanediol as a carbon source in *Escherichia coli*, *J. Bacteriol.* **126**:1166–1172.

Hacking, A. J., and Lin, E. C. C., 1977, Regulator changes in the fucose system associated with the evolution of a catabolic pathway for propanediol in *Escherichia coli*, *J. Bacteriol.* **130**:832–838.

Hacking, A. J., Aguilar, J., and Lin, E. C. C., 1978, Evolution of propanediol utilization in *Escherichia coli*: Mutant with improved substrate-scavenging power, *J. Bacteriol.* **136**:522–530.

Haldane, J. B. S., 1930, *Enzymes*, Longmans, Green and Co., London, pp. 80–83.

Heath, E. C., and Ghalambor, M. A., 1962, The metabolism of L-fucose. I. The purification and properties of L-fuculose kinase, *J. Biol. Chem.* **273**:2423–2426.

Heller, K. B., Lin, E. C. C., and Wilson, T. H., 1980, Substrate specificity and transport properties of the glycerol facilitator of *Escherichia coli*, *J. Bacteriol.* **144**:274–278.

Kluyver, A. J., and Schnellen, C., 1937, Uber die Vergarung von Rhamnose, *Enzymologia* **4**:7–12.

Krebs, H. A., and Kornberg, H. L., 1957, Energy transformations in living matter, *Ergeb. Physiol. Biol. Chem. Exp. Pharmakol.* **49**:212–298.

LeBlanc, D. J., and Mortlock, R. P., 1971a, Metabolism of D-arabinose: Origin of a D-ribulokinase activity in *Escherichia coli*, *J. Bacteriol.* **106**:82–89.

LeBlanc, D. J., and Mortlock, R. P., 1971b, Metabolism of D-arabinose: A new pathway in *Escherichia coli*, *J. Bacteriol.* **106**:90–96.

LeBlanc, D. J., and Mortlock, R. P., 1972, The metabolism of D-arabinose: Alternate kinases for the phosphorylation of D-ribulose in *Escherichia coli* and *Aerobacter aerogenes*, *Arch. Biochem. Biophys.* **150**:774–781.

Lerner, S. A., Wu, T. T., and Lin, E. C. C., 1964, Evolution of catabolic pathway in bacteria, *Science* **146**:1313–1315.

Lin, E. C. C., 1970, Evolution of catabolic pathways in bacteria, in: *Miami Winter Symposia*, Vol. 1, North-Holland, Amsterdam, pp. 89–102.

Lin, E. C. C., 1979, Importance of regulatory mutations in channeling the evolution of metabolic pathways in bacteria, in: *Genetics of Industrial Microorganisms* (O. K. Sebek and A. I. Laskin, eds.), American Society for Microbiology, Washington, D. C., pp. 274–279.

Lin, E. C. C., 1981, Conversion of reductases to dehydrogenases by regulatory mutations, in: *Trends in the Biology of Fermentations for Fuels and Chemicals* (A. Hollaender and R. Rabson, eds.), Plenum Press, New York, pp. 305–313.

Lin, E. C. C., Hacking, A. J., and Aguilar, J., 1976, Experimental models of acquisitive evolution, *BioScience* **26**:548–555.

Link, C. D. and Reiner, A. M., 1982, Inverted repeats surround the ribitol–arabitol genes of *E. coli*, *Nature* **298**:95–96.

Maleszka, R., Wang, P. Y., and Schneider, H., 1982, A ColEl hybrid plasmid containing *Escherichia coli* genes complementing D-xylose negative mutants of *Escherichia coli* and *Salmonella typhimurium*, *Can. J. Biochem.* **60**:144–151.

Mayo, J. W., and Anderson, R. L., 1968, Pathway of L-mannose degradation in *Aerobacter aerogenes*, *J. Biol. Chem.* **243**:6330–6333.

Mayo, J. W., and Anderson, R. L., 1969, Basis for the mutational acquisition of the ability of *Aerobacter aerogenes* to grow on L-mannose, *J. Bacteriol.* **100**:948–955.

McCarthy, B. J., and Bolton, E. T., 1963, An approach to the measurement of genetic relatedness among organisms, *Proc. Natl. Acad. Sci. USA* **50**:156–164.

Monod, J., 1947, The phenomenon of enzymatic adaptation, *Growth Symp.* **11**:223–289.

Mortlock, R. P., 1982, Metabolic acquisitions through laboratory selection, *Annu. Rev. Microbiol.* **36**:259–284.

Old, D. C., and Mortlock, R. P., 1977, The metabolism of D-arabinose by *Salmonella typhimurium*, *J. Gen. Microbiol.* **101**:341–344.

Oliver, E. J., and Mortlock, R. P., 1971a, Growth of *Aerobacter aerogenes* on D-arabinose: Origin of the enzyme activities, *J. Bacteriol.* **108**:287–292.

Oliver, E. J., and Mortlock, R. P., 1971b, Metabolism of D-arabinose by *Aerobacter aerogenes*: Purification of the isomerase, *J. Bacteriol.* **108**:293–299.

Ornston, L. N., and Ornston, M. K., 1969, Regulation of glyoxylate metabolism in *Escherichia coli* K-12, *J. Bacteriol.* **98**:1098–1108.

Ørskov, I., 1974, Genus VI. *Klebsiella* Trevisan 1885, 105 Nom. cons. Opin, 13, Jud. Comm. 1954, in: *Bergey's Manual of Determinative Bacteriology*, 8th ed. (R. E. Buchanan and N. E. Gibbons, eds.), Williams and Wilkins, Baltimore, pp. 321–325.

Power, J., 1967, the L-rhamnose genetic system in *Escherichia coli* K-12, *Genetics* **55**:557–568.

Richey, D. P., and Lin, E. C. C., 1972, Importance of facilitated diffusion for effective utilization of glycerol by *Escherichia coli*, *J. Bacteriol.* **112**:784–790.

Riley, M., and Anilionis, A., 1978, Evolution of the bacterial genome, *Annu. Rev. Microbiol.* **32**:519–560.

Ros J., and Aguilar, J., 1984, Genetic and structural evidence for the presence of propanediol oxidoreductase isoenzymes in *Escherichia coli*, *J. Gen. Microbiol.* **130**:687–692.

St. Martin, E. J., and Mortlock, R. P., 1976, Natural and altered induction of the L-fucose catabolic enzymes in *Klebsiella aerogenes*, *J. Bacteriol.* **127**:91–97.

St. Martin, E. J., and Mortlock, R. P., 1977, A comparison of alternate metabolic strategies for the utilization of D-arabinose, *J. Mol. Evol.* **10**:111–122.

Sanno, Y., Wilson, T. H., and Lin, E. C. C., 1968, Control of permeation to glycerol in cells of *Escherichia coli*, *Biochem. Biophys. Res. Commun.* **32**:344–349.

Sawada, H., and Takagi, Y., 1964, The metabolism of L-rhamnose in *Escherichia coli*. III. L-Rhamnulose-phosphate aldolase, *Biochim. Biophys. Acta* **92**:26–32.

Scangos, G. A., and Reiner, A. M., 1978, Ribitol and D-arabitol catabolism in *Escherichia coli*, *J. Bacteriol.* **134**:492–500.

Schwartz, N. B., and Feingold, D. S., 1972, L-Rhamnulose l-phosphate adolase from *Escherichia coli*. III. The role of divalent cations in enzyme activity, *Bioinorgan. Chem.* **2**:75–86.

Skjold, A. C., and Ezekiel, D. H., 1982a, Analysis of lambda insertions in the fucose utilization region of *Escherichia coli* K-12: Use of *fuc* and *argA* transducing bacteriophages to partially order the fucose utilization genes, *J. Bacteriol.* **152**:120–125.

Skjold, A. C., and Ezekiel, D. H., 1982b, Regulation of D-arabinose utilization in *Escherichia coli* K-12, *J. Bacteriol.* **152**:521–523.

Sridhara, S., and Wu, T. T., 1969, Purification and properties of lactaldehyde dehydrogenase in *Escherichia coli, J. Biol. Chem.* **244**:5233–5238.

Sridhara, S., Wu, T. T., Chused, T. M., and Lin, E. C. C., 1969, Ferrous-activated nicotinamide adenine dinucleotide-linked dehydrogenase from a mutant of *Escherichia coli* capable of growth on 1,2-propanediol, *J. Bacteriol.* **98**:87–95.

Takagi, Y., and Sawada, H., 1964a, The metabolism of L-rhamnose in *Escherichia coli.* I. L-Rhamnose isomerase, *Biochim. Biophys. Acta* **92**:10–17.

Takagi, Y., and Sawada, H., 1964b, The metabolism of L-rhamnose in *Escherichia coli.* II. L-Rhamnulose kinase, *Biochim. Biophys. Acta* **92**:18–25.

Taylor, A. L., and Trotter, C. D., 1967, Revised linkage map of *Escherichia coli, Bacteriol. Rev.* **31**:332–353.

Wilson, D. M., and Ajl, S., 1957a, Metabolism of L-rhamnose by *Escherichia coli.* I. L-Rhamnose isomerase, *J. Bacteriol.* **73**:410–414.

Wilson, D. M., and Ajl, S., 1957b, Metabolism of L-rhamnose by *Escherichia coli.* II. The phosphorylation of L-rhamnulose, *J. Bacteriol.* **73**:415–420.

Wilson, B. L., and Mortlock, R. P., 1973, Regulation of D-xylose and D-arabitol catabolism by *Aerobacter aerogenes*; *J. Bacteriol.* **113**:1404–1411.

Wimpenny, J. W. T., and Firth, A., 1972, Levels of nicotinamide adenine dinucleotide and reduced nicotinamide adenine dinucleotide in facultative bacteria and the effect of oxygen, *J. Bacteriol.* **111**:24–32.

Wu, T. T., 1967a, Growth of a mutant of *Escherichia coli* K-12 on xylitol by recruiting enzymes for D-xylose and L-1,2-propanediol metabolism, *Biochim. Biophys. Acta* **428**:656–663.

Wu, T. T., 1976b, Growth on D-arabitol of a mutant strain of *Escherichia coli* K12 using a novel dehydrogenase and enzymes related to L-1,2-propanediol and D-xylose metabolism, *J. Gen. Microbiol.* **94**:246–256.

Wu, T. T., 1978, Experimental evolution in bacteria, *Crit. Rev. Microbiol*, **6**:33–51.

Wu, T. T., Lin, E. C. C., and Tanaka, S., 1968, Mutants of *Aerobacter aerogenes* capable of utilizing xylitol as a novel carbon source, *J. Bacteriol.* **96**:447–456.

CHAPTER 6

# The Evolved β-Galactosidase System of Escherichia coli

BARRY G. HALL

## 1. Introduction

The overall objective of research in experimental evolution is to understand the variety of ways by which organisms can evolve new physiological functions. How does an organism that is already well adapted to its environment evolve a new function to cope with an altered environment? The synthetic theory of evolution does not really address the problem of the evolution of novel functions, nor does population genetics deal with the appearance of new functions.

Molecular biology has given us enormously powerful tools for studying the mechanisms of evolution at the molecular level, and application of biochemical and molecular techniques has already fundamentally altered the ways in which we think about evolution. By comparing protein and nucleic acid sequences we can construct molecular evolutionary trees, and we can deduce the probable sequences of primitive genes. We can integrate this information with paleontological data concerning the time of species divergence in order to estimate the rate at which nucleotide or amino acid substitutions are fixed in populations. Such considerations have led to the concept of the molecular evolutionary clock.

One of the major gaps in the field of molecular evolution is that while we can observe changes in protein sequence and can observe enormous genetic variability, we have been generally unable to directly relate those

---

BARRY G. HALL • Microbiology Section, Biological Sciences Group, University of Connecticut, Storrs, Connecticut 06268.

changes or that variability to fitness and changes in fitness. Although it is clear that mutations are the raw material of evolution and that selection is a major force in evolution, we know far too little about the detailed relationships between selective pressures and evolution. Evolutionary theory does not tell us what information we need in order to predict the evolutionary consequences of a particular environmental pressure applied to a particular organism.

It is a major goal of these studies to directly establish causal relationships between fitness changes and molecular changes by applying strong selective pressures to populations of microorganisms and monitoring the molecular changes that occur as those populations adapt. To accomplish that goal it is necessary to identify the molecular changes (nucleotide substitutions) that have occurred, to directly relate those substitutions to physiological–functional changes that occurred, and then to directly relate those functional changes to changes in fitness. We can then establish a series of rigorous, casual links between sequence changes and adaptive fitness by following the consequences of the changes from DNA to protein to physiological (cellular) level to interactions of the organism with its environment.

A decade ago Campbell et al. (1973) reported the isolation and characterization of a lactose-utilizing mutant derived from a lacZ deletion strain of E. coli K12. The deletion strain, which synthesized a functional lactose permease constitutively, had been streaked onto lactose-TTC (fermentation indicator) plates and incubated for several weeks. At the end of this period lactose-fermenting papillae were present on the surface of most colonies. Cells from one papilla were streaked onto lactose minimal agar, and when colonies appeared, the plates were again incubated for an extended period until papillae appeared on the surface of the colonies. These presumably represented faster growing lactose-positive cells, and these were again restreaked onto minimal lactose agar. After five such rounds of selection Campbell had obtained an isolate, designated EBG-5, which grew nearly as rapidly as wild type on lactose. The EBG-5 strain synthesized a new $\beta$-galactosidase enzyme constitutively. The new enzyme was named evolved $\beta$-galactosidase (EBG). The EBG enzyme was distinguished from the classical $\beta$-galactosidase by several criteria: (1) it had a different $K_m$ for the substrate ONPG (O-nitrophenyl-$\beta$-galactoside); (2) it had a stringent requirement for potassium and was inhibited by sodium, while the classical enzyme preferred sodium to potassium; (3) it had a larger native molecular weight than classical $\beta$-galactosidase; and (4) it failed to cross react with antiserum directed against classical $\beta$-galactosidase. The locus responsible for the EBG enzyme and for the lactose-positive phenotype was located on the opposite side of the map from the *lac* operon, and cotransduced about 5% with the *metC* locus. Based upon a failure to cotransduce with *serA*, it was estimated to be near *tolC*.

## 2. Development of the Evolved β-Galactosidase System as a Tool for Studying Evolution

The EBG system appeared ideally suited for use as a model system to study evolution. First, it allowed precise definition of the new function being selected: hydrolysis of the β-galactoside bond in the disaccharide lactose. Second, the lactose permease specified by the *lac* operon could be used to transport lactose into the cell, and the degradation products (glucose and galactose) were easily metabolized. Third, a wide variety of lactose analogues, useful as both substrates and inducers, had been developed for studying the *lac* operon and were commercially available. Fourth, the system offered an opportunity for comparing the newly evolved EBG enzyme with the well-characterized *lacZ* β-galactosidase.

The isolation of mutants as papillae on colonies is unusual, and is itself an important aspect of the validity and utility of this system as a model for studying evolution. The more any model system can resemble the real world, i.e., organisms in their natural environments, the more useful it is likely to be. The lactose-TTC plates used as the selective medium in these experiments mimic quite well the selective conditions one might expect to encounter in nature. The plates contain broth as a primary nutrient and lactose as a secondary, nonmetabolizable nutrient. Colonies grow initially at the expense of the broth, which is exhausted as colonies approach the size of about $10^{10}$ cells per colony. Rare spontaneous mutants in the population have an enormous selective advantage over the nonmutant cells in the colony, as they grow at the expense of the secondary resource, lactose, to form papillae on the surface of the colonies. All of the mutations studied are spontaneous mutations; thus we consider only that spectrum of mutations that would be expected to occur in natural settings. Finally, because the colonies are clonal, that is, each colony is derived from a single cell, the mutations must occur *after* the cells are on the selective plate. This means that two papillae occurring on the surfaces of two well-separated colonies must have arisen from independent mutations. It is thus possible to isolate 20 or so independent mutants from the surface of a single plate, and thus to study the variety of different mutations that can lead to lactose utilization.

From a *lacZ* deletion strain unrelated to that of Campbell, I initially selected 34 independent lactose-utilizing deriviatives (Hall and Hartl, 1974).

These isolates fell into two classes based upon growth rates in lactose: type I strains grew rapidly with no lag, and type II strains grew more slowly after a lag of several hours. Further experiments showed that type I strains synthesized EBG enzyme constitutively (as did Campbell's EBG-5 strain), but type II strains synthesized EBG enzyme only when induced by lactose. The EBG enzymes from type I and type II strains were indistinguishable from each other and from Campbell's EBG enzyme: and all cross-reacted with Campbell's anti-EBG antiserum. The EBG genes from both types mapped to a site near *tolC*. When EBG-constitutive mutants were selected from the regulated type II strains, they grew at exactly the same rate as the type I strains, confirming that the regulated and constitutive strains synthesized the same EBG enzymes.

The observation that the majority of the EBG$^+$ isolates were regulated led to the idea that the ancestral gene, which I termed $ebgA^0$, might likewise be regulated by lactose. This is in fact the case (Hartl and Hall, 1974). When the unevolved *lacZ* deletion strain DS4680A is grown in medium containing lactose, EBG$^0$ enzyme activity is induced. That activity is detected by assaying for the hydrolysis of the synthetic substrate ONPG. The observation that the induced cells still could not grow on lactose suggested that the unevolved enzyme hydrolyzed ONPG much more efficiently than it hydrolyzed lactose. The gene specifying the unevolved enzyme, EBG$^0$, mapped to the same locus near *tolC* as did the gene for EBG$^+$ enzyme that allowed lactose utilization.

The regulation of EBG enzyme synthesis was shown to be subject to negative control (Hall and Hartl, 1975) by a repressor that is the product of the *ebgR* gene, which is located very close to *ebgA*, the structural gene for the EBG enzyme.

A wide variety of β-galactoside compounds were tested as inducers of EBG enzyme synthesis, but only lactose was found to be an effective inducer (Table I). The thiogalactosides, which are powerful inducers of the *lac* operon, were noninducers of EBG enzyme synthesis (Hall and Clarke, 1977). Phenylgalactoside is a noninducer, but it does serve as a (poor) substrate for EBG enzymes, and was therefore used to select constitutive ($ebgR^-$) mutants. Such mutants synthesize 4.5–5% of their soluble protein as EBG enzyme (Hall, 1976a; Hall and Clarke, 1977). One such constitutive mutant was used as a source for the purification of the unevolved enzyme, EBG$^0$ enzyme. Characterization of that purified enzyme (Hall, 1976a) permitted accurate estimation of the level of synthesis of EBG enzyme in the unevolved strain. In the absence of inducer, the unevolved strain synthesizes about 0.003% of its soluble protein as EBG enzyme, i.e., 3–5 molecules/cell. Lactose induces that strain about 100-fold to give a specific synthesis of 0.3%, or about 6% of the maximal level

## Table I
Induction of Evolved β-Galactosidase Enzyme Synthesis in Various Strains

| Strain | ebgR Allele | Inducer | Specific synthesis[a] |
|---|---|---|---|
| DS4680A | $ebgR^+$ | None | 0.0022 |
| | | IPTG, TMG, galactose Meliboise, lactobionate Phenyl-galactoside | 0.001–0.0027 |
| | | Methyl-galactoside | 0.01 |
| | | Lactulose | 0.027 |
| | | Galactosyl-arabinose | 0.018 |
| | | Lactose | 0.26 |
| 1B1 | $ebgR^-$ | None | 4.2 |
| A4 | $ebgR^{+U}$ | None | 0.0030 |
| | | Lactulose | 0.056 |
| | | Lactose | 1.23 |
| 5A103 | $ebgR^{+L}$ | None | 0.002 |
| | | Lactulose | 1.03 |
| | | Lactose | 2.2 |

[a] 100 × specific activity of crude extract/specific activity of pure enzyme.

of synthesis found in $ebgR^-$ strains. Other β-galactoside sugars, such as lactulose, galactosyl-β-D-arabinose, or lactobionic acid, are not effective inducers, giving at best ten-fold induction above the basal level of expression (Hall and Clarke, 1977).

One of the most powerful aspects of experimental evolutionary studies is the potential for identifying and characterizing an unevolved gene and its product and comparing it with the evolved gene that has acquired a new function. The unevolved enzyme is a multimeric protein consisting of six identical 120,000-dalton subunits (based upon analytical ultracentrifugation and SDS gel electrophoresis). It is active toward ONPG, but virtually inactive toward lactose. The low specificity of $EBG^0$ enzyme for lactose is consistent with the observation that a constitutive ($ebgR^-$) mutant that synthesizes 5% of its soluble protein as $EBG^0$ enzyme is unable to utilize lactose as a carbon and energy source (Hall, 1976a; Hall and Clarke, 1977). The various $ebgA^+$ genes were shown to be alleles of $ebgA^0$ and to specify EBG enzymes with greatly increased activity toward lactose (Hall, 1976a). The details of the changes that increase lactase activity are discussed in Section 4 on kinetic analyses.

The unevolved lacZ deletion strain is the beginning point for these studies. We can summarize its relevant properties by saying that *E. coli* K12 strains have a gene, $ebgA^0$ that directs the synthesis of $EBG^0$ enzyme when cells are grown in the presence of lactose. The $EBG^0$ enzyme is a β-galactosidase, but not a lactase. Under sufficient selective pressure, the

$ebgA^0$ gene can evolve into an $ebgA^+$ gene specifying an enzyme with greatly enhanced lactase activity.

A critical step in this project was the determination of the number of mutations involved in evolving lactase activity. Two lines of evidence showed that mutations to $ebgA^+$ can be accounted for by a single mutation in the structural gene (Hall, 1977). Evidence discussed below shows that there are two distinct sites within *ebgA* where such mutations may occur. Other evidence (discussed below) has shown that although a single mutation in *ebgA* is sufficient to produce lactase activity, mutations in the regulatory gene *ebgR* are also required for the evolution of lactose utilization (Hall and Clarke, 1977).

Among the initial isolates, the homogeneity of growth rates on lactose suggested that the same mutation had occurred repeatedly. A second set of spontaneous mutants was isolated, and three of them, although constitutive for EBG enzyme synthesis, grew only 40% as fast as the other constitutives isolated (Hall, 1976a). These mutants proved to have evolved a second new function: growth on lactulose (galactosyl-$\beta$-D-fructose) (Hall, 1976a; Hall and Clarke, 1977). Genetic analysis clearly showed that the mutations conferring the new phenotype were in the *ebgA* gene.

I considered the possibilities that the new phenotype (slower growth on lactose, accompanied by growth on lactulose) might be an intermediate step on the way to rapid growth on lactose; or that the new phenotype might be a second step in which rapid growth on lactose had been sacrificed to a broader substrate range that included lactulose. Since there had been no selection for lactulose utilization, the second possibility seemed unlikely, but both alternatives were ruled out by showing that both phenotypes arose as the consequence of single mutations in the structural gene, mutations that occurred at the rate of $2.2 \times 10^{-9}$ per cell division (Hall, 1977). The two phenotypes were then alternatives to each other rather than sequential steps.

## 3. Evolution of Multiple Functions for Evolved $\beta$-Galactosidase Enzyme: An Evolutionary Pathway

The above observations led to two new questions: (1) What was the evolutionary potential of the *ebgA* gene, i.e., how many different new substrate specificities could be evolved by mutations in this gene? (2) could evolutionary pathways be demonstrated i.e., were there some new specificities that would require a series of mutations in the gene? Because lactose was the only effective inducer of EBG enzyme synthesis, I used strains that synthesized EBG enzyme constitutively to approach these questions. I began by selecting from an $ebgA^0$ strain one that could utilize β-methylgalactoside as a carbon source. The resulting mutation was specific for growth on β-methylgalactoside. The mutant was unable to utilize other β-galactosides (Hall, 1976b).

I next chose an array of commercially available β-galactoside sugars, and began by screening the existing set of $EBG^+$ strains to determine which sugars could be utilized. The strain synthesizing the unevolved enzyme $EBG^0$ was unable to utilize any of the sugars tested. The evolved strains fell into two classes: Class I strains grew rapidly on lactose, extremely slowly (doubling times >24 hr) on galactosyl-β-D-arabinose, and failed to grow on either lactulose (galactosyl-β-D-fructose) or on lactobionate (galactosyl-β-D-gluconate) (Table II). Class II strains grew well on lactulose, somewhat more slowly on lactose, extremely slowly on galactosyl arabinose, and failed to grow on lactobionate. Within a class the growth rates and patterns were very homogeneous. Among the isolates tested, all of which had been selected for lactose utilization, 90% were class I and 10% were class II. When I selected lactulose-utilizing deriviative from the $EBG^0$-constitutive strain, 100% of them grew well on lactulose, more slowly on lactose, extremely slowly on galactosyl arabinose, and not at all on lactobionate. These were originally called class III (Hall, 1978a), however, subsequent studies (Hall and Zuzel, 1980) have shown that these are indistinguishable from class II strains; thus, the class III designation has been dropped. All attempts to select galactosyl arabinose-utilizing or lactobionate-utilizing strains directly from the $EBG^0$ strain were unsuccessful.

Because class II strains utilized lactulose while class I strains did not, it was of interest to ask whether lactulose-utilizing mutants could be obtained from class I strains and whether they would be different from class II strains. When class I strains were subjected to selection for lactulose utilization, a new class arose. Class IV strains, which carry two mutations in the $ebgA$ gene, grow rapidly on lactose, slightly slower on lactulose, and at a moderate rate on galactosyl arabinose (Table II). This was the first class to thrive on galactosyl arabinose, and subsequent studies (detailed below) showed that class IV differs in several important respects from classes I and II.

All attempts to select lactobionate-utilizing mutants from the unevolved strain or from class I or class II strains were unsuccessful; however, lactobionate-utilizing mutants arose from class IV strains at exactly the frequency expected for single spontaneous mutations in $ebgA$. It was concluded that lactose and lactulose utilization can each arise as the consequence of a single mutation in ebgA, that effective galactosyl arabinose utilization requires two mutations in $ebgA$, and that lactobionate utilization requires three mutations in $ebgA$.

## Table II
### Properties of Evolved β-Galactosidase Enzymes and Growth of Strains Synthesizing Those Enzymes Constitutively[a]

| Class | Property | Lactose | Lactulose | Galactosyl arabinose | Lactobionate |
|---|---|---|---|---|---|
| EBG⁰ (n = 1) | $V_{max}$ | 620 | 270 | 52 | No detectable activity |
| | $K_m$ | 150 | 180 | 64 | |
| | Specificity | 4.0 | 1.5 | 0.81 | |
| | Growth rate | 0 | 0 | 0 | 0 |
| Class I (n = 3) | $V_{max}$ | 3566 | 69 | 185 | No detectable activity |
| | $K_m$ | 22 | 34 | 14 | |
| | Specificity | 160 | 2.1 | 12.7 | |
| | Growth rate | 0.45 | 0 | 0.03 | 0 |
| Class II (n = 3) | $V_{max}$ | 2353 | 1887 | 356 | No detectable activity |
| | $K_m$ | 59 | 26 | 25 | |
| | Specificity | 40 | 73 | 14.4 | |
| | Growth rate | 0.19 | 0.26 | 0.02 | 0 |
| Class IV (n = 9) | $V_{max}$ | 1461 | 430 | 737 | 105 |
| | $K_m$ | 0.82 | 7.9 | 3.0 | 15.4 |
| | Specificity | 1800 | 55 | 244 | 6.7 |
| | Growth rate | 0.37 | 0.18 | 0.13 | 0 |
| Class V (n = 1) | $V_{max}$ | 590 | 215 | 349 | 370 |
| | $K_m$ | 0.69 | 6.5 | 4.9 | 3.0 |
| | Specificity | 850 | 33 | 70 | 123 |
| | Growth rate | 0.18 | 0.10 | 0.07 | 0.20 |

[a] $n$, Number of purified enzymes from independent mutants; $V_{max}$ in units/mg of pure enzyme; $K_m$ in mM substrate; growth rates in hr$^{-1}$.

If we define evolution of a new enzymatic function as a mutation that alters an enzyme so that it degrades a resource rapidly enough to permit growth, then the wild-type EBG enzyme has the potential for evolving at least five new functions: lactose, lactulose, galactosyl arabinose, lactobionate, and methylgalactoside hydrolysis.

It was further concluded that the above series of spontaneous mutations constituted an evolutionary pathway. The pathway for the evolution of lactobionate utilization was originally thought to be "obligatory" and to consist of a class I mutation, selected on lactose, followed by a

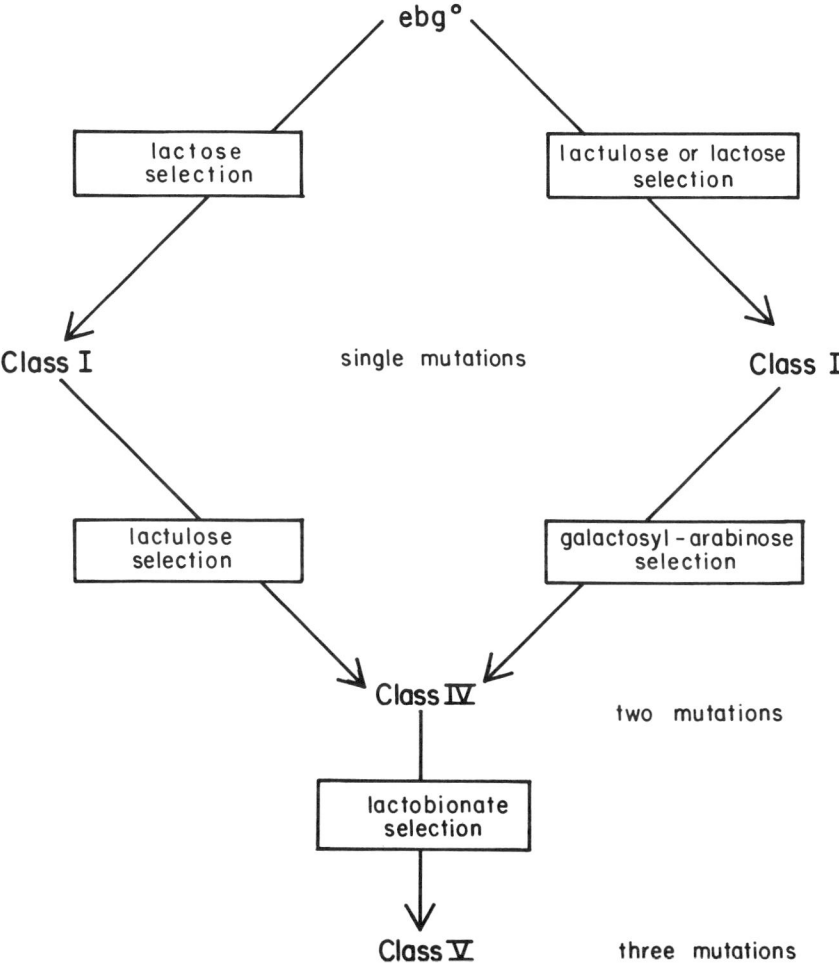

**Figure 1.** Pathway for evolution of lactobionate hydrolysis.

second mutation to class IV, selected on lactulose, followed by a third mutation to class V, selected on lactobionate (Hall, 1978a). Class II strains were considered "evolutionary dead ends" simply because I did not know what selective pressure to apply in order to evolve further substrate specificities. The realization that both class I and class II strains grew extremely slowly on galactosyl arabinose (gal-ara), but that class IV strains grew well on gal-ara, led me to wonder whether I could select gal-ara-utilizing derivitives of class II strains. The selection was carried out with two questions in mind: (1) how would the gal-ara-positive mutants compare with class IV strains in terms of growth rates on the various $\beta$-galactoside sugars? and (2) how would they compare with class IV strains in terms of their evolutionary potential for yielding lactobionate-utilizing derivatives? A series of spontaneous gal-ara-positive mutants were obtained from several class II strains. Surprisingly, they were phenotypically indistinguishable from the class IV strains described above (Hall and Zuzel, 1980a). Figure 1 diagrams the evolutionary pathways by which *lacZ* deletion strains of *E. coli* can evolve to grow on lactose, lactulose, galactosyl arabinose, and lactobionate. It is a branched, converging pathway. The validity of this proposed pathway is supported by both genetic analysis and kinetic analysis of purified EBG enzymes.

## 4. Kinetic Analysis of Evolved $\beta$-Galactosidase Enzymes

I purifed (average 92% pure) the EBG enzymes from the unevolved strain, from three class I strains, from three class II strains, from nine class IV strains, and from a class V strain. The Michaelis–Menton kinetic parameters $K_m$ and $V_{max}$ were determined for each enzyme on each of ten substrates, and the thermal stability of each enzyme was determined (Hall, 1981). Table II shows the pooled results of those studies for the utilizable substrates.

There were several reasons for this extensive biochemical analysis of the evolved enzymes. First, there was the matter of classification of the enzymes. It was important to verify that classes identified based upon growth properties were homogeneous with respect to the properties of the enzymes. This biochemical classification was intended to serve as a rigorous test of the model presented above for the evolutionary pathway. In particular, it was important to test the hypothesis that enzymes from class IV strains would be essentially alike independent of the pathway by which they evolved, or whether they were evolved by recombination between a class I and a class II *ebgA* allele. Likewise, enzymes from class I or class II strains that were obtained from crosses between class

IV and unevolved strains should be true class I or class II enzymes. Second, it was important to determine exactly how the specificities of the enzymes had changed during the evolutionary sequence. In particular, I wanted to determine how deliberate selection for enhancement of one substrate specificity affected other substrate specificities. Third, I wanted to know if there was a consistent pattern in the way that particular kinetic parameters altered during evolution. Fourth, there was the question of alteration in specificity for the nonreducing end of the substrate. All the selection had been carried out using β-galactoside sugars with different nongalactoside moieties. I asked whether activity toward C-5-substituted galactosides had altered during the evolutionary sequence. Finally, it is generally realized that a large part of the informational content of the amino acid sequence of a protein is involved with assuring correct folding and stability, rather than being directly involved with formation of an active site. Many mutations that do not directly affect the active site do reduce the stability of enzymes and lead to "temperature-sensitive" phenotypes. Thus, it is often assumed that there has been strong selection for stability, and that evolution will have generated an optimally stable sequence, implying that most mutations should lead to reduced thermostability. The availability of active enzymes carrying zero, one, two, or three substitutions allowed me to test this hypothesis.

It was evident that classification of these EBG enzymes based upon $K_m$, $V_{max}$, and the specificity $V_{max}/K_m$ for a variety of substrates is unambiguous and identical with classification based upon growth characteristics. The standard error of the mean for each parameter was less than 10% of the mean value of that parameter for all classes, substrates, and parameters; i.e., there was very high homogeneity within each class. Thus, the earlier conclusions concerning evolutionary pathways, and the conclusion that a class IV allele is simply one that carries a class I plus a class II mutation, are strongly supported.

It was also clear that selection for utilization of a particular sugar *always* increased the specificity of the enzyme for that sugar. The $V_{max}$ for the selected sugar always increased, and $K_m$ always decreased. Although the $V_{max}$ for the selected sugar always increased, the $V_{max}$ for other substrates often decreased; i.e., evolution was not the result of a general increase in the $V_{max}$ independent of substrate. There is a general correlation between substrate specificity and the growth rate upon that substrate; however, there is one particularly interesting exception to that rule. The specificity of class IV enzymes for lactose is more than tenfold higher than the specificity of class I enzymes for lactose, yet class IV strains grow only 80% as fast as class I on lactose. This example particularly illustrates the risk of considering *in vitro* data without taking physiological conditions into account. The difference in specificity $V_{max}/K_m$ is the consequence of class IV enzymes having a $K_m$ for lactose, 0.8 mM, that is 25-fold lower than that of class I enzymes, 22 mM. The internal physiological concentration of lactose is about 20 mM; consequently, class I enzymes are operating at about half of maximal velocity, or about 1800 nmole/min per mg. The class IV enzymes are operating essentially at a $V_{max}$ of about 1400 nmole/min per mg. Thus, increased specificity that comes from a substantially decreased $K_m$ is not biologically meaningful or particularly useful when the $K_m$ is significantly below the physiological substrate concentration. Thus, a class IV strain can improve its *in vivo* rate of lactose hydrolysis *only* by mutations that increase $V_{max}$. Cornish-Bowden (1976) has predicted on theoretical grounds that enzymes should evolve so that they have $K_m$ values near the physiological substrate concentration. These results are very consistent with that prediction.

The unevolved enzyme $EBG^0$ has detectable activity toward lactose, lactulose, and galactosyl arabinose, although the level of activity is insufficient for growth. Absolutely no activity of the unevolved enzyme for lactobionic acid could be detected, nor was activity detectable for class I or class II enzymes. Class IV enzymes exhibited detectable activity toward lactobionate, although the specificity was far too low to permit growth on lactobionate. Thus, while it could be argued that we have only enhanced activity for lactose, lactulose, and galactosyl arabinose, it is clear that the evolutionary steps first *created* a new lactobionate hydrolysis activity for EBG enzyme (class IV), then enhanced that activity to a level that permitted growth (class V).

I considered a series of synthetic substrates in which the glycones were substituted galactoses: activity toward *o*- and *p*nitrophenyl-$\beta$-D-fucosides (C-5 substituent, CHHH) and activity toward *o*- and *p*-nitrophenyl-$\alpha$-L-arabinosides (C-5 substituent, H) were compared with activity toward *o*- and *p*-nitrophenyl-$\beta$-D-galactosides (C-5 substituent, HCHOH). Although changes in the specificity for these glycones did occur and were consistent within a class, the changes in general paralleled the change in specificity for the galactosides. Thus, the ability of EBG enzymes to discriminate among these various glycone residues was largely unaffected by the mutations that drastically altered the specificities for various $\beta$-galactoside sugars.

The thermal stability of EBG enzyme was unaltered by the amino acid substitutions that afforded the enzymes new substrate specificities. This may simply reflect the ability of a rather large protein to withstand punishment, or it may be that the sequences that contribute to the shape of the active site are not themselves particularly important for the retention of overall conformation. In either case, it is clear that enzymes have not necessarily evolved so that their sequences are optimized for stability.

## 5. Evolution by Intragenic Recombination

The observation that class IV mutants could be derived from both class I and class II strains led me to hypothesize that a class IV *ebgA* gene might simply be one that carried a class I mutation and a class II mutation simultaneously. Were this true, when a class IV strain, independent of its evolutionary route, is crossed with an unevolved strain it should be possible to recover both class I and class II recombinants. Such recombinants were recovered from those crosses, and it was demonstrated that indeed a class IV allele simply consisted of a class I + a class II mutation. Mapping experiments showed that there is a site at the *ebgR* proximal end of the *ebgA* gene where class I mutations occur, and a second site, about one-third of the gene away, where class II mutations occur (Fig. 2). The presence of mutations in both sites results in the class IV phenotype (Hall and Zuzel, 1980a). The extent of the two sites and the constraints for effective mutations in those sites is not yet known. Class IV strains, regardless of the method of selection, can give rise to lactobionate-utilizing strains.

Effective utilization of galactosyl arabinose is a property of class IV strains. The "class IV = class I + class II" model predicts that class IV recombinants should be recovered from crosses of class I × class II strains. Gal-ara$^+$ recombinants were selected from such crosses, and they indeed proved to be class IV. No gal-ara$^+$ recombinants were recovered from crosses of class I × class I, or class II × class II strains (Hall and Zuzel, 1980a). I had therefore demonstrated that the new function "galactosyl arabinose utilization" could evolve by intragenic recombination. I suggested that similar events could occur in nature in a situation in which under different selective pressures, different alleles of the same

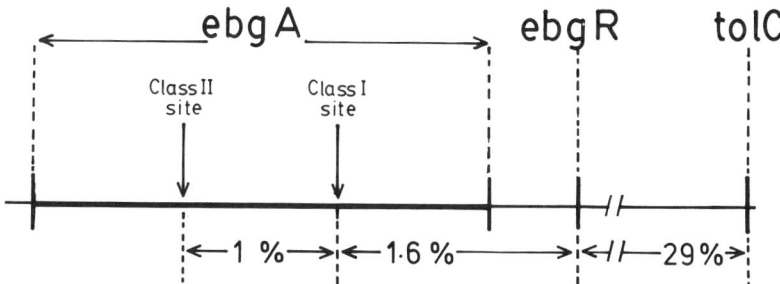

**Figure 2.** Map of *ebgA* gene. Distances are in percent recombination. Based upon recombination frequencies of other markers in this region, the class I and class II sites are about 1000 bp apart.

gene have diverged to give enzymes with somewhat different substrate specificities. Remixing of the populations would afford opportunities for intragenic recombination to generate a new allele with a substrate specificity present in neither parent. The new substrate specificity could permit the recombinant to exploit a new resource or ecological niche available to neither parent (Hall and Zuzel, 1980a).

## 6. Allolactose Synthesis: Another New Function for Class IV Enzyme

One of the properties of the EBG system that distinguishes it from the lactose system is the inability of EBG enzyme to convert lactose into an inducer of the *Lac* operon (Campbell *et al.*, 1973; Hall and Hartl, 1974). Lactose is not an inducer of the *Lac* operon, but is converted into the inducer allolactose by the transgalactosylation activity of the *lacZ* $\beta$-galactosidase (Jobe and Bourgeois, 1972). As a consequence, lactose does not induce synthesis of the lactose permease in *lacZ* deletion strains. Because the strains that I use are $lacI^+$, it has been necessary to include the gratuitous inducer IPTG in the medium to induce synthesis of the lactose permease (Hall and Hartl, 1974). I established early in this study that even $ebgR^-$ strains are dependent upon IPTG for lactose utilization (Hall and Hartl, 1974).

Following long-term selection for rapid growth on both lactose and lactulose, an unstable mutant population arose from the $ebgR^-$ class II strain 5A2. The unstable population segregated clones that were $LAC^+$ on medium lacking IPTG (Rolseth *et al.*, 1980). Genetic and biochemical analysis showed that the strain was still $lacI^+$ and that it still transported lactose via the lactose permease. The mutation permitting IPTG independent lactose utilization was mapped to *ebgA*, and it was concluded that the strain synthesized a mutant EBG enzyme that converted lactose into an inducer of the *lac* operon (Rolseth *et al.*, 1980).

Hydrolysis can simply be considered a special case of transgalactosylation of $\beta$-galactosidase enzymes in which the galactose residue is transferred to water. Most $\beta$-galactosidases are also capable of transferring the galactose moiety to alcohols or to other sugars (Huber *et al.*, 1976). In the case of *lacZ* enzyme acting on lactose, the transgalactosylation may either be "direct," in which case, following breakage between the $\beta$-1,4 bond between galactose and glucose, a $\beta$-1,6 bond is formed between the same molecules; or it may be "indirect," in which case the original glucose molecule is released from the enzyme, and the galactose is transferred to another glucose molecule or to any other suitable "acceptor" molecule (Huber et al., 1976). I have investigated the transgalactosylation activity of EBG enzymes, beginning with purified enzyme obtained from the IPTG-independent strain mentioned above (Hall, 1982a). The major question was whether the mutant EBG enzyme was converting lactose to allolactose, or whether it was synthesizing another inducer of the *lac* operon. Lactose was digested with purified mutant enzyme, and the digestion products were analyzed by both paper chromatography and by high-pressure liquid chromatography (HPLC). The only transgalctosylation product detected was allolactose, based upon comigration with true allolactose and upon color following spraying with a specific reagent that distinguishes 1,4 from 1,6 linkages. Kinetic analysis showed that 7% of the digested substrate was converted to allolactose, the remainder being digested to glucose + galactose.

When either lactose or ONPG was digested in the presence of 100 mM [$^{14}$C]glucose, the galactose moiety was transferred to glucose at about 10% of the rate at which it was transferred to water. The product of this indirect transgalactosylation was not allolactose. The two indirect transgalactosylation products detected failed to comigrate with either lactose or with allolactose, did not migrate as would be expected for either 1,3- or 1,2-linked products, and the products were not reducing sugars. The biological significance of these products is not clear; however, this is the first example of which I am aware of in which the products of direct and of indirect transgalactosylation are different.

When purified EBG enzyme from the IPTG-independent strain SJ60 was characterized kinetically on the ten substrates described above, it became clear that it was a class IV enzyme. That observation, together with the observation that strain SJ60 exhibited a typical class IV growth pattern, suggested that transgalactosylation to form allolactose might be a general property of class IV enzymes. Strains carrying all of the available class IV alleles, and strains carrying all of the well-characterized class I and class II alleles, were screened for their IPTG-indēpendent lactose phenotype on MacConkey lactose indicator plates. In the absence of IPTG all class I and all class II strains were Lac$^-$, while all class IV strains were Lac$^+$. Thus, *in vivo*, only class IV EBG enzyme is able to convert lactose into the *lac* operon inducer allolactose at a rate sufficient to induce synthesis of the lactose permease. When lactose was digested with purified EBG$^0$ enzyme and purified class I and class II enzymes, no transgalactosylation products were detected. When lactose was digested with any of several class IV enzymes, allolactose was synthesized at about 7% of the rate at which glucose + galactose were released. Efficient transgalactosylation is thus another new function unique to the double mutant class IV EBG enzyme (Hall, 1982a).

## 7. The Role of Regulatory Mutations in the Evolution of Lactose Utilization

Although the analysis of mutations affecting the *ebgA* gene has been carried out in constitutive (*ebgR$^-$*) strains, 90% of the lactose-utilizing mutants isolated from the unevolved (*ebgA$^0$ ebgR$^+$*) strain are inducible for EBG enzyme synthesis. It was initially assumed that those regulated strains carried mutations only in the *ebgA* gene. Based upon the properties of purified EBG enzymes, it was possible to measure accurately the specific EBG enzyme synthesis (the fraction of soluble protein that is EBG enzyme) in strains synthesizing those enzymes. Those measurements showed that, while they had the same basal (uninduced) level of synthesis, the *evolved* inducible strains had a specific synthesis of 1.2% during lactose induction; i.e., an induced level fourfold higher than the unevolved strain (see Table I). Genetic analysis demonstrated that the increased level of synthesis was the result of a mutation in the regulatory gene *ebg*R. The evolved repressor gene was designated *ebg*R$^{+U}$, the *U* standing for a mutation that has "upped" the sensitivity of the repressor to lactose (Hall and Clarke, 1977).

The question was *why* had the repressor coevolved with the structural gene for EBG enzyme?

Published properties of the Lac permease permitted calculation of the physiological internal lactose concentration (Winkler and Wilson, 1966). From that, and from data on the kinetics of lactose hydrolysis, it was possible to estimate the *in vivo* lactase activity of uninduced, induced, and constitutive cultures. Correlations between *in vivo* lactase activity and growth rates on lactose were very strong ($R = 0.99$), and showed that there is a threshold level of lactase activity below which lactose utilization does not occur (Hall and Clarke, 1977). The lactase activity of strains synthesizing even the most active EBG enzyme is below that threshold when synthesis is under control of the wild-type (*ebgR$^+$*) repressor. This meant that the structural and regulatory genes were coevolving in order to provide a sufficient concentration of EBG enzyme to permit growth. The data suggest that mutations in both *ebgA* and *ebgR* are absolutely necessary for growth on lactose. Both the catalytic efficiency and the concentration of EBG enzyme must be increased well above the levels provided by the unevolved strain.

This necessity means that *all* of the papillae are the result of spontaneous double mutations. The frequency of mutations to *ebgR$^-$* is about $10^{-8}$, and while the frequency of mutations of *ebgR$^{+U}$* has not been measured, it is expected to be even lower. The mutation rate to *ebgA$^+$* is $2 \times 10^{-9}$ (Hall, 1977). If these mutations were occurring independently, they would be expected at a frequency of about $10^{-18}$, yet they occur at a frequency that results in the appearance of papillae on the surfaces of most colonies that contain $10^{10}$ cells. The conditions are such that sequential mutations are highly unlikely (Hall, 1982c), and it is necessary to conclude that these spontaneous mutations are not independent of each other. It may well be that spontaneous mutations occur in closely linked clusters (much like nitrosoguanidine-induced mutations), and that most studies fail to detect these simply because they have not been sought.

## 8. Directed Evolution of a Repressor

Having shown that I can direct the evolution of an enzyme so that it can hydrolyze a series of new substrates (Hall, 1978a), I asked whether I could direct the evolution of a repressor so that it can recognize a new molecule as an inducer (Hall, 1978b). Beginning with a strain that carried a wild-type repressor gene ($ebgR^+$), but synthesized a class II EBG enzyme that hydrolyzes lactulose effectively, I selected spontaneous mutants that could grow on lactulose. The parental strain was unable to utilize lactulose because it is not an effective inducer of EBG enzyme synthesis (Hall and Clarke, 1977). As expected, the majority of the lactulose-utilizing mutants were $ebgR^-$, i.e., they synthesized EBG enzyme constitutively. Among the 3000 colonies screened, nine proved to be nonconstitutive and strongly induced by lactulose. Genetic analysis confirmed that the mutations in these strains lay in the $ebgR$ gene.

I had previously shown that selecton for EBG enzyme with enhanced lactase activity often resulted in an enzyme with increased activity toward other substrates (Hall, 1978a). Likewise, the evolved repressor that was sensitive to lactulose as an inducer also proved to have greatly increased sensitivity to lactose, methylgalactoside, and galactosylarabinose as inducers. The evolved repressor, designated $ebgR^{+L}$, was unchanged in its sensitivity to the synthetic galactoside IPTG, and its ability to repress enzyme synthesis in the absence of inducers was unaltered. This study represents the first deliberate attempt to evolve a repessor to respond to a specific new inducer. The results suggest that there are great similarities between the evolution of enzymes and the evolution of regulatory proteins.

## 9. The Fully Evolved EBG Operon

The evolutionarily mature operon might be defined as one that specifies a sufficient quantity of all the proteins necessary to carry out some metabolic function, and is expressed only in response to an environmental signal indicating the requirement for the function. The wild-type EBG operon is clearly not "matured" for the function "lactose utilization," and its evolutionary progress toward that function must overcome several barriers.

First, the EBG$^0$ enzyme is an ineffective lactase. That barrier can be overcome by either a class I or a class II mutation within *ebgA*. Second, the wild-type repressor specified by *ebgR*$^+$ does not allow enough EBG enzyme synthesis for growth on lactose. That barrier can be overcome by *ebgR*$^+U$ or *ebgR*$^+L$ mutations. Thus, *ebgA*$^+$ *ebgR*$^+L$ strains can respond appropriately to the environmental signal, lactose, but they can not utilize lactose because (in the absence of IPTG) synthesis of the lactose permease is not induced. The above results suggested that this barrier might be overcome by a second mutation in *ebgA* leading to class IV enzyme.

Strain 5A1 synthesizes a class II EBG$^+$ enzyme, but is Lac$^-$ because it is *ebgR*$^+$ (Hall and Clarke, 1977) Its descendent, strain 5A103, is Lac$^+$ in the presence of IPTG because the *ebgR*$^{+L}$ mutation specifies a repressor that is very sensitive to lactose as an inducer (Hall, 1978b), but it is lactose-negative in the absence of IPTG because class II enzymes do not make allolactose. The *ebgR*$^{+L}$ repressor is also sensitive to galactosyl arabinose as an inducer (Hall, 1978B); therefore, I selected a spontaneous gal-ara$^+$ mutant in the presence of IPTG. The mutant, strain 5A1032, was Lac$^+$ in the absence of IPTG. EBG enzyme from strain 5A1032 was characterized and shown to be a typical class IV enzyme. I demonstrated that in the absence of lactose, strain 5A1032 expresses the basal level of EBG enzyme and lactose permease; but that in the presence of lactose both EBG enzyme and lactose permease are synthesized at a rate that permits lactose utilization. I further showed that the failure of the parental strain 5A103 to grow on lactose is solely attributable to its failure to synthesize the lactose permease, since lactose induces synthesis of EBG enzyme in that strain as well as it does in strain 5A1032.

Strain 5A1032 thus meets the above definition of the evolutionarily mature operon. In the absence of the environmental signal (lactose), neither the hydrolytic enzyme nor the transport system is expressed at significant levels. In the presence of the signal both the enzyme and the transport system are synthesized at levels sufficient for growth. The system is thus both energetically conservative and able to respond appropriately to its environment.

Wilson (1977) has argued that the evolution of higher organisms involves the evolution of new regulatory mechanisms, particularly those that lead to coregulation of blocks of genes that were previously independent. In light of that concept, it is particularly interesting that the *ebg* operon has evolved so that it not only controls its own expression, but it controls the expression of another operon, the *lac* operon, one of whose products is required for the complete function "lactose utilization." This evolutionary sequence from an unevolved operon to a maturely evolved operon that responds appropriately to its environment required three steps: a structural gene mutation, followed by a regulatory gene mutation, followed by a second structural gene mutation (Hall, 1982b).

## 10. A Model for Evolution in Diploid Organisms

One of the purposes of a model system is to provide explanations for general phenomena. One of the strengths of *E. coli* as a model system is that utilization of this haploid organism greatly simplifies the analysis of the data obtained. On the other hand, it also reduces the utility of the system as a general model because it inhibits extension of the arguments to eukaryotic diploid organisms. Evolution of new metabolic functions in diploids might be subject to constraints not present in haploids. The most obvious of those constraints is the problem of heterozygosity.

When an advantageous mutation arises in a haploid organism it is perforce "homozygous," and therefore confers an immediate selective advantage upon its host. When such a mutation arises in a diploid organism, on the other hand, it is necessarily heterzygous; and the selective advantage of that mutation depends upon whether it is dominant, recessive, or codominant with respect to the wild-type allele. In order to explore this problem, I constructed a series of merdiploid strains that carried a wild-type and an evolved *ebgA* allele, or a wild-type and an evolved *ebgR* allele (Hall, 1980). I showed that heterozygosity at *ebgA* does not significantly affect the selective advantage of the evolved *ebgA* alleles. In all respects the various $ebgA^+$ alleles behaved as codominant with the unevolved $ebgA^0$ allele. On the other hand, heterozygosity at *ebgR* virtually eliminated the selective advantage of evolved *ebgR* alleles. As mentioned earlier, the unevolved repressor is not sufficiently sensitive to lactose as an inducer; thus the amount of $EBG^+$ enzyme synthesized is not enough for growth (Hall and Clarke, 1977). In merodiploids the unevolved repressor is completely dominant with respect to the evolved repressor; thus the heterozygotes synthesize the level of EBG enzyme characteristic of the unevolved repressor. This is not surprising, since the unevolved allele $ebgR^+$ can be considered as analogous to the *lac* "super repressor" alleles that specify *lac* repressor that is less sensitive to inducers than wild-type repressor and dominant with respect to $lacI^+$ alleles (Bourgeois and Jobe, 1970).

For all bacterial systems examined, regulatory mutations have been required for evolution of new metabolic functions (Clarke, 1978; Hall, 1982c). For systems under negative control it is expected that these critical regulatory mutations would be recessive, as was the case for the evolved repressor mutations. If mutations that allow regulatory proteins to recognize new metabolites as inducers are recessive for negatively controlled systems, then evolution of new functions for those systems would be very difficult in diploids since they would not be advantageous until rendered homozygous by recombination or by genetic drift. Such mutations are expected to be dominant for positively controlled systems, however; thus, evolution of new functions for *positively* controlled systems should be no more difficult in diploid than in haploid organisms. I suggested that this might explain the paucity of examples of simple negative control in higher eukaryotes, where negatively controlled systems may have been lost as evolutionary dead ends (Hall, 1980).

## 11. Future Perspectives

The stated goal of these studies was to establish causal links between molecular changes and changes in fitness. At this point we have been able to directly link changes in quantity and activity of EBG enzymes to changes in growth rates on several $\beta$-galactoside sugars. To complete the causal links, it will be necessary to determine the changes in DNA sequence that result in such specific alterations in enzyme specificity and repressor–inducer interactions. In addition, it will be necessary to examine the consequences of these changes at the population level in environments other than those where $\beta$-galactoside sugars are the primary nutrient. That is, we must ask whether gains in fitness under one condition have resulted in a loss of fitness under other biologically realistic conditions. These two questions look at the opposite ends of the chain of causal links—the ultimate source of variation, DNA changes, and the ultimate effect of those changes on the population. Those studies are presently underway.

## References

Bourgeois, S., and Jobe, A., 1970, Super repressors of the *lac* operon, in: *The Lactose Operon* (J. R. Beckwith and D. Zipser, eds.), Cold Spring Harbor Laboratory, Cold Spring Harbor, New York, pp. 325–341.

Campbell, J. J., Lengyel, J., and Langridge, J. 1973, Evolution of a second gene for $\beta$-galactosidase in *Escherichia coli*, *Proc. Natl. Acad. Sci. USA* **70**:1841–1845.

Clarke, P. H., 1978, Experiments in microbial evolution, in: *The Bacteria,* Vol. VI (L. N. Ornsto and J. R. Sokatch, eds.), Academic Press, New York, pp. 137–218.

Cornish-Bowden, A., 1976, The effect of natural selection on enzymic catalysis, *J. Mol. Biol.* **101**:1–9.
Hall, B. G., 1976a, Experimental evolution of a new enzymatic function: Kinetic analysis of the ancestral ($ebg^0$) and evolved ($ebg^+$) enzymes, *J. Mol. Biol.* **107**:71–84.
Hall, B. G., 1976b, Methylgalactosidase activity: An alternative evolutionary destination for the $ebgA^0$ gene, *J. Bacteriol.* **126**:536–538.
Hall, B. G., 1977, The number of mutations required to evolve a new lactase function in *Escherichia coli*, *J. Bacteriol.* **129**:540–543.
Hall, B. G., 1978a, Experimental evolution of a new enzymetic function. II. Evolution of multiple functions for *EBG* enzyme in *E. coli*. *Genetics* **89**:453–465.
Hall, B. G., 1978b, Regulation of newly evolved enzymes. IV. Directed evolution of the *EBG* repressor, *Genetics* **90**:673–681.
Hall, B. G., 1980, On the evolution of new metabolic functions in diploid organisms *Genetics* **96**:1007–1017.
Hall, B. G., 1981, Changes in the substrate specificities of an enzyme during directed evolution of new functions, *Biochemistry* **20**:4042–4049.
Hall, B. G., 1982a, Transgalactosylation activity of EBG β-galactosidase synthesizes allolactose from lactose, *J. Bacteriol.* **150**:132–140.
Hall, B. G., 1982b, Evolution of a regulated operon in the laboratory, *Genetics* **101**:335–344.
Hall, B. G., 1982c, Evolution on a petri dish: Using the evolved β-galactosidase system as a model for studying acquisitive evolution in the laboratory, in: *Evolutionary Biology*, Vol. 15 (M. K. Hecht, B. Wallace, and G. T. Prance, eds.), Plenum Press, New York, pp. 85–150.
Hall, B. G., and Clarke, N. D., 1977, Regulation of newly evolved enzymes, III. Evolution of the *ebg* repressor during selection for enhanced lactase activity, *Genetics* **85**:193–201.
Hall, B. G., and Hartl, D. L., 1974, Regulation of newly evolved enzymes. I. Selection of a novel lactase regulated by lactose in *Escherichia coli*, *Genetics* **76**:391–400.
Hall, B. G., and Hartl, D. L., 1975, Regulation of newly evolved enzymes. II. The *ebg* repressor, *Genetics* **81**:427–435.
Hall, B. G., and Zuzel, T., 1980, Evolution of a new enzymatic function by recombination within a gene, *Proc. Natl. Acad. Sci. USA* **77**:3529–3533.
Hartl, D., and Hall, B. G., 1974, A second naturally occurring β-galactosidase in *E. coli*. *Nature* **248**:152–153.
Huber, R. E., Kurz, G., and Wallenfels, K., 1976, A quantitation of the factors which affect the hydrolase and transgalactosylase activities of β-galactosidase (*E. coli*) on lactose, *Biochemistry* **15**:1994–2001.
Jobe, A., and Bourgeois, S., 1972, *Lac* repressor–operator interaction. VI. The natural inducer of the *lac* operon, *J. Mol. Biol.* **69**:397–408.
Rolseth, S., Fried, V., and Hall, B. G., 1980, A mutant *ebg* enzyme that converts lactose into an inducer of the *lac* operon, *J. Bacteriol.* **142**:1036–1039.
Wilson, A., 1977, Gene regulation in evolution, in: *Molecular Evolution* (F. Ayala, ed.), Sinauer Associates, Sunderland, Massachusetts, pp. 225–234.
Winkler, H. H., and Wilson T. H., 1966, The role of energy coupling in the transport β-galactosides by *Escherichia coli*, *J. Biol. Chem.* **241**:2200–2211.

CHAPTER 7

# Amidases of Pseudomonas aeruginosa

PATRICIA H. CLARKE

## 1. Introduction

### 1.1. Biochemical Activities of *Pseudomonas* Species

In choosing an enzyme of *Pseudomonas aeruginosa* as the starting point for studies in experimental evolution, we had in mind the well-known biochemical versatility of *Pseudomonas* species. Many catabolic pathways had been described in detail for species belonging to this genus, and biochemists interested in isolating strains that could attack unusual organic compounds had often found that the isolates obtained from enrichment cultures could be assigned to the *Pseudomonas* group. A very comprehensive survey of laboratory strains carried out 60 years ago by Den Dooren De Jong (1926) had pointed to *Pseudomonas putida* as a particularly versatile organism. He found that one of his strains could utilize about 80 of 200 test compounds as the sole carbon source for growth. Among the compounds used in Den Dooren De Jong's experiments were the sodium salts of bromosuccinic and $\alpha$- and $\beta$-bromopropionic acids. These compounds were laboratory chemicals, and it could be asked whether they were attacked by enzymes with very wide substrate specificities that could not distinguish between halogenated and nonhalogenated substrates or whether there were specific enzymes that had evolved to act on organic compounds that occur rarely if at all in nature. Such questions become particularly pertinent when it is realized that during the last century the

---

*PATRICIA H. CLARKE* • Department of Biochemistry, University College London, London WC1E 6BT, England.

activities of the chemical industry have released many synthetic chemicals into the natural environment. Some of these are deliberately introduced as pesticides or herbicides and others are released casually as unwanted byproducts of the manufacture of organic chemicals. In view of the biochemical versatility of the pseudomonads, one hypothesis that could be put forward was that as pseudomonads had already evolved a wide variety of enzymes to metabolize the very many organic compounds that are synthesized by plants and animals, they might still have the potential of evolving enzymes with novel substrate specificities. When some of the earlier records of strain isolation were examined it appeared that the methods used were ideally suited for selecting new variants from the bacterial strains already present in the soil or water inoculum. For example, in a later survey of the nutritional capabilities of the aerobic pseudomonads Stanier *et al.* (1966) listed many strains of *Pseudomonas* species that had been isolated by enrichment culture from polluted soil or water, including one that had been isolated "from clay that had been suspended in kerosene for 3 weeks." This would appear to be a good method for selecting strains able to grow well on hydrocarbons.

From these general observations on pseudomonads it was thought worthwhile to find out if changes in the substrate specificity of a selected enzyme could be detected by selecting for mutants with altered growth phenotype.

### 1.2. Choice of Enzyme System

Before deciding on an appropriate enzyme it was useful to make a list of the characteristics that would be most helpful for the investigation. First, it should be a catabolic enzyme that could be made rate-limiting for growth and, if possible, it should lead straight into a known catabolic pathway. A catabolic enzyme of a pseudomonad would be likely to be an inducible enzyme and it would be useful if the inducer specificity were to be somewhat different from the substrate specificity of the enzyme itself. Second, it would be essential to explore the enzyme kinetics in some detail, and the kinetics of enzyme synthesis might also be of relevance. For these purposes it would be useful to have an enzyme that could be assayed rapidly, since a very large number of enzyme assays would be needed. The third requirement was that the original substrate should be one for which analogues could be readily obtained either because they were commercially available or because they could be synthesized without too much difficulty.

The choice of the species to be used presented no difficulty. At the time this work started Bruce Holloway had begun genetic studies with

## 1.3. Growth of *Pseudomonas aeruginosa* on Acetamide

Acetamide (Fig. 1) had some obvious attractions as a starting substrate. Amides of the homologous aliphatic series could be compared with acetamide and in addition one could consider amides with double bonds or substituent groups, such as hydroxyls or halogens, in the side chain or the substitution of one or of both hydrogens of the amide nitrogen or the substitution of the carbonyl oxygen by sulfur. Many of these compounds were commercially available and others could be prepared relatively easily. Hydrolysis of acetamide produces acetate, which can be metabolized via the tricarboxylic acid cycle, together with ammonia to

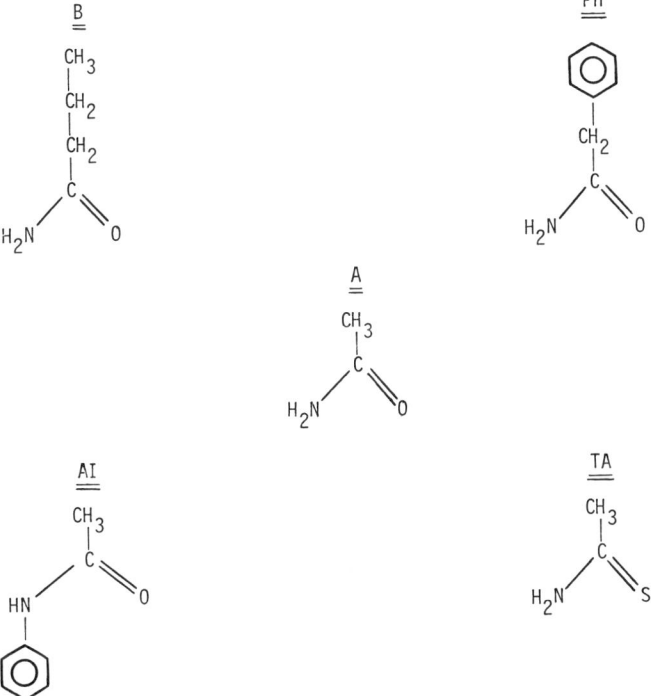

**Figure 1.** Acetamide and some of the related amides used for investigating amidase evolution. A, Acetamide; B, butyramide; Ph, phenylacetamide; AI, acetanilide; TA, thioacetamide.

provide a nitrogen source for growth. Although Den Dooren De Jong (1926) had reported that some bacteria could utilize acetamide, there was at that time no detailed information on the properties of bacterial aliphatic amidases. We tested several strains of *P. aeruginosa* for growth on acetamide as the sole source of carbon and nitrogen and selected strain PAC1 (at that time NCTC8602) for further study (Kelly and Clarke, 1960). Later, Stanier *et al.* (1966) reported that all their *P. aeruginosa* strains (29 were tested) grew on acetamide. Bühlmann *et al.* (1961) used a test medium containing glucose, acetamide, and an indicator and found that strains of *P. aeruginosa* produced an alkaline reaction. Mossel and Van Zadelhoff (1971) examined 500 *P. aeruginosa* isolates and found that only nine were unable to produce an alkaline reaction on a selection medium containing acetamide and an indicator. Growth on acetamide, presumed to be dependent on the activity of an aliphatic amidase, thus appears to be a characteristic of *P. aeruginosa* strains (see also Section 6.5).

### 1.4. The Wild-Type Amidase of *Pseudomonas aeruginosa* PAC1

Kelly and Clarke (1960, 1962) found that the range of aliphatic amides that could be used for growth by *P. aeruginosa* was very limited and could be related to the substrate and inducer specificities of an inducible aliphatic amidase. The two- and three-carbon amides acetamide and propionamide were the only ones that could provide the sole source of carbon and nitrogen. Bacteria grown on acetamide had higher specific activities for amide hydrolysis than those grown on propionamide. On the other hand, bacteria grown with either acetamide or propionamide hydrolyzed propionamide at a faster rate than acetamide. Thus, the early investigations showed that *P. aeruginosa* was able to synthesize an inducible aliphatic amidase with narrow substrate specificity. Further, a comparison of a number of other amides as potential inducers or substrates confirmed that the inducer specificity was not the same as the substrate specificity (Table I).

Among the amides tested in these experiments, only two substituted amides were hydrolyzed at an appreciable rate, glycollamide and acrylamide. Both these carry side chains that are similar in size to acetamide and propionamide, and this was the first indication of the limitations imposed by the length of the amide side chain. More sensitive assays showed later that butyramide and lactamide could be hydrolyzed at a low rate by the wild-type enzyme. However, a larger number of amides had good inducing activity and included lactamide (but not butyramide) and the N-substituted amides $N$-methylacetamide and $N$-acetylacetamide. The latter finding was useful in that it meant that experiments on the kinetics of amidase induction could be carried out under conditions of gratuity with the enzyme not being required for the growth of the culture. The poor substrate activity of lactamide meant that it was hydrolyzed very slowly by growing cultures, and for experiments on the kinetics of induction N-methylacetamide, N-acetylacetamide, or lactamide were used as gratuitous inducers with cultures in the exponential growth phase. A few amides were found to be neither inducers nor substrates, but appeared to prevent amidase induction by inducing amides. These included cyanoacetamide, thioacetamide, and butyramide. Formamide, which showed weak inducing activity, reduced the rate of enzyme induction by the nonsubstrate inducing amide N-acetylacetamide. These properties were an advantage in the selection of various classes of mutants. Fluoroacetamide was both substrate and inducer, but inhibited growth and was used later to select amidase-negative mutants (Clarke and Tata, 1973). The wide variety of amides available and the subtle response of the amidase system to small changes in amide structure provided very powerful tools for mutant selection. Mutants could be selected because they grew better than the parental strains on a particular medium and others could be selected because they were resistant to growth inhibition by certain amides added to an otherwise satisfactory medium. Table II indicates some of the amides used for mutant selection at different stages of these studies.

A general outline of the inducible amidase of *P. aeruginosa* was apparent from the pioneering studies of M. Kelly. As particular aspects were studied in more detail, the properties of the enzyme and its regulatory control became clearer and it was possible to devise a medium for selecting a particular class of mutant. The following sections describe some of the methods used and the ways in which biochemical and genetic studies helped to exploit this enzyme system. All the mutants retained in our collection were assigned PAC strain numbers according to the accepted system for bacterial nomenclature, but since it is often helpful to be able to recognize a particular strain by its phenotype, the following account also includes, where appropriate, the series number used on first isolation. For example, C1 (PAC101) denotes the first constitutive strain isolated and B6 (PAC351) denotes a mutant producing an altered enzyme capable of hydrolyzing butyramide at a rate sufficient for this amide to be used as a growth substrate. Some of the earlier work on the amidase of *P. aeruginosa* has been the subject of reviews (Clarke, 1974, 1978, 1980).

## 2. Amidase Regulatory Mutants

We were able to isolate amidase mutants with altered enzymes and also amidase mutants with altered regulation of enzyme synthesis. These had a variety of amide growth phenotypes. Since the isolation of the

## Table I
### Comparison of Amides as Substrates, Inhibitors, Inducers, and Amide Analogue Repressors[a]

| Amide | | Substrate | Inducer | Enzyme inhibitor | Amide analogue repressor |
|---|---|---|---|---|---|
| Formamide | $HCONH_2$ | + | + | NT | + |
| Acetamide | $CH_3CONH_2$ | ++ | ++ | NT | NT |
| Propionamide | $CH_3CH_2CONH_2$ | +++ | ++ | NT | NT |
| Butyramide | $CH_3CH_2CH_2CONH_2$ | Trace | − | NT | + |
| iso-Butyramide | $CH_3{>}CHCONH_2$ $CH_3$ | − | − | NT | NT |
| Valeramide | $CH_3(CH_2)_3CONH_2$ | − | NT | NT | NT |
| Hexamide | $CH_3(CH_2)_4CONH_2$ | − | NT | NT | NT |
| N-Methylformamide | $HCONHCH_3$ | − | +++ | − | NT |
| N-Ethylformamide | $HCONHC_2H_5$ | − | + | NT | NT |
| N-Methylacetamide | $CH_3CONHCH_3$ | − | +++ | − | NT |
| N-Ethylacetamide | $CH_3CONHC_2H_5$ | − | ++ | NT | NT |
| N-Acetylacetamide | $CH_3CONHCOCH_3$ | − | +++ | NT | NT |
| N-Phenylacetamide | $CH_3CONHC_6H_5$ | − | − | − | + |
| N-Dimethylacetamide | $CH_3CON(CH_3)_2$ | − | − | − | − |

## AMIDASES OF PSEUDOMONAS AERUGINOSA

| Compound | Formula | | |
|---|---|---|---|
| N-Methylpropionamide | CH₃CH₂CONHCH₃ | − | NT |
| N-Ethylpropionamide | CH₃CH₂CONHC₂H₅ | − | NT |
| Cyanoacetamide | CH₂CNCONH₂ | − | + |
| Iodoacetamide | CH₂ICONH₂ | NT | NT |
| Glycine amide | CH₂NH₂CONH₂ | − | + |
| Sarcosine amide | CH₂NH(CH₃)CONH₂ | − | + + |
| Glycollamide | CH₂OHCONH₂ | + | + + |
| Acrylamide | CH₂=CHCONH₂ | NT | NT |
| β-OH-propionamide | CH₂OHCH₂CONH₂ | − | + |
| Lactamide | CH₃CHOHCONH₂ | + + + | NT |
| Pyruvate | CH₃COCONH₂ | NT | NT |
| Methylcarbamate | CH₃OCONH₂ | + + | NT |
| Urea | NH₂CONH₂ | − | NT |
| Benzamide | C₆H₅CONH₂ | NT | NT |
| Thioacetamide | CH₃CSNH₂ | − | + |

*a* Strain PAC1 was grown in acetamide medium and the washed suspensions used to test for substrate and inhibitor activities. Cultures were grown in succinate–minimal salt medium with amides added to test for inducer activity. N-Methylacetamide was present as inducer to test for amide analogue repression. The relative activities are shown as $+++$, $++$, $+$, and trace; $-$, no activity detected; NT, not tested. [Data are from Kelly and Clarke (1962).]

## Table II
### Interactions of Amides with Amidase of *Pseudomonas aeruginosa* Strain PAC1

| Amide | Characteristics |
| --- | --- |
| Formamide | Poor substrate, poor inducer; used in S/F medium for isolation of constitutive and formamide-inducible mutants |
| Acetamide | Good substrate and inducer; supports good growth |
| Propionamide | Good substrate and inducer; supports good growth |
| Butyramide | Very poor substrate, represses amidase synthesis; used to isolate regulator and altered enzyme mutants |
| Lactamide | Very poor substrate but good inducer; used for gratuitous enzyme induction; used with succinate for isolation of catabolite repression-resistant mutants |
| $N$-Acetylacetamide | Good inducer, used for gratuitous amidase induction, hydrolyzes spontaneously on prolonged incubation at 37°C |
| Valeramide | Not a substrate, represses amidase induction; used to isolate altered enzyme mutants |
| Phenylacetamide | Not a substrate, represses amidase induction; used to isolate altered enzyme mutants |
| $N$-Phenylacetamide (acetanilide) | Not a substrate, represses amidase induction; used to isolate altered enzyme mutants |
| Fluoracetamide | Substrate and inducer, but inhibits growth due to lethal synthesis; used to isolate amidase-negative mutants |
| Acrylamide | Inducer and substrate but inhibits growth |
| Urea | Inhibits amidase activity; used to isolate altered enzyme mutants |

altered enzyme mutants was often dependent on particular changes in regulation, it is convenient to start with some of the regulatory mutants. The mutants described in this section are those in which amidase synthesis has become (1) constitutive, (2) altered in inducer specificity, or (3) resistant to catabolite repression.

### 2.1. Isolation of Mutants from Succinate/Formamide Medium

Formamide is a poor substrate and a poor inducer of amidase. *Pseudomonas aeruginosa* is unable to grow on formate, so the most that formamide can provide is the nitrogen source for growth. The wild-type strain grows very poorly on succinate/formamide (S/F) medium, but mutant colonies appear within a few days either spontaneously or at higher frequency from cultures treated with chemical mutagens. It had been predicted that if constitutive mutants arose, they would be able to outgrow the wild type since they would not be dependent on induction by formamide. Succinate provides the carbon source, but exerts additional selection pressure, since it is a strong catabolite repressor of amidase synthesis (see Section 2.2). Several classes of mutants can be isolated from S/F plates (Fig. 2): Magnoconstitutive strains synthesize high amidase levels in the absence of inducers; semiconstitutive mutants have lower specific activities, but may produce higher levels if an inducing amide is present; formamide-inducible strains are altered in specificity of induction, so that the rate of induction by formamide now reaches or surpasses that by acetamide. These regulator mutants all produce wild-type amidase and it could be concluded that they carried mutations in a regulator gene *amiR*. Transductional analysis showed that the *amiR* mutations were closely linked to the structural gene for the enzyme, *amiE* (Brammar *et al.*, 1967).

### 2.1.1. Response of Constitutive Mutants to Butyramide

Butyramide is unavailable as a growth substrate for the wild-type strain for two reasons. It is a very poor amidase substrate and has no inducing power. However, it also has the property of preventing induction by inducing amides and this could be interpreted as being due to competition for an amide-binding site of a regulatory protein. When the constitutive strains isolated from S/F plates were examined it was found that some grew on butyramide plates while others were unable to do so. Studies on the kinetics of synthesis by the constitutive mutants revealed that the addition of butyramide to some strains growing in a minimal salt medium containing pyruvate as carbon source resulted in almost immediate ces-

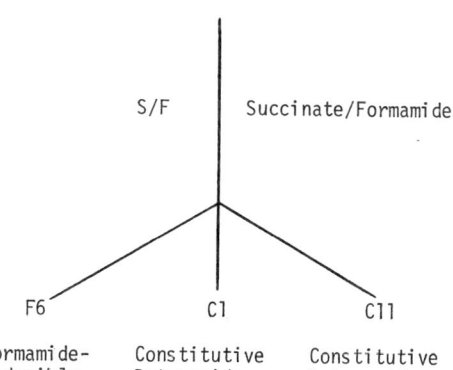

**Figure 2.** Examples of regulatory mutants isolated from wild-type *Pseudomonas aeruginosa* PAC1 on succinate/formamide (S/F) plates.

sation of amidase synthesis. These butyramide-sensitive strains were those that were unable to grow on butyramide plates. Those strains that were resistant to repression by butyramide during growth in a minimal salt medium were able to grow on plates with butyramide as carbon and nitrogen source. The butyramide-resistant strains produced high amidase levels, but since this was the wild-type enzyme with little activity on butyramide, the rate of growth at the early stages was limited by the slow rate of butyramide hydrolysis (J. E. Brown and Clarke, 1970) (see also Section 4.1).

### 2.2.2. Amide Analogue Repression

We termed this effect "amide analogue repression" and interpreted it as due to the binding of butyramide (or another amide analogue) at the inducer-binding site. Amidase synthesis is under positive control by the *amiR* protein and it is assumed that when an inducing amide is bound, the wild-type regulator becomes changed to the active conformation (Farin and Clarke, 1978). Butyramide and other amide analogue repressors (Tables I and II) prevent amidase synthesis by competing with inducing amides, and it is assumed that when butyramide is bound at the inducer-binding site of the wild-type regulator protein the change to the active conformation does not take place. With the constitutive strains we can distinguish between those that are butyramide-sensitive and those that are butyramide-resistant. The regulator proteins of both classes are active in the absence of amides, but the regulator protein of the resistant strains may have little or no affinity for butyramide, while those of the sensitive strains might bind butyramide and then undergo conformational change to the inactive form. Studies on the effects of different concentrations of butyramide on amidase synthesis by constitutive strains in pyruvate–minimal salt medium showed that they ranged from being highly sensitive to being almost insensitive to butyramide repression. This could be expressed in terms of the repression constant $K_{Rep}$ in an analogous way to the induction constant $K_{Ind}$ (J. E. Brown and Clarke, 1970).

This is expressed graphically in Figs. 3 and 4. Figure 3 illustrates the effect of adding 20 mM butyramide to a culture of the constitutive strain C11 growing in succinate minimal medium. Following the addition of butyramide, the rate of amidase synthesis decreased almost immediately. Strain C11 was the parent of many of the altered enzyme mutants and this sensitivity to repression by butyramide (and other amide analogue repressors) was an important factor. By measuring the rates of amidase synthesis of constitutive strains in the presence of different concentrations of butyramide, it was possible to relate the extent of repression to the

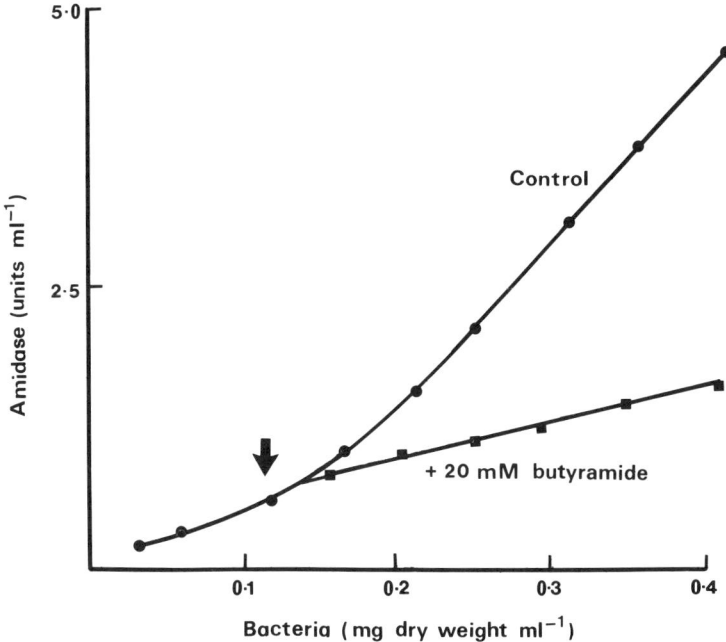

**Figure 3.** Repression of amidase synthesis by butyramide. The constitutive mutant C11 was grown in succinate–minimal salt medium, and at the time indicated by an arrow the culture was divided into two parts. (●) control culture; (■) 20 mM butyramide added (J. E. Brown and Clarke, 1970).

butyramide concentration. In Fig. 4 it can be seen from the double reciprocal plot that strain C11 was much more sensitive to butyramide repression than strain L10. The resistance of strain L10 to amide analogue repression was also an important factor in selection of further mutants.

Jayaraman et al. (1966) had previously observed that certain $I^-$ constitutive mutants of the *lac* operon were repressed by ONPF (2-nitrophenyl-$\beta$-D-fucoside). The *lac* operon is under negative control by the *lacI* protein, so that these results could be interpreted as a change from an inactive to an active repressor conformation as a result of binding ONPF. The interactions between the low-molecular weight effectors (inducers and corepressors) and the regulator proteins allow changes in protein conformation that can either stimulate or prevent transcription of structural genes. Among the amidase-constitutive mutants were others in which amidase synthesis could be decreased in rate (i.e., slightly repressed) by amides such as *N*-acetylacetamide that are inducers for the

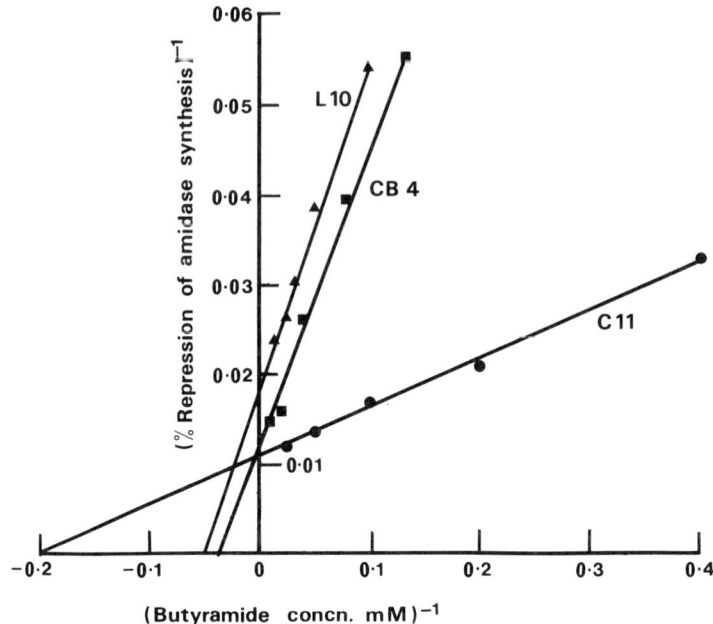

**Figure 4.** Mutant strains L10, CB4, and C11 were grown in succinate–minimal salt medium and divided into six parts when the cultures had reached an early stage of exponential growth. One part was kept as control and various concentrations of butyramide added to the other five. The differential rates of amidase synthesis were measured and are shown here as double reciprocal plots of percentage repression against butyramide concentration. Strains L10 and CB4 are able to grow with butyramide, since they are much more resistant to butyramide repression than is strain C11 (J. E. Brown and Clarke, 1970).

wild-type strains. Such changes in regulatory phenotype strongly suggest that new characteristics may evolve in nature as the result of mutations in regulatory genes.

### 2.2. Isolation of Mutants from Succinate/Lactamide Medium

Lactamide is a poor substrate and a very good inducer, while succinate is a strong catabolite repressor. The wild-type strain grows well on lactamide as sole carbon and nitrogen source, but when succinate is present the catabolite repression of amidase synthesis severely affects the rate of growth. At a ratio of about 20:1 succinate–lactamide the wild type grows slowly on S/L plates, but within the faint background large colonies begin to appear after a few days. Several different classes of mutant are able to grow well on this medium (see Section 4.1), but the usual isolate is catabolite-repression resistant. From the wild-type strain it is possible to isolate both inducible and constitutive catabolite-resistant mutants. The latter invariably carry mutations in the amidase regulator gene *amiR* as well as the mutations conferring resistance to catabolite repression by succinate. Many other carbon compounds repress amidase synthesis and a variety of media can be constructed in which succinate, malate, glucose, etc., provide the carbon source and lactamide the nitrogen source. The key to this selection procedure is the high inducing and poor enzyme substrate activity of lactamide. The bacteria require very high enzyme levels to overcome the poor enzyme activity, and the mutation to catabolite resistance appears to be the most probable response. The mutants isolated from these media are not a homogeneous group and their mutations have not been found to be linked to the *amiR* and *amiE* genes. (Smyth and Clarke, 1975a). Section 3.4 describes promoter mutants in which resistance to catabolite repression is tightly linked to the *amiE* gene.

### 2.3. Isolation of Mutants with Altered Inducibility

In Section 2.1 it was mentioned that some of the mutants isolated from succinate/formamide plates were altered in the specificity of induction and had become inducible by formamide at rates approaching that for acetamide in the wild-type strain. For a mutation to altered inducibility to be of value it would be essential for the amidase being induced to have a significant level of activity on the particular amide in question. Mutants with altered inducibility did not appear in any of the other selection methods, but Section 6.2 describes the route by means of which a butyramide-inducible mutant was obtained. The identification of mutants with altered inducibility requires the screening of large numbers of possible candidates, many of which will have grown by scavenging the products of hydrolysis carried out by other colonies on the plate. Further, it is probable that mutations to give altered inducer-binding sites are less frequent than those conferring constitutivity, so that the latter would predominate.

## 3. Amidase-Negative Mutants

### 3.1. Isolation of Acetamide-Negative mutants

Mutants with an acetamide-negative phenotype were first isolated by selecting for minute colonies on plates containing acetamide as the major carbon source with trace amounts of succinate to allow the amidase mutants to grow large enough to be identified (Skinner and Clarke, 1968). This method gave mutants defective in acetate metabolism as well as amidase-negative mutants with defects in the structural gene *amiE*. With fluoroacetamide added to a minimal medium containing pyruvate and ammonium salts it was possible to isolate large numbers of acetamide-negative mutants by positive selection for fluoracetamide resistance (Clarke and Tata, 1973).

### 3.2. Mutations in the *amiE* Gene

The absence of acetamidase activity may be due to point mutations in the structural gene *amiE* or to deletions of part or all of the sequence. A further possibility is that the substrate specificity may be changed so that the original amide substrate can no longer be hydrolyzed at a rate sufficient for growth to occur. In Section 4 some altered enzyme mutants are described that have lost the ability to hydrolyze acetamide but can hydrolyze other amides. This acetamide-negative phenotype made it possible to use transduction for fine structure mapping for both amidase-defective and certain altered-amidase mutants (Betz *et al.*, 1974).

### 3.3. Mutations in the *amiR* Gene

One reason for discarding the view that amidase might be under negative control of transcription was that among more than $10^3$ constitutive strains examined there were none that were constitutive at higher temperatures and inducible at a lower temperature. Some *lacI* mutants produce thermolabile repressors and these strains are constitutive for $\beta$-galactosidase when grown at higher growth temperatures ($>37°C$), but require the presence of an inducer for $\beta$-galactosidase synthesis at lower temperatures (Sadler and Novick, 1965). It was thought possible that such mutants might occur for *P. aeruginosa* amidase but could not be detected by the methods employed. An alternative explanation was that the regulator protein exerted a positive control on transcription, and if so, the prediction was that it should be possible to obtain amidase-negative mutants with defects in the regulator gene. Also, it should be possible to obtain mutants that had an amidase-negative phenotype at elevated temperatures but were amidase-positive at lower temperatures. In the latter case it would be essential to demonstrate that the temperature sensitivity was not due to a mutation in the structural gene for the enzyme resulting in a thermolabile amidase protein. F. Farin screened a large number of acetamide-negative mutants obtained by selection for fluoracetamide resistance (Farin and Clarke, 1978). Among these were some derived from the constitutive strain C11 (PAC111) producing high amidase activity but sensitive to butyramide repression. Some of these were found to be acetamide-negative at 37°C but able to grow on acetamide at a reasonable rate at 30°C. Further analysis of two of these mutants showed that amidase was synthesized when they were grown at 28°C but ceased almost immediately when the temperature was raised to 43°C. The amidase produced by these strains was as thermostable as the wild type and retained 100% of its activity for longer than 15 min at 60°C, so that a thermosensitive mutation in the *amiE* gene could be ruled out. Figure 5 shows the effect of a shift in temperature from 29°C to 43°C on the synthesis of amidase by mutant RTS1 growing in a pyruvate–minimal salt medium. At the same time as the shift up in temperature, other portions of the culture were given either 50 mM butyramide or 10 mM succinate. It can be seen that mutant RTS1 had retained the sensitivity to amide-analogue repression by butyramide and to catabolite repression by succinate that had been characteristic of the parent strain C11.

Similar results were obtained with an inducible strain. Strain F6

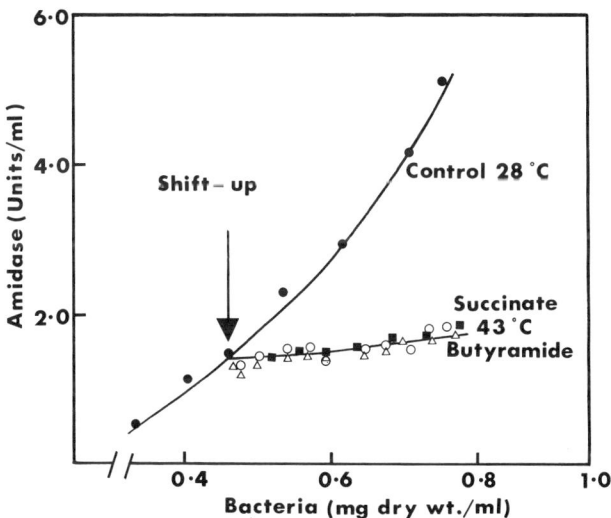

**Figure 5.** Effect of a temperature shift from 28 to 43°C on the rate of amidase synthesis by the temperature-sensitive mutant RTS1 compared with the addition of butyramide as amide analogue repressor and succinate as catabolite repressor. The culture was grown at 28°C in lactate–minimal salt medium until the early exponential growth phase and divided into four parts at the time indicated by the arrow. (●) Control culture; (■) shift up to 43°C; (△) 50 mM butyramide added; (○) 10 mM succinate added (Farin and Clarke, 1978).

(PAC153) had been isolated on S/F plates and found to be altered in inducer specificity. This strain grows well on S/F plates because it is readily induced by formamide. Among fluoracetamide-resistant mutants isolated from F6 were some that were acetamide-negative at 37°C and constitutive for amidase synthesis at 30°C. The second mutation that had conferred constitutivity to strain F6 segregated out in a cross with an amidase-negative (*amiE*) mutant to give a constitutive stable recombinant. It appeared that in this case it was necessary for the two mutations to be present: *amiR*43, the original formamide-inducible mutation of F6, and *amiR*221, the constitutive mutation, together resulting in a thermolabile regulator protein. This is an example of the cumulative effect of mutations on protein properties that we were to observe frequently among the altered enzyme mutants.

Further evidence for positive control by the *amiR* protein came from analysis of revertants of amidase-negative mutants derived from a strain with a wild-type *amiR* gene and a mutation *amiE*16 in the structural gene conferring altered substrate specificity (Fig. 6). Strain IB10 (PAC398) was a recombinant in a cross between an altered enzyme mutant and an amidase-negative mutant (see Sections 4.3 and 6.3) and FIB29 was an acetamide-negative mutant derived from IB10 by fluoracetamide selection. The acetamide-positive revertants from FIB29 were tested for their regulatory phenotype and for the substrate specificity of the enzyme. The hypothesis was that some of the acetamide-negative mutants would have mutations in the *amiE* gene and that these would revert at more than one site to give a variety of amidase proteins but would retain the wild-type *amiR* and remain inducible. On the other hand, if the amidase-negative phenotype was due to a mutation in the *amiR* gene, then the *amiE*16 phenotype would be retained but the revertants would exhibit a range of different regulatory phenotypes. The latter prediction was fulfilled for FIB29 and FIB32, while other acetamide-negative strains isolated from IB10 had mutations in the enzyme structural gene as expected. These findings indicated that amidase synthesis required the product of the regulatory gene *amiR*.

### 3.4. Promoter Mutations

Another possible site for amidase-negative mutations is the *amiE* promoter. Since the number of down-promoter mutants was expected to be much less than either *amiE* or *amiR* amidase-negative mutants, it was necessary to devise a more complex selection procedure. It was difficult to identify the sites of down-promoter mutations other than by intensive

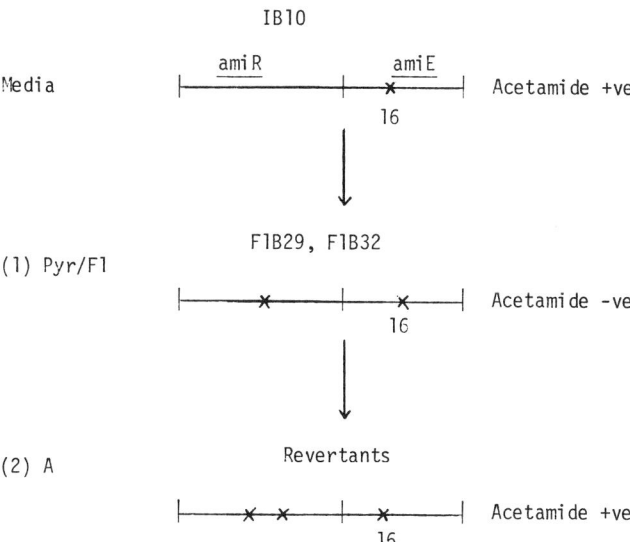

**Figure 6.** Derivation of amidase-negative mutants with lesions in the regulator gene *amiR*. Strain IB10 (wild-type *amiR* gene, producing B amidase) was the parent of acetamide-negative FIB mutants. Some FIB mutants gave revertants with an inducible phenotype, indicating that the acetamide-negative mutation had been in the enzyme structural gene. FIB29 and FIB32 gave revertants with a variety of regulatory phenotypes and all retained the *amiE*16 mutation (Farin and Clarke, 1978).

gene mapping, and the methods available at that time were not sensitive enough for that. P. F. Smyth identified promoter mutants by looking for catabolite-resistance mutations very closely linked to the *amiE* gene (Smyth and Clarke, 1975b). The starting strain was the constitutive mutant C1 (PAC101), which is resistant to butyramide repression. This character was important for identification of mutants at a later stage of the selection procedure. Fluoracetamide-resistant mutants were selected from strain C1, and 116 leaky mutants that grew slightly on acetamide but not at all on lactamide were tested for reversion on succinate/butyramide (S/B) plates. The rationale for this procedure was that growth on S/B plates would require both resistance to butyramide repression and resistance to catabolite repression. If the C1 mutation for butyramide resistance was maintained, the second mutation might be in the promoter region, giving a down-promoter mutant with a low rate of amidase synthesis. From these amidase-defective mutants the selection on S/B plates might be expected to yield up-promoter mutants with both a high rate of amidase synthesis and resistance to catabolite repression. Transductional analysis could show whether or not the mutation conferring resistance to catabolite repression was tightly linked to the *amiE* gene. Only seven of the 116 fluoracetamide-resistant mutants derived from strain C1 gave revertants on S/B plates.

The S/B revertants from each of these seven leaky mutants were tested for transductional linkage to the *amiE* gene. Six sets gave 99–100% linkage and these were presumed to be up-promoter mutants. This strategy for finding promoter mutants was based on the expectation that both the mutation resulting in a low rate of amidase synthesis and the subsequent mutation giving up-promoter properties would occur in the promoter region. This was not the only way in which such catabolite-resistant mutants could arise, since the mutations might have occurred in the *amiR* gene. In that case the linkage would be expected to be somewhat less (Brammar et al., 1967). Moreover, the presumed up-promoter mutants obtained in this experiment still retained the regulatory features associated with the original constitutive mutation in the *amiR* gene of the strain C1. Another way in which the leaky mutants could have acquired the ability to grow on S/B plates would have been a mutation in *amiE* to produce an altered enzyme that could hydrolyze butyramide. Smith and Clarke (1975b) found that such mutants could be isolated directly from the wild type by mutagenesis and selection on S/B plates (see also Section 4.1). However, all the presumed up-promoter mutants were tested for substrate specificity and all produced wild-type enzyme.

The selection of these promoter mutants was dependent on making use of several amides that had different effects on the amidase system. First, the constitutive strain C1 was isolated from S/F plates (containing the weak inducer formamide), since it was able to synthesize amidase without induction. Second, a large number of fluoracetamide-resistant mutants were isolated from strain C1 by positive selection, since this amide inhibits growth of strains with an active amidase. Leaky amidase mutants could be identified by slight growth on the good substrate acetamide and complete absence of growth on the poor substrate lactamide. Butyramide, with succinate, provided a very strong selection system since it is hydrolyzed very slowly by the wild-type enzyme, and a mutation giving resistance to catabolite repression, together with the original C1 mutation giving resistance to amide analogue repression, is essential for growth to take place. Of the six up-promoter mutants obtained by this selection procedure, strain PAC433 was found to produce very high amidase activities and was subsequently used as the source of amidase DNA for cloning experiments (Section 6.4). Further studies on the amidase promoter are being directed to a comparison of DNA sequences.

## 4. Mutants with Altered Enzymes

The isolation of mutants producing altered enzymes with new substrate specificities was achieved by the straightforward method of selecting for mutants that could use the novel amide as a source of carbon or nitrogen or both. Success depended on starting with a strain with appropriate regulatory properties.

### 4.1. Butyramide-Utilizing Mutants: B Group

The first of the altered enzyme mutants was strain B6 (PAC351) isolated on butyramide plates from strain C11 (PAC111) (J. E. Brown *et al.*, 1969). The parent strain C11 was a constitutive strain with a high rate of amidase synthesis, but it was unable to grow on butyramide since it was very sensitive to amide analogue repression. The B group of butyramide-utilizing mutants were isolated following mutagenesis with NMG ($N$-methyl-$N'$-nitro-$N$-nitrosoguanidine) and were found to produce an amidase with altered substrate specificity and altered electrophoretic mobility. Strain B6 retained sensitivity to amide analogue repression by cyanoacetamide, but it was not possible to test directly for repression by butyramide since it was so rapidly hydrolyzed by the novel enzyme. The constitutive strain C11, with its potential for synthesizing high levels of amidase and its sensitivity to butyramide repression, was a fortunate choice as a potential parent for mutants with new substrate specificities. Later, Smyth and Clarke (1975b) found that mutants producing B amidases could be isolated directly from the wild-type strain by selection on succinate/butyramide plates (Section 3.4). This required a mutation in the regulator gene to give constitutivity and a mutation in the structural gene *amiE* to produce an altered enzyme. Another system in which a double mutational event leads to a novel growth phenotype is in the selection of *ebg* mutants of *E. coli* K12. Mutations in both the regulator gene *ebgR* and the structural gene *ebgA* are needed to get mutants able to utilize lactose for growth (Hall and Hartl, 1974).

Another class of *P. aeruginosa* amidase mutants can also be isolated from butyramide plates (Fig. 7). These are the constitutive mutants that are resistant to butyramide repression (CB group) and are similar in properties to the butyramide-resistant mutants isolated by selecting for constitutive mutants on succinate/formamide plates (Section 2.1). The butyramide-utilizing regulatory mutants can be isolated from the wild-type inducible strain and also from butyramide-sensitive constitutive mutants, such as strain C11. The two classes of butyramide-utilizing mutants, novel enzyme or high levels of a poor enzyme, represent two different solutions

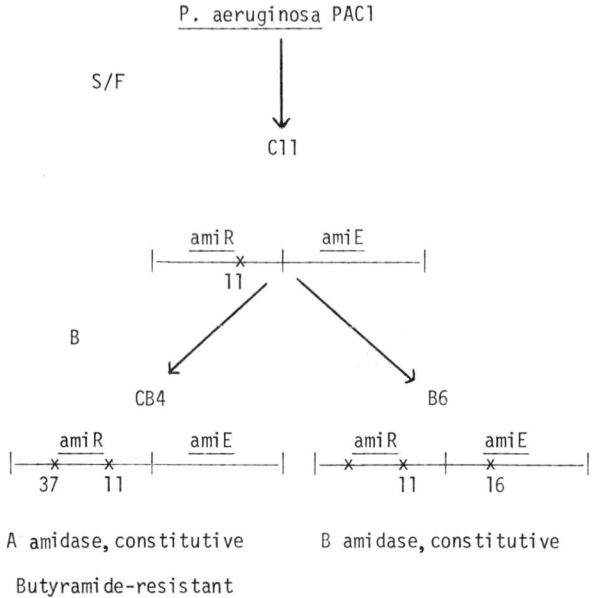

**Figure 7.** Derivation of butyramide-utilizing mutants. The constitutive mutant C11 gave rise to two classes of butyramide-utilizing mutants. Strain CB4 has an additional mutation in the *amiR* gene conferring resistance to butyramide repression, and B6 produces an altered amidase (J. E. Brown and Clarke, 1970).

to the problem of growth on a poor substrate. The wild-type enzyme, in this case A amidase, may have a very low affinity for the poor substrate or a low rate of activity (high $K_m$, low $V_{max}$) or both. If the amount of amidase synthesized in unit time is expressed as $E$ and the rate of butyramide hydrolysis per enzyme molecule in unit time as $B$, then there is a critical value for $P$ in the equation $EB = P$ that determines whether or not growth can occur (Clarke and Lilly, 1969). With the B group mutants producing the altered amidase the significant factor is the change in $B$, the rate of hydrolysis of butyramide in unit time. For the CB group of regulatory mutants the significant change is in $E$, the rate of amidase synthesis in unit time. In both cases $P$ reaches the threshold value for growth. Even so, with some of the CB mutants the rate of hydrolysis of butyramide at the early stages of growth is so slow that it becomes rate-limiting (J. E. Brown and Clarke, 1970). Similar observations were made by Hall and Clarke (1977) for the growth of some of their *ebg* mutants on lactose.

It is theoretically possible for altered enzyme mutants to appear on

S/L plates. Since lactamide is a very poor substrate and succinate represses the rate of amidase synthesis, a mutant producing an amidase with a high rate of lactamide hydrolysis would be expected to have a growth advantage. None has been recognized among the many mutants examined. Strain B6 is sensitive to catabolite repression by succinate, but produces B amidase with higher activity toward lactamide as well as butyramide (Fig. 7). Strain B6 grows well on succinate/lactamide plates by virtue of its altered enzyme rather than by any change in sensitivity to catabolite repression.

### 4.2. Valeramide-Utilizing Mutants: V Group

Valeramide does not support the growth of any of the constitutive strains carrying the wild-type *amiE* gene or the B group of mutants. J. E. Brown *et al.* (1969) isolated a number of valeramide-utilizing mutants from strain B6. These were a heterogeneous group and carried at least two mutations in the structural gene representing sequential enzyme evolution. Several produced unstable enzymes and two of them had lost the ability to grow on acetamide. This indicated that the substrate specificity could be altered by mutation in such a way that the optimum shifted to amides with longer aliphatic side chains. Mutant V9 was used for enzyme studies and also as the parent strain for further mutant selection (Section 5.1).

### 4.3. Phenylacetamide-Utilizing Mutants: Ph Group

#### 4.3.1. PhB Amidases

Phenylacetamide has a very bulky side chain and since the wild-type amidase is limited to amides with small side chains, this presented a considerable challenge for the evolution of new substrate specificities. The product of hydrolysis of phenylacetamide is phenylacetate, which, unlike the aliphatic carboxylic acids, cannot be utilized as a carbon source by *P. aeruginosa*. The selection media for the phenylacetamide-utilizing mutants contained phenylacetamide as nitrogen source and succinate as the carbon source. Strain B6 yielded 16 phenylacetamide-utilizing mutants that appeared to produce identical enzymes, and PhB3 (PAC377) (*amiR*11 *amiE*16, 67) was selected for more detailed studies (Betz and Clarke, 1972). It has completely lost the ability to grow on acetamide and could be used as donor or recipient in transductional crosses with other acetamide-negative strains. From one of these crosses a transductant was obtained that had a wild-type *amiR*$^+$ gene together with the *amiE*16 mutation originally present in strain B6. This was an indication that a mutation could be retained during subsequent evolutionary events. There was no direct method whereby such a strain could be selected, since, although it produced the altered B amidase, it was unable to grow on butyramide because butyramide could not induce amidase synthesis. Later, this strain, IB10, was used to select $amiR^-$ mutants (Section 3.3) and butyramide-inducible mutants (Section 6.3).

### 4.3.2. PhV Amidases

One of the valeramide-utilizing mutants, strain V9, gave rise to two different phenylacetamide-utilizing mutants (Fig. 8): PhV1 had lost the ability to grow on acetamide, but PhV2 grew on acetamide rather slowly. Both were considered to have three mutations in the *amiE* gene.

The phenylacetamide-utilizing mutants PhB3, PhV1, and PhV2, together with the valeramide-utilizing mutant V9, constitute the B6 family of altered enzyme mutants and illustrate the way in which the substrate specificity of an enzyme can be progressively changed by sequential mutation (Sections 5 and 6). It was assumed that the change in substrate specificity could be related to changes in the conformation of the enzyme protein around the substrate-binding site. It was possible that there were severe limitations on the amino acid substitutions that would allow the bulky phenylacetamide molecule to be accommodated at the active site while still retaining good catalytic activity. On the other hand, it was thought that it might be possible to arrive at an amidase with activity for phenylacetamide by a totally different series of mutational events.

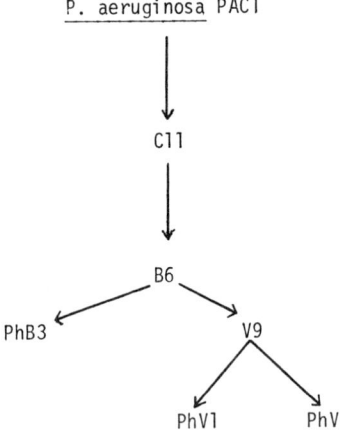

**Figure 8.** The B6 family of mutants producing amidases with altered substrate specificities. The novel substrates utilized by these mutants are B6, butyramide; V9, valeramide; PhB3, PhV1, and PhV2, phenylacetamides (see also Fig. 9).

### 4.3.3. PhF Amidases

The butyramide-resistant group of constitutive mutants had the characteristic of producing high amidase levels in the presence of an amide that acted as an analogue repressor for other strains. Phenylacetamide represses the synthesis of amidase in strain B6 but has little effect on the CB group of mutants. One of these butyramide-resistant mutants, CB4 (PAC128), gave rise to phenylacetamide-utilizing mutants in a single mutational selection (Fig. 9). PhF1 has the ability to grow slowly on acetamide. It is thought to have a single mutation in the structural gene and the enzyme is thermolabile.

### 4.3.4. PhA Amidases

Another route to the evolution of a phenylacetamidase from an acetamidase was via a catabolite-repression-resistant mutant. Strain L10 (PAC142) was isolated from succinate/lactamide plates and is constitutive and resistant to both catabolite repression and butyramide repression. Some acetamide-negative mutants were isolated from this strain with a view to examining cross-reacting proteins with little or no amidase activity. One of these LAm1 (PAC326), gave the phenylacetamide-utilizing

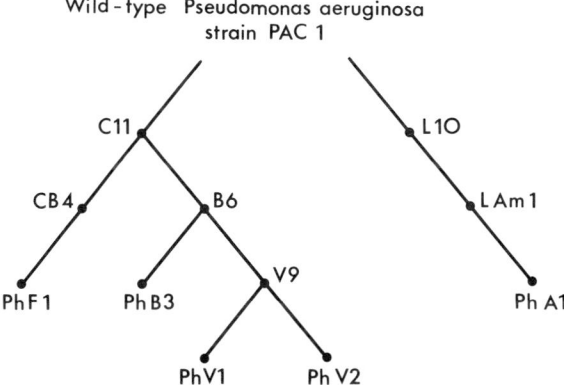

**Figure 9.** Family tree of the evolution of phenylacetamidases. The constitutive mutant C11 was derived by a single mutational step from wild-type PAC1. Strain C11 gave rise to the butyramide-utilizing mutant B6 and by successive mutational steps to the valeramide-utilizing mutant V9 and the phenylacetamide-utilizing mutants PhB3, PhV1, and PhV2. The phenylacetamide-utilizing mutant PhF1 was obtained in a single mutational step from the regulatory mutant CB4 (see Fig. 7). Strain L10 is a derepressed mutant obtained directly from PAC1 and successive mutational events gave the amidase-defective mutant LAm1 and the phenylacetamide-utilizing mutant PhA1 (Betz and Clarke, 1972).

mutant PhA1, which is assumed to have two mutations in the *amiE* gene. This mutant retained the ability to grow slowly on acetamide and produced a very thermolabile amidase.

### 4.4. Acetanilide-Utilizing Mutants: AI Group

4.4.1. AI3 Amidase

The wild-type strain does not utilize N-substituted acetamides, although some hydrolase activity can be measured with the purified wild-type amidase (Section 5.2.1). The hydrolysis of acetanilide (*N*-phenylacetamide) releases aniline and acetate. Aniline is not a good nitrogen source for PAC1 and it seemed preferable to present acetanilide as a carbon source and to provide ammonium sulfate as the nitrogen source. Using this medium, it was possible to isolate acetanilide-utilizing mutants from strain L10 (PAC142). This strain was chosen as a possible parent of the acetanilide-utilizers since it produces large amounts of amidase in minimal media. Strain L10 grown in succinate–minimal salt medium produces about 10% of its total soluble protein as amidase. A single mutational event gave rise to mutant AI3, which produces a mutant amidase in high yield. Strain L10 is able to grow on butyramide because it is constitutive and resistant to butyramide repression. The acetanilide mutant AI3 has retained the ability to grow on acetamide and succinate/formamide but is no longer able to utilize butyramide. Since it does not appear to be altered in regulation, the inability to grow with butyramide must be related to a difference in the conformation around the substrate-binding sites. There is only a single mutation and a single amino acid change is involved (P. R. Brown and Clarke, 1972) (Section 5.1.2).

4.4.2. Urea-Resistant Amidases

Gregoriou *et al.* (1977) found that strain AI3 was unable to grow on plates containing acetanilide as the carbon source and urea as the nitrogen source. Urea inhibits both wild-type and AI3 amidases (Section 5.2.3). The AIU group of mutants were isolated from acetanilide/urea plates and produced amidases that were less sensitive to urea inhibition. Unlike the parent strain AI3, they were unable to grow on succinate/formamide plates, suggesting that there had been a change in enzyme structure leading to a difference in substrate specificity. Some, but not all, of the AIU mutant amidases were less thermostable than AI3 amidase, but the cell extracts from all AIU strains gave complete immunological cross reactions with antiserum to purified AI3 amidase, indicating that there had been no major alterations in the enzyme structure. The only significant difference that could be detected for this class of mutant was a decreased rate of reaction with urea and thus a decreased sensitivity to urea inhibition. One of the AIU amidases had a different pH activity profile from that of AI3 amidase, but others were very similar to AI3.

Hydroxyurea could not be used as a nitrogen source, but P. R. Brown et al. (1978) found that it inhibited growth of strain AI3 in media containing either acetanilide or acetamide as the carbon source. Mutants isolated from plates containing acetanilide/$NH_4$/hydroxyurea produced amidases that were much less sensitive to inhibition by urea and hydroxyurea. These mutants were unable to grow on acetamide or on succinate /formamide plates and the enzymes did not hydrolyze either of these amides. This shift in substrate specificity resembled that of some of the Ph group of mutants in that the new amidases had lost the ability to attack some of the original substrates of the wild-type enzyme.

Mutants isolated from strain AI3 on acetamide/hydroxyurea plates were selected for their ability to utilize acetamide as a carbon source in the presence of hydroxyurea. Their growth properties were unlike those of the mutants isolated from strain AI3 on acetanilide/urea or acetanilide/$NH_4$/hydroxyurea plates. All retained the ability to grow on succinate/formamide as well as acetamide and none grew on acetanilide as carbon source. With this group of mutants we see a shift back in substrate specificity to a pattern more like that of the wild-type enzyme. Cell extracts had acetamide hydrolase activities that were very similar to those for strain L10, and under the same growth conditions these were tenfold higher than those for strain AI3. The ability to grow on acetamide/hydroxyurea plates was related to the increased rate of acetamide hydrolysis, since the amidases produced by these strains were more sensitive to hydroxyurea inhibition than that of strain AI3. Strain L10, from which AI3 had been derived, was also able to grow on acetamide/hydroxyurea plates and this could be related to the high level of wild-type amidase produced by this derepressed mutant.

With the AI3 family of mutants we can see that growth on amides in the presence of urea and hydroxyurea can be achieved by the production of novel enzymes. With acetanilide as carbon source the mutant enzymes were altered in amide substrate specificity and were less sensitive to inhibition by urea and hydroxyurea. With acetamide as carbon source the mutation regained the high specific activity toward acetamide as substrate and this property, together with the derepressed regulatory system, allowed growth in the presence of hydroxyurea in spite of high sensitivity to inhibition by this compound. This is an interesting use of enzyme inhibitors by P. R. Brown and colleagues to select novel amidase mutants. Media used for the isolation of amidase mutants are listed in Table III.

## Table III
## Amide Media Used for Mutant Selection

| | Medium[a] | | | | | |
|---|---|---|---|---|---|---|
| | Carbon source | Nitrogen source | Class of mutants isolated | Mutant[b] | Parent | Reference |
| S/F | Succinate | Formamide | Constitutive | C11 | PAC1 | Brammar *et al.* (1967) |
| | | | Butyramide-sensitive | C1 | PAC1 | Brammar *et al.* (1967) |
| | | | Butyramide-resistant Semiconstitutive | C5 | PAC1 | Brammar *et al.* (1967) |
| | | | Inducible Formamide-inducible | F6 | PAC1 | Brammar *et al.* (1967) |
| A | Acetamide | — | Revertant Amidase-negative | — | — | Skinner and Clarke (1978) |
| B | Butyramide | — | Constitutive Butyramide-resistant | CB4 | C11 | J. E. Brown and Clarke (1970) |
| | | | Altered enzyme | B6 | C11 | J. E. Brown *et al.* (1969) |
| | | | Butyramide-inducible | BB1 | IB10 | Turberville and Clarke (1981) |
| S/B | Succinate | Butyramide | Constitutive Butyramide-resistant Catabolite repression-resistant | SB1 | PAC1 | Smyth and Clarke (1975b) |

| | Media | Characteristics | SB62 PAC433 | PAC1 PAC432 | PAC1 | Reference |
|---|---|---|---|---|---|---|
| L | Lactamide | Altered enzyme | | | | Smyth and Clarke (1975b) |
| | | Promoter | | | | Smyth and Clarke (1975b) |
| | | Revertant | | | | Betz and Clarke (1972) |
| | | Amidase-negative | | | | |
| S/L | Succinate | Catabolite repression-resistant | L10 | — | PAC1 | Smyth and Clarke (1975a) |
| | Lactamide | Constitutive | | | | |
| | | Inducible | L11 | PAC1 | | Smyth and Clarke (1975a) |
| | | Constitutive | PAC148 | C1 | | Smyth and Clarke (1975a) |
| V | Valeramide | Altered enzyme | V9 | B6 | | J. E. Brown et al. (1969) |
| S/Ph | Succinate Phenylacetamide | Altered enzyme | PhB3 | B6 | | Betz and Clarke (1972) |
| Pyr/F | Pyruvate (NH$_4$)$_2$SO$_4$ (fluoracetamide) | Amidase-negative | PhV1 | V9 | | Betz and Clarke (1972) |
| | | Enzyme defective | — | — | | Clarke and Tata (1973) |
| | | Regulator defective | TS1 | C11 | | Farin (1976) |
| | | Promoter | PAC432 | C1 | | Smyth and Clarke (1975b) |
| AI | Acetanilide (NH$_4$)$_2$SO$_4$ | Altered enzyme | AI3 | L10 | | Brown and Clarke (1972) |
| AIU | Acetanilide Urea | Resistant to urea inhibition | AIU | AI3 | | Gregoriou et al. (1977) |
| AI/OHU | Acetanilide (NH$_4$)$_2$SO$_4$ (hydroxyurea) | Resistant to urea and hydroxy urea inhibition | OUCH | AI3 | | P. R. Brown et al. (1978) |

$^a$The composition of the media was based on known characteristics of the amidase as inducers and substrates as shown in Tables I and II. Examples are given of a few mutants isolated from these media. The series number of the mutants is given in some cases to indicate the most important character. Many revertants and amidase-negative mutants have been isolated and comprise various different subclasses.

## 5. Properties of Wild-Type and Mutant Amidases

### 5.1. Enzyme Structure

#### 5.1.1. Molecular Weight and Subunits

The wild-type amidase A can be readily purified from extracts of the high-producing strains C11 and L10. After precipitation of nucleic acids with streptomycin the extracts are heated at 60°C for 15 min to remove extraneous protein, then fractionated with ammonium sulfate, and finally subjected to column chromatography on DEAE–Sephadex. Similar methods were used for the purification of the mutant amidases, but for the thermolabile enzymes the heat treatment was reduced to 5 min and omitted entirely in some cases (Betz and Clarke, 1972; Paterson and Clarke, 1979).

Molecular weight determinations by sedimentation equilibrium gave a value of about 200,000 and the subunit molecular weight determined by electrophoresis in SDS–polyacrylamide gels was calculated as about 35,000 by P. R. Brown *et al.* (1973). Subsequently this was revised to 41,000 by Auffret (1976) for the wild-type enzyme and confirmed by Gregoriou and Brown (1979), who obtained a value of 41,600 ± 1100 for AI3 amidase. From the most recent data for subunit molecular weights the molecular weight of the amidase hexamer is considered to be about 240,000. The molecular weights indicate a structure of six subunits and since only one $NH_2$-terminal and only one COOH-terminal amino acid were detected, it was concluded that the subunits were identical. Further evidence for a structure of six identical subunits came from analysis of the number of bands produced on SDS–polyacrylamide gels when the enzyme protein had been reacted with dimethylsuberimidate (Davies and Stark, 1970). All the mutant enzymes that have been purified have properties that can be correlated with the structure of the wild-type enzyme, although some of the V group and Ph group mutants produce less stable enzymes. Whereas the wild-type enzyme retains its hexameric structure and loses little of its activity during purification, some of the less stable mutant enzymes are both heat-labile and cold-labile and give poor yields. The B amidase produced by strain B6 is as thermostable as the wild type and gives complete immunological cross reaction to antisera raised against A amidase. The V and Ph amidases cross react with diffuse bands, which is consistent with the relatively labile nature of these mutant enzymes. Figure 10 compares the effect of heating at 60°C on the activities of preparations of Ph amidases (Betz and Clarke, 1972).

The AI3 amidase gives complete cross reactions to antisera raised against A or B amidases, indicating that all three enzymes are antigenically very similar. The AIU mutants isolated by Gregoriou *et al.* (1977) gave complete cross reactions with antiserum to AI3 amidase, indicating that

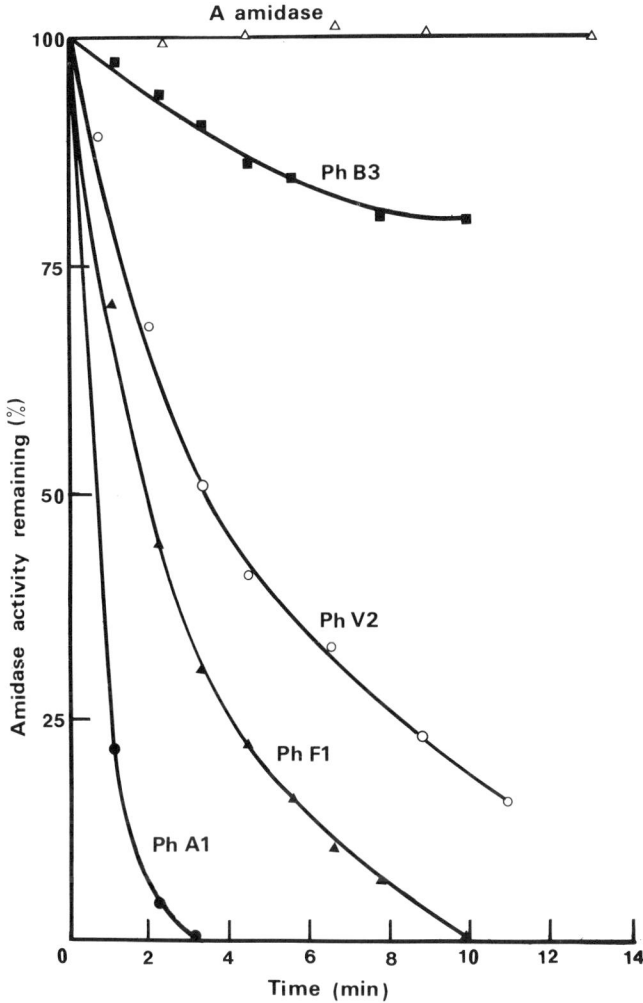

**Figure 10.** Thermal stability of wild-type A amidase compared with amidases produced by the phenylacetamide-utilizing mutants PhB3, PhV2, PhF1, and PhA1 (Betz and Clarke, 1972).

there had been no significant changes around the antigenic determinants. The AI3 amidase is slightly less thermostable than PhB3 amidase and loses about 40% of its activity on heating at 60°C, others were similar to AI3 in thermal stability, again suggesting that the mutations conferring resistance to inhibition by urea were not identical.

### 5.1.2. Amino Acid Sequences

The primary structure of the wild-type enzyme was investigated by A. D. Auffret and R. P. Ambler in the University of Edinburgh (Auffret, 1976). Strain L10 (PAC142) was used as the enzyme source since it provides large amounts of the wild-type enzyme. Some of the mutant enzymes were also included in the sequence studies. The other approach to this problem was to compare the peptides obtained from digestion of the amidases with proteolytic enzymes to see if it was possible to detect peptide differences between wild-type and mutant enzymes.

P. R. Brown and Clarke (1972) found a peptide difference between the AI3 amidase and A amidase. Following digestion with trypsin + chymotrypsin, the peptides were separated by electrophoresis and chromatography, and among the neutral peptides of the AI3 enzymes was one that did not appear in the digests of the wild-type enzyme, while the latter gave a peptide that was absent from the AI3 enzyme. From the sequences of the two related hexapeptides it could be concluded that in the AI3 enzyme an isoleucine residue had been substituted for a threonine present in the wild type (Table IV). Auffret (1976) was able to identify this peptide within the sequence of the amidase gene. No other amino acid substitutions could be found in the A13 enzyme and this is consistent with the physicochemical properties of this enzyme.

The family of mutant amidases derived from strain B6 (Fig. 8) appeared to have retained the original mutation *amiE*16 throughout the subsequent evolutionary steps. This was confirmed by finding that the substitution of a serine residue by a phenylalanine residue could be detected in peptides obtained by thermolysin digestion of purified amidases from strains B6, V9, and PhV1. This substitution gave an additional site for chymotrypsin hydrolysis, and digests of B6, PhB3, V9, and PhV1 amidases gave an additional peptide that was absent from digests of the wild-type enzyme (Paterson and Clarke, 1979). From sequence analysis this extra peptide could be identified with the $NH_2$-terminal sequence of the protein (A. D. Auffret and R. P. Ambler, unpublished results) and the *amiE*16 mutation could be assigned to amino acid 7 from the $NH_2$-terminus (Table V).

**Table IV.**
Amino Acid Sequences of Peptides[a]

| Source of peptide | Sequence |
|---|---|
| A Amidase | Ser-Leu-Thr-Gly-Glu-Arg |
| AI3 Amidase | Ser-Leu-Ile-Gly-Glu-Arg |

[a] Peptides obtained after trypsin + chymotrypsin digestion of purified amidases (P. R. Brown and Clarke, 1972).

## Table V
### Amino Acid Substitution at *amiE*16

| | Sequence |
|---|---|
| Chymotryptic peptide from B amidase[a] | Met-Arg-His-Gly-Asp-Ile-<u>Phe</u> |
| NH$_2$ terminus of A amidase[b] | Met-Arg-His-Gly-Asp-Ile-Ser-Ser-Ser-Asn- |

[a]The substitution of Phe for Ser provided an additional site for chymotryptic digestion and this peptide was found in digests of amidases from B6, PhB3, V9, and PhV1.
[b]Data from R. P. Ambler (personal communication).

The substitutions in AI3 and B6 enzymes have very little effect on enzyme stability. The significant change is in substrate specificity and in both cases the amino acid substitution results in the loss of a potential hydrogen bond, which could result in a slight increase in the flexibility of the protein. Neither substitution is likely to be directly concerned with the catalytic site, but the minor effect on the conformation of the enzyme protein is probably enough to account for the change in substrate specificity. The side chains of isoleucine and phenylalanine are more hydrophobic than the equivalent amino acids of the wild-type amidase. In the case of phenylalanine in the amidases of the B6 family the side chain is considerably larger than that of the serine that it replaces. Detailed studies with purified enzymes showed that the wild-type enzyme has low activity for both butyramide and acetanilide, and the respective amino acid changes are presumed to give slightly different conformational effects that allow these more bulky molecules to the accommodated more readily at the active sites of their respective enzymes.

### 5.2. Catalytic Activities

5.2.1. Amidase Substrates

The physiological reaction of the wild-type and mutant amidases is the hydrolysis of amides, but *in vitro* the enzymes are able to carry out a number of other reactions (Table VI). The wild-type amidase hydrolyzes short-chain esters such as methyl acetate, ethyl acetate, and *n*-propyl acetate at about 2% of the rate for the hydrolysis of acetamide. In addition to the hydrolase reactions, the enzyme catalyzes the transfer of the acyl moiety of the amide and ester substrates to hydroxylamine. Related aliphatic acids can also take part in a transferase reaction to form acylhydroxamates (McFarlane *et al.*, 1965). All these reactions are considered to involve the formation of an acyl intermediate at the active site of the enzyme. Hollaway and Ticho (1979) showed that there was direct competition between ethyl acetate and acetamide in amidase hydrolase re-

**Table VI.**
Amidase Reactions[a]

(1) Amide hydrolysis
$R \cdot CO \cdot NH_2 + H_2O = R \cdot CO_2^- + NH_4^+$

(2) Transferase activity (amide as substrate)
$R \cdot CO \cdot NH_2 + NH_2 \cdot OH + H^+ = R \cdot CO \cdot NH \cdot OH + NH_4^+$

(3) Transferase activity (acid as substrate)
$R \cdot CO_2^- + H^+ + NH_2 \cdot OH = R \cdot CO \cdot NH \cdot OH + H_2O$

(4) Ester hydrolysis
$R \cdot CO_2 \cdot R' + H_2O = R \cdot CO_2^- + H^+ + R' \cdot OH$

(5) Transferase activity (ester as substrate)
$R \cdot CO_2 \cdot R' + NH_2 \cdot OH = R \cdot CO \cdot NH \cdot OH + R' \cdot OH$

[a]Mutants selected for changes in substrate specificity with respect to reaction (1), amide hydrolysis, were also altered with respect to reaction (2), the transferase reaction.

actions. With purified enzymes it was possible to test for low activity on amides and related compounds that were not growth substrates and were not attacked at significant rates by cell extracts of cultures grown under inducing conditions (Kelly and Kornberg, 1964; J. E. Brown et al., 1969). Woods et al. (1979) found that acetohydrazide, N-methylacetamide, and N-methylpropionamide could act as substrates in both hydrolase and transferase reactions with purified preparations of the wild-type enzyme. The low activity toward minor substrates was often only picked up when comparisons were made between wild-type and mutant enzymes. For example, the hydrolysis of butyramide by the wild-type enzyme at about 2% of the acetamide rate was not noticed until the constitutive butyramide-resistant mutants had been isolated. The affinity of the wild-type amidase for butyramide is so low that a very high level of enzyme is needed to reach the threshold rate of butyramide hydrolysis required for growth. J. E. Brown et al. (1969) reported apparent $K_m$ values for wild-type and B6 amidases at 500 and 74 mM, respectively.

In a similar way, it had not been suspected that acetanilides could be attacked by the wild-type enzyme, but P. R. Brown et al. (1978) detected low specific activities with extracts of the depressed strain L10, obtaining values about 4% of those for the mutant AI3 enzyme. It was found that acetanilides containing substituents in the para position, including paracetamol and 4-nitroacetanilide, were good substrates for the AI3 amidase (P. R. Brown and M. J. Smyth, unpublished data).

Tables VII and VIII present data obtained by P. R. Brown and Clarke (1972) and Betz and Clarke (1972) for hydrolase and transferase reactions with wild-type and mutant amidases. With the wild-type enzyme Hollaway

## Table VII
Comparison of A, B, and AI Amidases: Apparent Michaelis Constants

| | | Apparent $K_m$ value (mM) | | |
|---|---|---|---|---|
| Enzyme reaction | Substrate | AI Amidase | A Amidase | B Amidase |
| Amide hydrolase | Acetamide | 30 | 1 | 2 |
| | Propionamide | 22 | 21 | 6 |
| Acyl transferase | Acetamide | 53 | 19 | 15 |
| | Propionamide | 208 | 55 | 10 |

*a*Measurements were made with purified enzymes. [Data are from P. R. Brown and Clarke (1972).]

*et al.* (1980) obtained a $K_m$ value for acetamide in the hydrolase reaction of 0.8 mM, but for ethyl acetate the $K_m$ value was >0.4 M. The changes leading to acquisition of a new amide growth phenotype appear to be related to a decrease in the apparent $K_m$ value for the novel substrate, but the $V_{max}$ may also be altered. Comparing the A and B amidases, a decrease in $K_m$ from 500 to 70 mM might still seem insufficient to get reasonable enzyme activity, but this is accompanied by a corresponding increase in $V_{max}$. It is interesting that the $K_m$ values for phenylacetamide with the Ph amidases are 1–3 mM, but as this amide is not very soluble, it may be that altered enzymes with less affinity for phenylacetamide would be useless to the bacteria and therefore not selected. The mutant *ebg* enzymes studied by Hall (1981) also differed from the wild type and from each other with respect to both $K_m$ and $V_{max}$ values.

### 5.2.2. Amidase Assays

The most widely used assay for amidase activity has been the transferase reaction with acetamide or another amide as substrate (Brammar and Clarke, 1964). Hydrolase activity has been measured by several methods. Kelly and Clarke (1962) used the Conway method for ammonia estimation, and a ninhydrin reaction was used by J. E. Brown *et al.* (1969) and Betz and Clarke (1972). Woods *et al.* (1979) used a modified Berthelot method. Gregoriou and Brown (1979) coupled the hydrolase reaction to glutamate dehydrogenase and measured the rate of NADH oxidation. For rapid reactions and for detecting amidase activity in gels the Nessler reaction has been used (Kelly, 1960 .

The hydrolysis of acetanilides by AI3 amidase led to the development of a continuous spectrophotometric assay with 4-nitroacetanilide as substrate following the release of 4-nitroaniline. This method was used by P. R. Brown *et al.* (1978) and Gregoriou and Brown (1979) for kinetic studies

## Table VIII
### Kinetic Constants of Mutant Amidases of *Pseudomonas aeruginosa* in the Acyl Transferase Reaction[a]

| Strain | Apparent $K_m$ (mM) | | | $V_{max}$ (μmole/min per mg protein) | | |
|---|---|---|---|---|---|---|
| | Acetamide | Butyramide | Phenylacetamide | Acetamide | Butyramide | Phenylacetamide |
| Wild type | 16 | 500 | NA | 1800 | 19 | 0 |
| B6 | 15 | 77 | NA | 1500 | 190 | 0 |
| PhB3 | NA | 180 | 2 | 0 | 2 | 2 |
| PhV1 | NA | 200 | 3 | 0 | 5 | 2 |
| PhF1 | 120 | 38 | 1 | 12 | 18 | 5.8 |

[a] Measurements were made with purified enzyme preparations. NA, No activity. Strains PhB3, PhV1, and PhF1 grow on phenylacetamide. [Data are from Betz and Clarke (1972).]

on AI3 amidase and the enzymes that had become less sensitive to inhibition by urea and hydroxyurea.

Hollaway et al. (1980) devised a continuous assay for amidase using a pH-stat to follow the proton release associated with ester or amide hydrolysis and also used a continuous spectrophotometric assay in which enzyme activity was measured by color change with cresol red as indicator dye.

### 5.2.3. Amidase Inhibitors

The inhibition of amidase activity has been studied by several investigators with a view to understanding something about the mechanism of action of this enzyme. Kelly and Clarke (1962) recorded that iodoacetamide and urea were inhibitors of enzyme activity. Iodoacetamide could be considered both as an acetamide analogue and as a thiol-reacting agent. Further studies with other thiol reagents showed that PCMB (*p*-chloromercuribenzoate) and DTNB [5',5'-dithio-bis(2-nitrobenzoate)] also inhibited enzyme activity (McFarlane, 1967). Acetamide gave partial protection from inhibition by thiol reagents and it seemed plausible that the thiol reagents were preventing the formation of an acyl intermediate at the active site. Each subunit contains eight cysteines, and studies with stopped-flow techniques showed that six thiol groups, i.e., one for each subunit, reacted very rapidly with DTNB, whereas a further 12 reacted more slowly. Since the presence of acetamide did not protect the fast-reacting thiol groups from the reaction with DTNB, it was very unlikely that they were involved in substrate binding (M. R. Hollaway, unpublished results). During the course of amidase purification it had been found that the presence of compounds with reduced thiol groups, such as mercaptoethanol or dithiothreitol, was necessary to maintain enzyme activity. The hexameric structure appears essential for amidase activity and the role of the fast-reacting thiol groups is probably that of maintaining the oligomer intact.

Another possibility for the formation of acyl intermediates would be a serine residue at the active site. The wild-type enzyme was not inhibited by diisopropylphosphofluoridate and related compounds, but it must be remembered that only relatively small molecules can be accommodated at the active site, so that the participation of a serine residue cannot be completely ruled out. Alt et al. (1975) reported that an acetanilidase from *P. acidovorans* was strongly inhibited by organophosphorus compounds as well as by thiol reagents and have some evidence for the presence of a serine at the active site.

Urea inhibits hydrolase activity at lower concentrations than it does transferase activity, and the presence of hydroxamate in the transferase reaction is presumed to be responsible for this protection. With AI3 amidase Gregoriou and Brown (1979) obtained clear evidence that urea could bind to the active site of the enzyme and that acetamide could protect from urea inactivation. Urea was not hydrolyzed by AI3 amidase, and bound [$^{14}$C]urea was released unchanged from the inhibitor–enzyme complex. The molar ratio for urea bound to the inactive enzyme was found to be 0.9/1, which provides evidence that the hexameric enzyme is inhibited when a single inhibitor molecule is bound to an active site of the enzyme. They found that hydroxyurea and cyanate exhibited similar time-dependent inhibition of AI3 amidase and concluded that these compounds also interacted with the active site of the enzyme. Findlater and Orsi (1973) reported that acetaldehyde, formaldehyde, and acetaldehyde-ammonia produced time-dependent inhibition of the wild-type amidase, and Gregoriou and Brown (1979) found that they behaved in a similar way with AI3 amidase.

Hollaway *et al.* (1980) selected chloracetone as a potential site-directed inhibitor of amidase, since chloroketones had proved to be useful in studying the mechanism of action of proteases and a number of other enzymes. Using stopped-flow measurements, they observed very rapid reaction with chloracetone and showed that acetamide could protect from chloracetone inhibition. The concentration dependence of protection by acetamide gave a value for an apparent dissociation constant that was similar to the $K_m$ value for this substrate. This study indicated that chloracetone reacted with the enzyme in two stages, the first involving the reversible formation of an enzymically inactive species and the second the formation of an inactive enzyme–inhibitor complex from which activity could not be recovered. It was concluded that the second enzyme–inhibitor complex was the result of covalent bond formation between a group in the enzyme and chloracetone in a similar manner to the formation of the enzyme–substrate complex during the catalytic reaction.

## 6. Amidase Genes and Enzymes

### 6.1. Gene Mapping

*Pseudomonas aeruginosa* PAC strains can be transduced by several of the bacteriophages used by Holloway and colleagues with strain PAO. Bacteriophage F116 (Holloway and Monk, 1959) was used to transduce amidase-negative mutants with lysates prepared from amidase-positive constitutive strains, and the high transduction frequencies obtained indicated close linkage of genes *amiE* and *amiR* (Brammar et al., 1967). Acetate-negative mutations, including those affecting acetic thiokinase and isocitrate lysase, gave no transductional linkage to the amidase genes (Skinner and Clarke, 1968). Interrupted mating using PAO strains carrying FP2 as donors and amidase-negative strains as recipients indicated that the amidase genes were located in the later region of the PAO chromosome. More exact mapping, and the identification of linked auxotrophic markers, gives a map position at about 50 min (Holloway and Crockett, 1982). The drug resistance plasmid R68.45 (Haas and Holloway, 1976), which has chromosome-mobilizing ability with PAO, is also able to act as a sex factor in PAC strains. Values obtained for linkage in conjugal crosses indicate that the amidase genes of PAC are linked to genes *argFG* and to *met* and *nal* genes of strain PAC in an analogous way to the loci of strain PAO (R. Sambanthamurthi and P. H. Clarke, unpublished results).

## 6.2. Mutation

It was not always possible to estimate with accuracy the frequency of the mutations that I have described. Regulatory mutants could be readily isolated from S/F and S/L plates at a frequency of $10^{-6}$–$10^{-5}$. However, on media that allow some growth the number of mutant colonies appearing increased remarkably after a few days, since the total population on the plate could grow by utilizing the products of the first mutants and thus more mutants could appear. With the selection of constitutive mutants from the wild-type strain on butyramide plates there was no possibility of background growth, and, in addition, it was essential that the constitutive mutation should also confer resistance to butyramide repression. On butyramide medium, constitutive mutants were obtained at frequencies of $10^{-8}$–$10^{-7}$. For the altered-enzyme mutants the procedure was quite simple. If one or more colonies appeared spontaneously on the selective medium, they were picked off at once. The frequency with which these were found was $10^{-9}$ or less. If none were found, and there was a reasonable case for expecting that they ought to exist, then the culture was subjected to chemical mutagenesis or to ultraviolet irradiation. The treatment used was not related to the desired phenotype and was merely to increase the rate of mutation. No attempts were made to obtain specific base changes. The phenylacetamide-utilizing mutants obtained from strain B6 represent the most diverse group in terms of treatment. J. L. Betz obtained 16 PhB mutants that exhibited very similar features. Two were spontaneous, one was obtained after UV irradiation, four after treatment with EMS (ethyl methanesulfonate), and five after treatment with NMG (*N*-methyl-*N'*-nitro-*N*-nitrosoguanidine). The underlying philosophy was that if a particular type of mutation could happen it would happen and all we needed to do was to find it. When no new mutants appeared from one parental line, the next move was to try another parental line to see if it might have that particular evolutionary potential.

### 6.3. Role of Recombination

Betz et al. (1974) used transductional analysis for fine structure mapping of the *amiE* gene using mutations in the *amiR* gene as outside markers since no other linked markers were available. From crosses with strain PhB3 (phenylacetamide-positive, acetamide-negative) it was possible to isolate recombinants carrying only the *amiE*16 mutation in the structural gene, thus confirming that this mutation had been retained. Four classes of recombinants could be recognized; inducible A enzyme ($amiR^+$ $amiE^+$), constitutive A enzyme ($amiR11$ $amiE^+$), constitutive B enzyme ($amiR11$ $amiE16$), and inducible B enzyme ($amiR^+$ $amiE16$). A representative of the last class was strain IB10, and although this strain produced B amidase, it was unable to grow on butyramide, since butyramide did not induce an amidase synthesis. Cultures grown on acetamide produced amidase with the substrate specificity of the B enzyme (Sections 3.3 and 4.1). Turberville and Clarke (1981) used strain IB10 to select butyramide-utilizing mutants. Most of those isolated were constitutive, but BB1 (PAC181) had acquired a butyramide-inducible phenotype (Fig. 11). This strain was of particular interest since it had a novel growth phenotype that depended on the induction of an altered enzyme by its novel substrate. In this case we can trace several different genetic events. The wild-type strain underwent a series of mutational events to produce strain PhB3 ($amiR11$, $amiE16$, 67). Recombination between PhB3 and strain Am7 ($amiR^+$ $amiE7$) gave IB10 ($amiR^+$ $amiE16$), with the final step involving another mutation to give BB1 ($amiR206$ $amiE16$). Most of the studies in experimental enzyme evolution have concentrated on mutational events, but it is very reasonable to suppose that recombination may also play a part in the evolution of strains with new growth phenotypes. In the *ebg* system of *E. coli* Hall and Zuzel (1980) found that the *ebg* β-galactosidase could gain novel substrate specificity by recombination within the structural gene *ebgA*.

### 6.4. Alignment of *amiE* Gene and Protein

Strain PAC433 is an up-promoter mutant isolated by Smyth and Clarke (1975b) which produces high amidase activities. This strain was used as the source of the *ami* DNA cloned into bacteriophage lambda by Drew et al. (1980). Although the amidase genes are poorly expressed in *E. coli*,

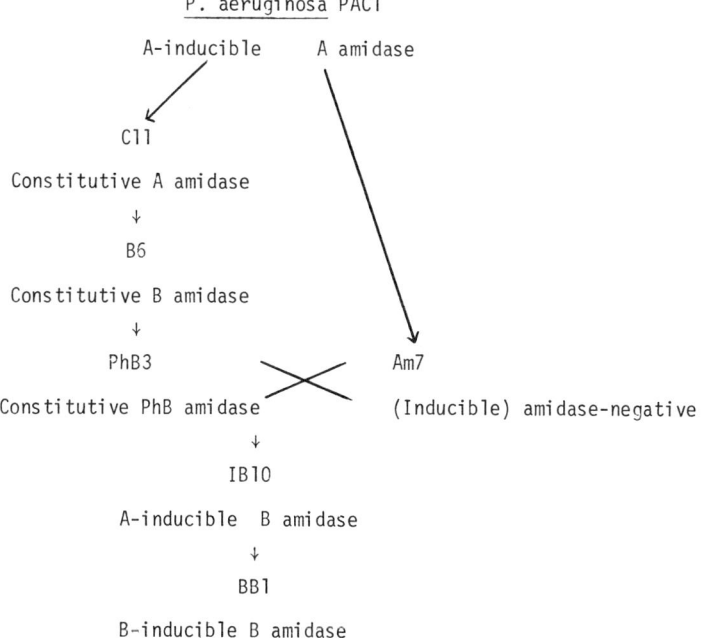

**Figure 11.** Derivation of a mutant strain producing an altered amidase induced by its novel substrate. Strain IB10 producing B amidase was obtained by recombination between strain PhB3 and the amidase-defective mutant Am7. A further mutational event gave strain BB1 producing B amidase and inducible by butyramide (Turberville and Clarke, 1981).

it was possible by inserting $Q^-S^-$ mutations into the bacteriophage genes to get high amidase activities in *E. coli* cultures infected by λ–*ami* hybrids. This made it possible to demonstrate that the enzyme produced under these conditions was identical in physicochemical properties with the *P. aeruginosa* A amidase. Further studies were carried out with sequences subcloned into plasmid pBR322. Since the $NH_2$-terminal amino acid sequence was known, it was possible to locate the most probable site of the corresponding gene sequence from the known target sites of restriction enzymes. Restriction mapping gave defined DNA sequences at known distances from the *Hind*III site that marked one end of the *P. aeruginosa* DNA. The target sequences of DNA were "translated" into amino acids for each of the three possible reading frames and then matched to the $NH_2$-terminal sequence of the amidase protein. This procedure located the *amiE* gene at about 250 bp from the *Hind*III site (Clarke *et al.*, 1981). This was confirmed by DNA sequencing. From the point of view of evolution of enzyme specificity it was of particular interest to identify the codon for serine in the wild-type *amiE* gene that had undergone change to give the phenylalanine replacement in B amidase. At this site the serine codon is one that allows a mutation to a phenylalanine codon by a single base change, while the two adjacent serine codons belong to a different coding group (W. J. Brammar, unpublished results).

### 6.5. Amidase Gene Capture

One of the striking observations made about the degradative capabilities of pseudomonads during the last decade is that many genes for catabolic pathways are carried on transmissible plasmids. One way in which novel enzymes could be spread through a bacterial population would be if the chromosomally located genes could be picked up by a plasmid with a broad host range. PAC 174 ($lys^-$ *amiR*11) carrying plasmid R68.44 was used successfully to isolate R' *arg* and R' *trp* plasmids by selecting for prototrophy for these markers in *E. coli* (Hedges and Jacob, 1977; Hedges et al., 1977). We were not able to obtain R'*ami* plasmids by this method, but they can be obtained by indirect selection. Morgan (1982) obtained a number of R' plasmids carrying genes located near *ami* on the chromosome of PAO by selecting for prototrophy in the corresponding auxotrophs of *P. putida* strain PPN. We have obtained R' plasmids carrying *ami* genes of strain PAC in a similar manner.

Many of the R' plasmids isolated in *P. putida* retain their transfer functions and can be transferred back into *P. aeruginosa* or into other bacterial species such as *E. coli*. The *P. aeruginosa* biosynthetic genes are poorly expressed in *E. coli* (Hedges et al., 1977; Clarke and Laverack, 1983). This is also the case for amidase genes. The *E. coli* strain W4100 carrying wild-type amidase genes on the R' plasmid grows very slowly on media containing acetamide as nitrogen source and either glucose or succinate as carbon source. It has proved possible to select faster growing mutants, and although the molecular basis of this adaptation is not yet known, it offers further possibilities for analyzing the evolution of new enzyme functions. The spread of genetic information through microbial populations by plasmid transfer presents a new dimension for the evolution of new metabolic activities.

### 6.5. How Many More Amidases?

The inducible aliphatic amidase of *P. aeruginosa* has fulfilled the demands made upon it to evolve new substrate specificities. In some cases it has been possible to trace the changes to single mutational events with the substitution at a single site of one amino acid for another. It may be asked whether similar chameleon attributes are to be found in other enzymes. Hartley (1974) suggested that it might be an unusual enzyme in that it could readily undergo structural changes that would give it new properties. It was from this point of view that he suggested that a multimeric enzyme might have the catalytic site at the junction between subunits. He suggested that the mutation of a single surface residue within the interface between subunits could allow a new subunit interaction that would change the quarternary structure without appreciable effect on tertiary structure. In the model proposed the subunits were each considered to have a substrate-binding site. In the studies with chloracetone as a site-directed inhibitor Hollaway *et al.* (1980) found that the inhibition was not monophasic and considered that this could result from a preexisting asymmetry in the active sites or information transfer between active sites whereby a change at one site could follow the modification of another within the same molecule, or an effect on the reactivity at the active site after modification at other loci. Gregoriou and Brown (1979) found that one molecule of the hexamer bound only one molecule of urea and they concluded that either the binding of the single molecule of urea renders the other sites inactive, or that the quarternary structure of the enzyme is such that only one catalytic site is available. The availability of mutant enzymes with altered response to inhibitors, in addition to the amidases that are able to act on several different amides, offers many opportunities for exploring these problems.

A more mundane answer to the question as to whether or not amidase is an enzyme particularly susceptible to evolutionary change is to say that it is not a special case except insofar as excellent selection systems are available for picking up mutants by using selective media. If this simple explanation were true, then similar numbers of altered enzyme mutants should be obtainable for other catabolic enzymes for which a variety of substrates could be made rate-limiting for growth. In this respect one has only to point to the diversity of *ebg* β-galactosidases selected by B. G. Hall (this volume, Chapter 6) and to encourage attempts to evolve new substrate specificities with other enzymes.

Another question that can legitimately be asked concerns the relevance of laboratory studies in experimental evolution to evolution in the natural enviroment. It was interesting to find that aliphatic amidases occur in other *Pseudomonas* species (Clarke, 1972). Among the aerobic pseudomonads studied by Stanier *et al.* (1966) were several species that produced amidases that gave some immunological cross reactions with antiserum to wild-type amidase from *P. aeruginosa* strain PAC1. Complete cross reactions occurred with all strains of *P. aeruginosa*, but as we have seen from the studies with B and AI amidases, this does not exclude a single amino acid change (or perhaps several more). The comparison with the aliphatic amidase of *P. putida* A87 was more interesting, and although it gave a strong cross reaction, there was marked spurring, indicating structural differences. R. P. Ambler and A. D. Auffret (unpublished results) have located 20 or so amino acid differences between the amidase of *P. putida* A87 and *P. aeruginosa* A amidase, most of them being conservative. The substrate specificities were similar and strongly suggest a common ancestor. Several of the *Pseudomonas* species also posses the ability to grow on phenylacetamide (Betz and Clarke, 1973). These natural phenylacetamides are quite different from the aliphatic amidases and bear no relationship to the phenylacetamides derived by experimental evolution in *Pseudomonas aeruginosa*. Equally, the amidase hydrolyzing acetanilide described by Alt *et al.* (1975) in a strain of *P. acidovorans* has no resemblance to the acetanilidase of *P. aeruginosa* AI3.

There seems no reason to suppose that the evolutionary potential of *P. aeruginosa* amidase has yet been exhausted, and the isolation of new mutants may be of value in attempting to understand the mechanism of action of this and similar enzymes. Together with the altered-enzyme mutants, it has been possible to obtain a range of regulatory mutants, and although one might consider that a derepressed mutant might be at a disadvantage in the natural environment, we have shown that it is also possible to obtain mutants that can be induced by their novel substrates. The metabolic versatility of pseudomonads that is so evident in nature has appeared in a dynamic way in these laboratory studies.

ACKNOWLEDGMENTS. It is a pleasure to record my appreciation of the work of my students and research associates who have contributed to amidase evolution at University College London. Bruce Holloway (Monash University) has been a constant advisor on genetic questions; Richard Ambler and Tony Auffret (University of Edinburgh) have advised on protein studies and have been sequencing the amidase protein; Bill Brammar (University of Leicester) collaborated on cloning the *amiE* gene and is currently sequencing it; Paul Brown and his colleagues (Kings College, London) have recently developed the new family of acetanilide mutants.

## References

Alt, J., Heymann, E., and Krisch, K., 1975, Characterization of an inducible amidase from *Pseudomonas acidovorans* AE1, *Eur. J. Biochem.* **53**:357–369.

Auffret, A. D., 1976, Structural Studies of a Bacterial Amidase, Ph.D. Thesis, University of Edinburgh.

Betz, J. L., and Clarke, P. H., 1972, Selective evolution of phenylacetamide-utilizing strains of *Pseudomonas aeruginosa*, *J. Gen. Microbiol.* **73**:161–174.
Betz, J. L., and Clarke, P. H., 1973, Growth of *Pseudomonas* species on phenylacetamide, *J. Gen. Microbiol.* **75**:167–177.
Betz, J. L., Brown, J. E., Clarke, P. H., and Day, M., 1974, Genetic analysis of amidase mutants of *Pseudomonas aeruginosa*, *Genet. Res. Camb.* **23**:335–359.
Brammar, W. J., and Clarke, P. H., 1964, Induction and repression of *Pseudomonas aeruginosa* amidase, *J. Gen. Microbiol.* **37**:307,–319.
Brammar, W. J., Clarke, P. H., and Skinner, A. J., 1967, Biochemical and genetic studies with regulator mutants of the *Pseudomonas aeruginosa* 8602 amidase system, *J. Gen. Microbiol.* **47**:87–102.
Brown, J. E., and Clarke, P. H., 1970, Mutations in a regulator gene allowing *Pseudomonas aeruginosa* 8602 to grow on butyramide, *J. Gen. Microbiol.* **64**:329–342.
Brown, J. E., Brown, P. R., and Clarke, P. H., 1969, Butyramide-utilizing mutants of *Pseudomonas aeruginosa* 8602 which produce an amidase with altered substrate specificity, *J. Gen. Microbiol.* **57**:273–285.
Brown, P. R., and Clarke, P. H., 1972, Amino acid substitution in an amidase produced by an acetanilide-utilizing mutant of *Pseudomonas aeruginose*, *J. Gen. Microbiol.* **70**:287–298.
Brown, P. R., Smyth, M. J., Clarke, P. H., and Rosemeyer, M. A., 1973, The subunit structure of the aliphatic amidase from *Pseudomonas aeruginosa*, *Eur. J. Biochem.* **34**:177–187.
Brown, P. R., Gregoriou, M., and Tata, R., 1978, Relationship between mutant amidases of *Pseudomonas aeruginosa* and hydroxyurea as an inhibitor, *Mol. Gen. Genet.* **165**:213–219.
Bühlmann, X., Visher, W. A., and Bruhlin, H., 1961, Identification of apyocyanogenic strains of *Pseudomonas aeruginosa*, *J. Bacteriol.* **82**:787–790.
Clarke, P. H., 1972, Biochemical and immunological comparison of aliphatic amidases produced by *Pseudomonas* species, *J. Gen. Microbiol.* **71**:241–257.
Clarke, P. H., 1974, The evolution of enzymes for the utilisation of novel substrates, in: *Evolution in the Microbial World* (M. J. Carlisle and J. J. Skehel, eds.), Cambridge University Press, Cambridge, pp. 183–217.
Clarke, P. H., 1978, Experiments in microbial evolution, in: *The Bacteria*, Vol. 6 (L. N. Ornston and J. R. Sokatch, eds.), Academic Press, New York, pp. 137–218.
Clarke, P. H., 1980, The utilization of amides by microorganisms, in: *Microorganisms and Nitrogen Sources* (J. W. Payne, ed.), Wiley, Chicester, England.
Clarke, P. H., and Laverack, P. D., 1983, Expression of the *argF* genes of *Pseudomonas aeruginosa* in *Pseudomonas aeruginosa*, *Pseudomonas putida* and *Escherichia coli*, *J. Bacteriol.* **154**:508–512.
Clarke, P. H., and Lilly, M. D., 1969, The regulation of enzyme synthesis during growth, in: *Microbial Growth* (P. M. Meadow and S. J. Pirt, eds.), Cambridge University Press, Cambridge, pp. 113–159.
Clarke, P. H., and Tata, R., 1973, Isolation of amidase-negative mutants of *Pseudomonas aeruginosa* by a positive selection method using an acetamide analogue, *J. Gen. Microbiol.* **75**:231–234.
Clarke, P. H., Drew, R. E., Turberville, C., Brammar, W. J., Ambler, R. P., and Auffret, A. D., 1981, Alignment of cloned *amiE* gene of *Pseudomonas aeruginosa* with the N-terminal sequence of amidase, *Biosci. Rep.* **1**:299–307.
Davies, G. E., and Stark, G. R., 1970, Use of dimethyl suberimidate, a cross-linking reagent in studying the subunit structure of oligomeric proteins, *Proc. Natl. Acad. Sci. USA* **66**:651–656.

Den Dooren de Jong, L. E., 1926, *Bijdrage tot de kennis van het mineralisatie proces*, Nijgh and Van Ditmar, Rotterdam.
Drew, R. E., Clarke, P. H., and Brammar, W. J., 1980, The construction *in vitro* of derivatives of bacteriophage lambda carrying the amidase genes of *Pseudomonas aeruginosa*, *Mol. Gen. Genet.* **177**:311–320.
Farin, F., and Clarke, P. H., 1978, Positive regulation of amidase synthesis in *Pseudomonas aeruginosa*, *J. Bacteriol.* **135**:379–392.
Findlater, J. D., and Orsi, B. A., 1973, Transition-state analogs of an aliphatic amidase, *FEBS Lett.* **35**:109–111.
Gregoriou, M., and Brown, P. R., 1979, Inhibition of the aliphatic amidase from *Pseudomonas aeruginosa* by urea and related compounds, *Eur. J. Biochem.* **96**:101–108.
Gregoriou, M., Brown, P. R., and Tata, R., 1977, *Pseudomonas aeruginosa* mutants resistant to urea inhibition of growth on acetanilide, *J. Bacteriol.* **132**:377–384.
Haas, D., and Holloway, B. W., 1976, R factor variants with enhanced sex factor activity in *Pseudomonas aeruginosa*, *Mol. Gen. Genet.* **144**:243–251.
Hall, B. G., 1981, Changes in the substrate specificities of an enzyme during directed evolution of new functions, *Biochemistry* **20**:4042–4049.
Hall, B. G., and Clarke, N. D., 1977, Regulation of newly evolved enzymes. III Evolution of the *ebg* repressor during selection for enhanced lactase activity, *Genetics* **85**:193–201.
Hall, B. G., and Hartl, D. L., 1974, Regulation of newly evolved enzymes I. selection of a novel lactase regulated by lactose in *Escherichia coli*, *Genetics* **16**:391–400.
Hall, B. G., and Zuzel, T., 1980, Evolution of a new enzymatic function by recombination within a gene, *Proc. Natl. Acad. Sci. USA* **77**:3529–3533.
Hartley, B. S., 1974, Enzyme families, in: *Evolution in the Microbial World* (M. J. Carlisle and J. J. Skehel, eds.), Cambridge University Press, Cambridge, pp. 151–182.
Hedges, R. W., and Jacob, A. E., 1977, *In vitro* translocation of genes of *Pseudomonas aeruginosa* on to a promiscuously transmissible plasmid, *FEMS Lett.* **2**:15–19.
Hedges, R. W., Jacob, A. E., and Crawford, I. P., 1977, Wide ranging plasmid bearing the *Pseudomonas aeruginosa* tryptophan synthetase genes, *Nature* **267**:283–284.
Hollaway, M. R., and Ticho, T., 1979, A competition time-course method for following enzymic reactions applied to the hydrolysis of acetamide catalysed by an aliphatic amidase, *FEBS Lett.* **106**:185–188.
Hollaway, M. R., Clarke, P. H., and Ticho, T., 1980, Chloracetone as an active-site-directed inhibitor of the aliphatic amidase from *Pseudomonas aeruginosa*, *Biochem. J.* **191**:811–826.
Holloway, B. W., 1955, Genetic recombination in *Pseudomonas aeruginosa*, *J. Gen. Microbiol.* **13**:572–581.
Holloway, B. W., and Crockett, 1982, Chromosome map of *Pseudomonas aeruginosa* PA0, *Genet. Maps* **2**:165–167.
Holloway, B. W., and Monk, M., 1959, Transduction in *Pseudomonas aeruginosa*, *Nature* **184**:1426–1427.
Jayaraman, K., Müller-Hill, B., and Rickenberg, H. V., 1966, Inhibition of the synthesis of β-galactosidase in *Escherichia coli* by 2-nitrophenyl-β-D-fucoside, *J. Mol. Biol.* **18**:339–343.
Kelly, M., 1960, An Investigation of the Action of *Pseudomonas aeruginosa* on Amides and Related Compounds, Ph.D. thesis, University of London.
Kelly, M., and Clarke, P. H., 1960, Amidase production by *Pseudomonas aeruginosa*, *Biochem. J.* **74**:21P.
Kelly, M., and Clarke, P. H., 1962, An inducible amidase produced by a strain of *Pseudomonas aeruginosa*, *J. Gen. Microbiol.* **27**:305–316.

Kelly, M., and Kornberg, H. L., 1964, Purification and properties of acyltransferases from *Pseudomonas aeruginosa*, *Biochem. J.* **93**:557–566.

McFarlane, N. D. D., 1967, Studies on *Pseudomonas aeruginosa* Amidase, Ph.D. Thesis, University of London, England.

McFarlane, N. D., Brammar, W. J., and Clarke, P. H., 1965, Esterase activity of *Pseudomonas aeruginosa* amidase, *Biochem. J.* **95**:24C–25C.

Morgan, A. D., 1982, Isolation and characterization of *Pseudomonas aeruginosa* R' plasmids constructed by means of interspecific mating, *J. Bacteriol.* **149**:654–661.

Mossel, D. A. A., and Van Zadelhoff, C. 1971, The selective detection of *Pseudomonas aeruginosa* in water, *J. Gen. Microbiol.* **69**:XIV.

Paterson, A., and Clarke, P. H., 1979, Molecular basis of altered enzyme specificities in a family of mutant amidases from *Pseudomonas aeruginosa*, *J. Gen. Microbiol.* **114**:75–85.

Sadler, J., and Novick, A., 1965, The properties of repressor and the kinetics of its action, *J. Mol. Biol.* **12**:305–327.

Skinner, A. J., and Clarke, P. H., 1968, Acetate and acetamide mutants of *Pseudomonas aeruginosa* 8602, *J. Gen. Microbiol.* **50**:183–194.

Smyth, P. F., and Clarke, P. H., 1975a, Catabolite repression of *Pseudomonas aeruginosa* amidase: The effect of carbon source on amidase synthesis, *J. Gen. Microbiol.* **90**:81–90.

Smyth, P. F., and Clarke, P. H., 1975b, Catabolite repression of *Pseudomonas aeruginosa* amidase: Isolation of promoter mutants, *J. Gen. Microbiol.* **90**:91–99.

Stanier, R. Y., Palleroni, N. J., and Doudoroff, M., 1966, The aerobic pseudomonads; A taxonomic study, *J. Gen. Microbiol.* **43**:159–271.

Turberville, C., and Clarke, P. H., 1981, A mutant of *Pseudomonas aeruginosa* PAC with an altered amidase induced by the novel substrate, *FEMS Microbiol. Lett.* **10**:87–90.

Woods, M. J., Findlater, J. D., and Orsi, B. A., 1979, Kinetic mechanism of the aliphatic amidase from *Pseudomonas aeruginosa*, *Biochim. Biophys. Acta* **567**:225–237.

CHAPTER 8

# Structural Evolution of Yeast Alcohol Dehydrogenase in the Laboratory

## CHRISTOPHER WILLS

## 1. Introduction

Because our approach to the problems of molecular evolution in the laboratory is rather different from that of most of the other workers contributing to this volume, I would like to begin this review of our recent work on yeast alcohol dehydrogenase with a historical note.

Our investigation of this system was sparked by a controversy that has animated population genetics and population biology in general since the time of Darwin, but most especially during the last decade and a half. This controversy centers on the relative roles of natural selection and of random genetic drift in producing the evolutionary changes we observe. Since the thorough working out of the consequences of random drift by Wright (1931), most of the controversy had until recently centered around the relative importance of selection and random events in the evolution of morphological differences or of striking genetic polymorphisms like the chromosomal inversion systems in *Drosophila* [see Wills (1981) for review]. However, with the discovery that enzyme polymorphisms are extremely widespread in nature (Harris, 1966; Lewontin and Hubby, 1966), a whole new area of controversy opened up. Because the connection between the phenotype detected by electrophoresis (the overall charge and shape of an enzyme or other protein molecule) and its effect on the fitness of the organism was not at all obvious, the immediate reaction on

---

*CHRISTOPHER WILLS* • Department of Biology, University of California at San Diego, La Jolla, California 92093.

the part of some biologists with a theoretical bent was to assume that differences detected at this level were primarily selectively neutral (Kimura, 1968). As a corollary of this, amino acid substitutions during the course of evolution were also assumed by some workers to be primarily due to mutation to selectively equivalent alleles and their subsequent fixation (King and Jukes, 1969).

A long, acrimonious, and ultimately frustrating debate ensued. While it could be shown that some genetic polymorphisms in man and a few other organisms were indeed under the influence of or maintained by selective forces, it would be quite beyond the resources of the biological community to investigate a sample statistically large enough to convince the neutralists, as they came to be known. Attempts to distinguish mathematically between the relative influences of selection and drift were also doomed to frustration, since it gradually became apparent that any distribution of allele frequencies could be explained by either selection or drift with the appropriate manipulation of other population parameters, such as size, subdivision of the population in the past and the present, and migration (Lewontin, 1974).

Nonetheless, in my opinion, the preponderance of evidence is that selection has played the larger role in maintaining molecular polymorphisms and causing molecular evolution (Wills, 1973, 1978). This is particularly so because the new techniques for sequencing DNA allow new measures of the influence of selection to be made. The nonrandom use of codons (Berger, 1977; Sanger *et al.*, 1977) and the influence of adjacent codons on each other (Lipman *et al.*, 1982) are only some of the new and fascinating pieces of evidence that have been found.

At no point in the controversy have the neutralists and selectionists been in conflict about the paramount role of selection in evolution. The controversy has been over the degree to which small phenotypic changes are influenced by selection. The neutralists have backed off somewhat and now admit that much of the variation they formerly thought neutral may be weakly selected against (Ohta, 1973). On the other hand, the selectionists will be hard pressed to find a selective explanation for all the polymorphisms that have been found at the level of the DNA (estimated to be as many as ten million in man).

Our original interest in developing the yeast alcohol dehydrogenase system was to investigate the question of the selective effect of amino acid substitutions. We considered that if a model system could be found that resulted in selective pressures approximating those found in nature, then some important evolutionary questions could be asked.

The first was: what proportion of a protein molecule is "seen" by selection? The hemoglobin molecule is by far the best investigated in this regard, first because its function and its tertiary and quaternary structures are known in such detail, and second because a large number of hemoglobinopathies have been traced to particular amino acid alterations. This last dimension of investigation is a particularly important one, since it illuminates the function of the molecule far more fully than any amount of study of the unaltered molecule could do. Half the amino acids of the $\alpha$ and $\beta$ chains have been shown to have an important function in the molecule, either in its physiological functions or in maintaining the tertiary structure (Goodman *et al.*, 1975). It may be that the function of the remaining half has simply not been perceived by the relatively crude methodologies currently available. It seems probable, therefore, that (especially if the hemoglobinopathies are any indication) most amino acid substitutions at these positions would have some impact on the functioning of the molecule and therefore on the Darwinian fitness of its carrier. To investigate these matters more thoroughly, it was desirable to have a model system different from hemoglobin. Would a similarly large proportion of residues be found to be important in function when another molecule was investigated? Could any patterns be seen that would give indications of the function of various parts of the molecule, as was the case with hemoglobins?

The second question had to do with the method of obtaining mutants. The hemoglobinopathies caused by amino acid substitutions have been discovered because they are manifested in clinical symptoms or because they result in a charge change on the molecule. In only one certain and two possible cases (hemoglobins S, C, and E) has the substitution resulted in a molecule that confers a higher fitness on the carrier under some conditions. It would, we considered, be most instructive to investigate a subset of possible mutations, *all* of which confer an increase in fitness on the organism under certain defined experimental conditions. What would the properties of this subset be? Is it possible to produce a similar adaptation in the organism by more than one type of change in the target molecule?

A third question, which we have yet to answer satisfactorily, has to do with the repeatability of the evolutionary process. If a highly restrictive environmental pressure is applied, will a small number of mutations be selected for repeatedly, or is the number of evolutionary pathways open to the organism large even under such a restrictive set of conditions?

In the hope of answering some of these questions, about 10 years ago we turned to the alcohol dehydrogenase of the yeast *Saccharomyces cerevisiae*. We were already using the organism in population genetics studies, something was known about the physiology and distribution of the two cytoplasmic and one mitochondrial enzymes (Lutstorf and Megnet, 1968), and the enzyme was plentiful in the cell and easily purified.

At about the time we began working on the system, Michael Ciriacy in Darmstadt began an extensive series of publications on the structural and regulatory genes of the ADH system (Ciriacy, 1975a,b, 1976, 1979). His efforts and ours complemented each other in the succeeding years, and together led to the recent exploration of the molecular biology of the system by Young and his collaborators, which will be dealt with later in this chapter.

## 2. The Biochemistry and Regulation of Yeast Alcohol Dehydrogenase

The alcohol dehydrogenases of *Saccharomyces cerevisiae* form a complex and highly regulated system. There are two isozymes found in the cytoplasm and coded for by nuclear genes. A third isozyme is sequestered in the mitoclondria, and is probably coded for by a nuclear gene as well, although the evidence is not firm.

The synthesis and function of the two cytoplasmic isozymes is firmly tied into the state of the cell's metabolism. ADH-I migrates more slowly on horizontal starch activity gel electrophoresis than does ADH-II, and is also largely constitutive, although there are circumstances under which ADH-I can be repressed (Denis *et al.*, 1981). The greatest amount of repression we have discovered occurs when the cells are grown on minimal medium (salts plus vitamins) with pyruvate as a sole carbon source. Varying amounts of repression can be achieved with other nonfermentable substrates as carbon sources.

ADH-I is the isozyme chiefly responsible for the production of ethanol from acetaldehyde under anaerobic or aerobic growth with high sugar levels in the medium. This appears to occur primarily because the high $K_m$ for ethanol of this isozyme ($2.4 \times 10^{-2}$ M) (Wills, 1976a) ensures that the concentration of ethanol in the cell can reach fairly high levels before the reaction acetaldehyde → ethanol slows appreciably. As a consequence, it is the structure and function of this isozyme that are of the greatest interest from the standpoint of understanding and perhaps improving ethanol production in yeast.

The other isozyme, ADH-II, migrates more rapidly at pH 8.9 on horizontal starch gel electrophoresis. It is sufficiently closely related to the first that there is a high degree of amino acid identity (Wills and Jörnvall, 1979), but nonetheless the two isozymes only form a limited amount of hybrid. Although the isozymes are tetramers, only a single hybrid band is seen on gels either before or after dissociation and reassociation with urea (C. Wills, unpublished results). Thus there is circumstantial evidence that the isozymes dimerize readily with themselves, but cross-isozyme dimers do not form. This isozyme is synthesized only in cells with functioning mitochondria, so that it is not synthesized under conditions of anaerobic growth or when the cell is petite (that is, when the mitochondria are nonfunctional so that the cell grows by means of anaerobic glycolysis). However, the isozyme is also repressed by glucose (although what this means at the physiological level is not at all clear), so that it is not synthesized when the cell is grown on minimal medium with glucose as a sole carbon source and with moderate aeration, or when it is grown on complete medium with high levels of glucose.

The regulation of ADH-II has been shown to be at the transcriptional level (Denis et al., 1981). Its $K_m$ for ethanol is one-tenth that of ADH-I. Thus, when it is induced, it is capable of rapid oxidation of ethanol present in the environment. Our current investigation of the mechanism by which this enzyme is induced suggests that certain mitochondrial transport systems are involved in the induction.

Very little is known about the mitochondrial ADH, except that it is entirely sequestered within the mitochondrion, migrates very slowly on electrophoresis at pH 8.9, and has the same native and subunit molecular weight as do the cytoplasmic enzymes. Its $K_m$ for ethanol and for NAD are very similar to those for ADH-I (C. Wills and T. Martin, unpublished results).

Mutants in structural and regulatory ADH genes can readily be obtained by allyl alcohol selection (Megnet, 1967). Allyl alcohol, itself harmless to the cell, is rapidly oxidized by ADH to the poisonous aldehyde acrolein. A sufficiently high concentration of this latter compound is capable of killing the cell, although the precise mechanism of acrolein sensitivity remains to be discovered. Using this and other selective techniques, Ciriacy (1975a,b, 1976, 1979) formalized the genetics of the structural genes and discovered an entire hierarchy of regulatory genes for this system. The regulatory genes discovered so far are all involved in the regulation of ADH-II.

The structural genes for the three enzymes are designated as follows: *ADC1*, the structural gene for the "constitutive" *ADH-I, ADR2*, the structural gene for the highly regulated ADH-II; and *ADM*, the putative structural gene for the mitochondrial ADH. This last enzyme appears on an activity gel as five distinct bands, but a single nuclear mutation is enough to remove all five bands. Thus, the *ADM* gene may be the structural gene, or may be a gene that activates the enzyme inside the mitochondrion. In the latter case, the five bands may be formed by the random association of the products of two different structural genes, or of two different post-transcriptionally modified subunits. Attempts to isolate electrophoretic mobility mutants of the enzyme have so far failed, although E. T. Young (personal communication) may have succeeded in cloning at least one of the structural genes.

Several regulatory genes have been investigated by Ciriacy. *ADR3$^C$* mutations occur at closely linked points upstream from the structural gene, and result in *cis*-acting constitutive mutations. Most of the constitutive mutations isolated by a variety of means have been shown to be due to the insertion of Ty1-like elements (Williamson *et al.*, 1981). These and other constitutive mutants for ADH-II continue to make the enzyme even as petites, so there is no absolute requirement for mitochondrial activity.

Because of the isolation of the *ADR3$^C$* mutants, it was possible to determine the nature of the control exerted by another unlinked locus at which both constitutive and repressed mutations have been isolated. These mutations, *ADR1$^C$* and *adr1*, are semidominant and recessive, respectively, and almost certainly must produce a regulatory protein that interacts with a region at or near the *ADR3* site in order to turn on the structural gene *ADR2* [see Englesberg and Wilcox (1974) for the criteria for a positive regulatory system].

Further levels of regulation were found. The constitutive mutations at *ADR1* can be masked epistatically by a pleiotropic unlinked mutation *ccr1*, which also prevents derepression of a number of other glycolytic enzymes. Other unlinked mutations with similar properties, *ccr2* and *ccr3*, have been found but have not been characterized as thoroughly. Finally, Ciriacy looked for derepressed mutants of a strain that was *ccr1 ADR1$^C$*, and found some that could be localized to yet another locus, *ADR4*. These latest mutations in the hierarchy were, surprisingly, found to be recessive.

Beier and Young (1982) have succeeded in cloning the *ADR3* region upstream from the ADH-II structural gene *ADR2*. Deletion of the *ADR3* region released the structural gene from glucose repression, and insertion of the regulatory region upstream from the ADH-I structural gene placed this gene under glucose regulation in turn.

The regulation of this system is thus well understood at the genetic level, though not at all understood at the biochemical level. The majority of the work in our laboratory is now directed toward understanding the biochemistry and physiology of this system.

The enzyme itself is a tetramer with a molecular weight of 141,200. It consists of four subunits with 347 amino acids each. ADH-I has been completely sequenced (Jörnvall, 1977a), and despite a limited amount of material, we were able to determine 85% of the primary structure of ADH-II. The two enzymes were found to be 95% identical over the sequenced regions (Wills and Jörnvall, 1979a). We also found that there was differential blocking of the amino termini of the two isozymes. Later we were able to separate the two isozymes from a single cellular preparation and show that ADH-II is only about 60% blocked by an acetyl group, while ADH-I is at least 90% blocked (Jörnvall et al., 1980). The physiological significance of this differential blocking has not yet been determined, but it suggests that the two isozymes are synthesized under conditions of an excess and a shortage of acetyl-CoA, respectively.

Because the production of ethanol is of such central importance to the physiology of the yeast cell, it is not surprising that alterations in this enzyme's function affect a number of important cellular processes.

As was mentioned earlier, it is possible to obtain both functional and regulatory mutants that decrease or eliminate ADH activity in the cell by growing cells in the presence of allyl alcohol. This three-carbon alcohol with a double bond is itself harmless to the cell, but can be oxidized to the poisonous aldehyde acrolein. A number of other double-bond alcohols, such as crotyl alcohol, can be used to isolate similar mutants. Selection schemes based on the same principle have been used to isolate ADH-negative mutants of *Drosophila* (Sofer and Hatkoff, 1972).

It is possible using this technique to make mutants in all the three structural genes, leading to a strain that lacks detectable levels of ADH activity (Wills and Phelps, 1975; Ciriacy, 1975a). Interestingly, such a strain does not lack *all* activity, since it can still grow on ethanol as a sole carbon source and careful measurements on mutants resistant up to 0.5 M allyl alcohol show a slight residual ADH activity of less than 1% of the wild type.

These mutants were used in genetic studies by Ciriacy and by ourselves. We also investigated the physiology of the mutants and found a striking effect of the presence of ADH activity on survival of the cells as petites. It was found that if cytoplasmic ADH activity were present, the cells could survive as petites. If it were absent, they could not. The presence or absence of mitochondrial ADH had no effect.

Our initial experiments were performed using ethidium bromide, which interferes with mitochondrial DNA synthesis. We found that the induction curve of petites produced with this compound when cytoplasmic ADH was present exactly matched the killing curve produced when the enzymes were absent. Later, M. Ciriacy (personal communication) improved on this technique by using antimycin A, which reversibly blocks electron transport in the mitochondrion at the level of cytochrome *b*. Treatment with low levels of this inhibitor—usually 1 ppm—produces a reversible petite phenocopy in wild-type cells, but kills cells lacking cytoplasmic ADH.

It seemed obvious that the killing of petite cells lacking cytoplasmic ADH was due to the failure of anaerobic glycolysis. The last stage in this pathway is the production of ethanol from acetate, and if this is prevented *and* mitochondrial activity is interfered with, the cell has no available source of energy. However, it was not immediately obvious why cells with an intact mitochondrial ADH should be affected in the same way as cells with no ADH activity.

A study of the pathways involved suggested that since glycolysis was largely blocked in cells without a cytoplasmic ADH, reducing equivalents somehow had to be transferred into the mitochondrion. The result of this transfer would be the continuation of the early steps in glycolysis even though the later steps would be blocked. Most particularly, it would be essential for the cell to generate $NAD^+$ in order to permit the conversion of glyceraldehyde-3-phosphate to 1,3-diphosphoglycerate. The only pathway open to the cell for transferring reducing equilavents into the mitochondrion, generating $NAD^+$, and permitting this step to proceed was the glycerophosphate-dihydroxyacetone phosphate shuttle.

This shuttle transfers reducing equivalents unidirectionally into the mitochondrion. Because the mitchondrial glycerophosphate dehydrogenase is a flavoprotein, and is thus able to feed electrons directly into the electron transport chain at the level of coenzyme Q, this shuttle is absolutely dependent on a functioning electron transport chain. Once this is disrupted, the shuttle ceases to operate.

One of the byproducts of the glycero phosphate–dihydroxyacetone phosphate shuttle is an excess of glycerol phosphate in the cytoplasm, which can be converted to glycerol and secreted by the cell. Indeed, cellular glycerol production was discovered early in the century by Neuberg and Reinfurth (1918), who poisoned the ADH system (as is now realized) with the addition of bisulfite. Glycerol production from yeast actually contributed to Germany's war effort during the World War I. We quickly found that our ADH-negative yeast also produced large quantities of glycerol, indicating their dependence on this pathway for energy production.

One observation remained to be explained. This was that cells that possessed an intact gene for the mitochondrial ADH were also killed by ethidium bromide or antimycin A. It might be thought that the mitochondrial ADH should be able to generate enough $NAD^+$ to enable glycolysis to continue. In fact, of course, it had been known for some time that the mitochondrial membrane is impermeable to $NAD^+$, so that the generation of $NAD^+$ within the mitochondrion would be unable to aid a cell that could not transfer it to the cytoplasm.

This observation, incidentally, provided two pieces of information of general biochemical interest. The first was the genetic demonstration of mitochondrial impermeability to $NAD^+$, which precisely reinforced the earlier data obtained from isolated mitochondrial preparations. The second was that the mitochondrial membrane of the petite cell is also impermeable to $NAD^+$. Incidentally, it would be most instructive to look for petite mutants that are capable of survival with only the mitochondrial ADH present; such mutants should have a defective inner mitochondrial membrane.

This rather complex set of pathways is summarized in Fig. 1.

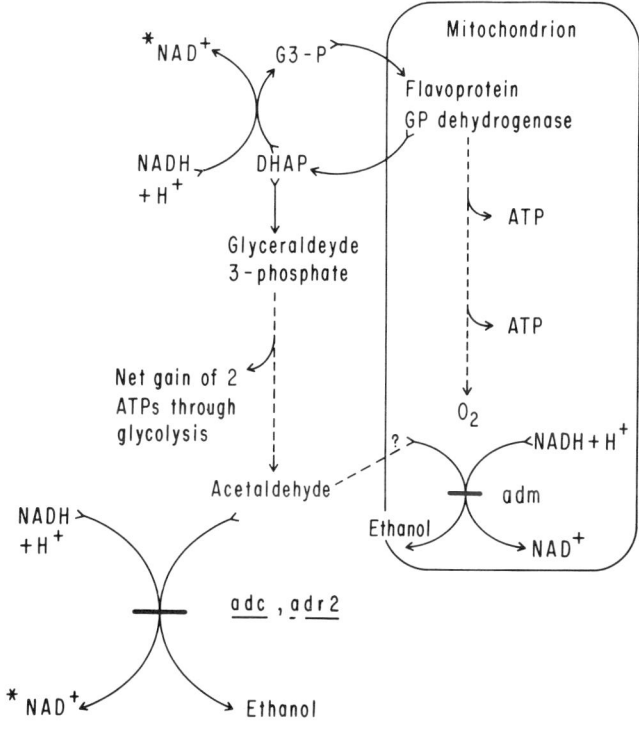

**Figure 1.** The energy pathways involved in sugar metabolism in yeast. The blocks caused by the three ADH structural gene mutations are shown as solid lines. When oxidative phosphorylation is blocked, the cell deficient in cytoplasmic ADH activity is unable to generate $NAD^+$ through the G3-P–DHAP shuttle. Any $NAD^+$ generated inside the mitochondrion by mitochondrial ADH activity is unable to pass into the cytoplasm, and the cell dies. [Modified from Fig. 7 of Wills and Phelps (1975).]

This information could be used to proceed in two directions. It was used by Young's group to clone the structural genes for ADH-I and ADH-II, and more recently to clone the linked regulatory region *ADR3*. The cloning of the structural genes was accomplished by screening for transformants of an ADH-negative strain capable of growth in the presence of antimycin A.

We have proceeded in a somewhat different direction, and used the system to provide tentative answers to some of the questions posed at the beginning of this chapter. We reasoned that if an ADH-negative strain was incapable of surviving as a petite, then a petite strain that produced ADH-I in the cytoplasm could be used as a tool to alter the gene itself. In particular, it could be subject to selection pressures that would normally select for mutations inactivating the structural genes for ADH-I. Such strains would not survive, and the only strains that would survive would be those in which the structural gene was altered but still functional.

We have employed several selective schemes in order to generate functional mutants of ADH-I, but only one has yielded results. This is selection in the presence of allyl alcohol, the compound that generated the ADH-negative mutants in the first place. Selection involving a variety of inhibitors of ADH activity, such as pyrazole and sodium thiosulfate, resulted in mutants, but none of them could be traced to the *ADC* structural gene. In the case of allyl alcohol selection of petite strains, however, a suprisingly high proportion of both spontaneous and induced mutations could be traced to the structural gene. In some experiments the proportion was as high as 40%.

We initially reported (Wills, 1976b) that the majority of electrophoretic mobility mutants selected for by this procedure migrated more slowly at pH 8.9 than did the wild-type enzyme. In subsequent experiments, however, a number of mutants were found that migrated more rapidly than the wild type.

A preliminary electrophoretic screen proved to be the most efficient way to pick up structural gene mutations. Most of the mutants were selected at fairly low levels of allyl alcohol, 5–10 mM, on complete medium. A few, however, were produced in a turbidostat, a modified chemostat in which the cell density is monitored continuously and kept constant (Northrop, 1954).

One of the questions posed at the beginning of this chapter was whether a sufficiently specific selective pressure could be applied that the mutants that result will represent only a small number of the possible changes that can occur in a particular structural gene. We are attempting to answer this question by refining the allyl alcohol selective procedure still further, though this work has not been completed. We are isolating temperature-sensitive mutants with both low and high permissive temperature and osmotic-remedial mutants in an attempt to see if certain electrophoretic mobility classes are represented more often than expected among these subcategories of functional mutants.

## 3. The Mechanism of Allyl Alcohol Resistance

While it is straightforward to isolate allyl alcohol-resistant mutants in petites and to demonstrate that the resistance segregates in subsequent generations along with altered electrophoretic mobility, it has not been as easy to determine the nature of the resistance at the biochemical and physiological levels. The discovery of the nature of the resistance and our relation of this to the kinetic differences seen in the mutant molecules have increased our understanding of the function of ADH in the yeast cell. It is an excellent illustration of the fact that even a protein with a well-defined function can be shown to have other important functions in the cell.

We quickly discovered that our first expectation about the nature of the resistance was incorrect. We had expected that the mutant enzymes would still be able to oxidize ethanol, their normal substrate, but be unable, or at least less able, to oxidize allyl alcohol. The amount of acrolein produced in the mutant cells as a result of enzyme action would be reduced. Such an alteration should be manifested, for example, as a greatly increased $K_m$ for allyl alcohol relative to that for ethanol. If a higher concentration of allyl alcohol was needed for the mutant enzyme to catalyze the oxidation reaction at half maximum velocity, then this would explain the resistance to this compound.

In fact, the kinetics of a variety of mutant enzymes did not bear out this expectation. While none of the mutant enzymes had as high a turnover number as did the wild-type enzyme and some showed substantial alterations in $K_m$, only one showed a substantial change in the *ratio* of the $K_m$ values for ethanol and allyl alcohol.

We were faced with an apparent paradox. Alterations in the enzymes had been selected for, yet these alterations apparently had little effect on the rate at which the poisonous acrolein was produced by the cell. In order to solve this dilemma, we had to consider the dynamics of the reaction itself.

We were able to show, using labeled allyl alcohol, that this compound was not metabolized further by the cell. There was no significant transfer of the label to any of the cellular components (Wills and Phelps, 1978). Therefore, the concentration of acrolein in the cell should reach a steady state as a result of the action of the alcohol dehydrogenase alone—no other enzyme acts on allyl alcohol in the cell. It would therefore be a function of the concentration of allyl alcohol in the cell, which in turn would be a direct function of the concentration of this compound in the medium.

Given this circumstance, there is only one factor that can be altered in the mutant cells. This is the relative concentration of reduced and oxidized NAD. The relative concentrations of these cofactors for the ADH reaction have a strong influence on the equilibrium concentrations of allyl alcohol and acrolein, as expressed by the following equation:

$$K'_{eq} = \frac{[NADH][acrolein]}{[NAD][allyl\ alcohol]}$$

where the equilibrium constant refers to the ratio of these compounds at a particular pH. At pH 7.0, this ratio is approximately $10^{-3}$, so that even at equilibrium there is a large preponderance of the harmless alcohol.

Obviously, however, if the ratio of NADH to NAD in the cell were altered, such that the concentration of NADH were increased, then this in turn would decrease the relative concentration of acrolein to allyl alcohol at equilibrium. It must be remembered that while the reaction involving acrolein and allyl alcohol will move with time to some approximation of equilibrium, the enzyme in the living cell is primarily catalyzing the ethanol–acetaldehyde reaction and is also responsible for the reduction of other long-chain aldehydes to their alcohols. These latter reactions will not normally reach their equilibrium.

We have not been able to determine the permeability of yeast cells to allyl alcohol, but it may be presumed that over time the interior of cells immersed in medium containing allyl alcohol at a given concentration will approximate the external concentration. Further, the mechanism of action of acrolein on the cell is unknown, though it may combine with proteins through addition reactions (Rando, 1974). Whatever the mechanism, it seems likely that acrolein disappears rather slowly once it has been formed. The time required for these secondary chemical reactions should be far greater than that required for enzymatically catalyzed reactions.

The ratio of NAD to NADH in the cell will therefore be primarily a function of normal metabolic processes in the cell. If it is possible for a mutant enzyme to alter this ratio markedly, then it is a safe assumption that the ratio is at least partially under the control of the enzyme itself.

In a series of investigations (Wills, 1976a; Wills and Phelps, 1978; Wills and Martin, 1980), we were able both to confirm that the resistance to allyl alcohol could be traced to an alteration in the NADH–NAD ratio, and that ADH plays an important role in governing this ratio. The effect of the mutant enzymes on the ratio was demonstrated in a number of ways: by direct measurement, using sensitive cycling techniques (Kato *et al.*, 1973), and by manipulation of the environment of the cells. For example, it was found that addition of ethanol to the medium increased the level of resistance of the mutant cells, while addition of acetaldehyde decreased it. This is what would be expected of the ADH reaction were governing the ratio in the cell.

Changes in the ratio were substantial, up to 2- to 3-fold. One of the most intriguing observations to come from this work, in addition to the light it cast on the mechanism of action of the mutant ADHs, was that when cells possessed both ADH-I and ADH-II the ratio was strongly buffered against the effects of environmental manipulation. The ratio could be altered markedly by addition of ethanol or acetaldehyde when the strain possessed only the wild-type ADH-I or ADH-II (though the magnitude of the shift was not as great as in the mutant cells). But if a cell possessed both enzymes, the same environmental manipulation had little effect on the ratio. These observations provided strong evidence that the ADH system is under complex and sensitive control and as a result regulates important aspects of cell metabolism.

All the mutants we have examined so far, either at the kinetic or the biochemical level, appear to owe their resistance to allyl alcohol to a shift in the NADH–NAD ratio. In this particular instance, a specific selective pressure has repeatedly resulted in a set of mutants that owe their survival to a particular biochemical mechanism. As will be apparent in the next section, however, there appear to be a large number of ways in which this shift in the ratio can be affected.

## 4. Amino Acid Substitutions in the Mutant ADHs

Five mutant ADHs have been investigated at the primary structure level, using a combination of standard techniques and HPLC peptide separation (Wills *et al.*, 1982). Three of the mutant substitutions have been determined precisely, and two determined approximately. One of the mutant enzymes (S-AA-1) may be a double mutant, but more work is needed to confirm this possibility.

The positions of the mutants are given in Fig. 2. Because the tertiary structure of ADH-I is not yet known, we have been forced to employ the structure of horse liver ADH in order to obtain an approximate idea of the position of the substitutions. Even though there is only 25% amino acid identity between the two molecules, there is a great deal of evidence

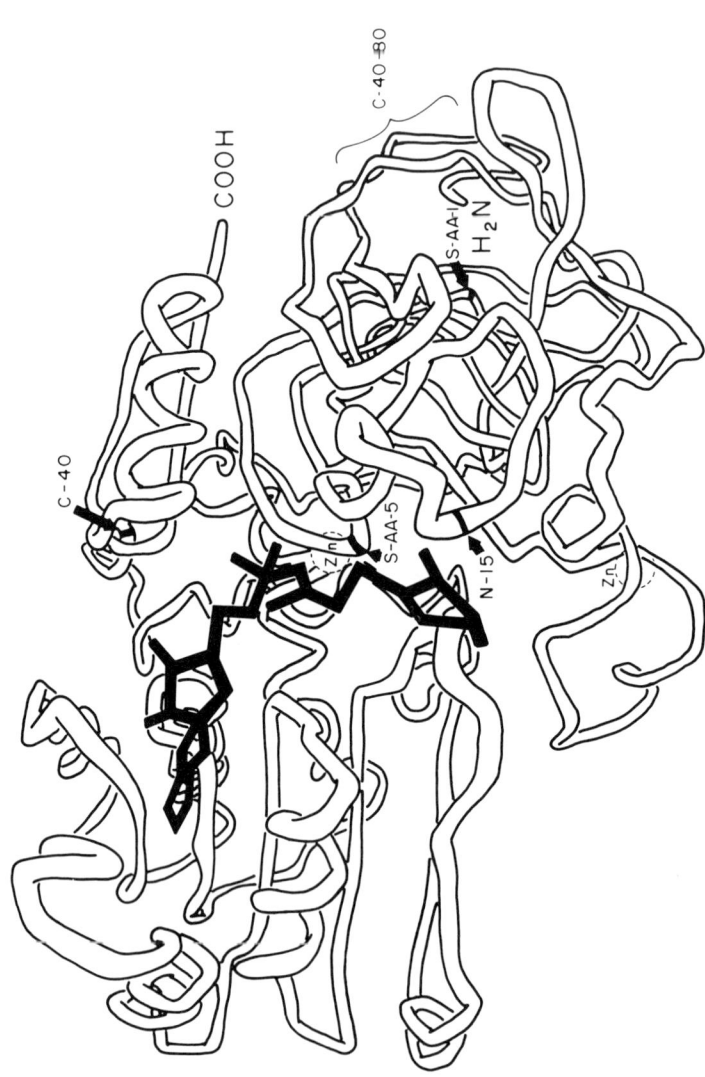

**Figure 2.** Positions of functional yeast ADH mutants mapped to their approximate positions on the tertiary structure of the horse liver ADH monomer. The cofactor-binding domain to the left and the substrate-binding domain to the right are separated in the figure by the bound cofactor molecule, shown in black. The cleft through which the substrate enters is at the bottom of the molecule in this diagram. [Taken with permission from Fig. 9 of Wills et al. (1982).]

that the tertiary structures are highly similar (Jörnvall, 1977b). In many instances, pairs of amino acid differences have been found between the two enzymes that would have the effect of preserving the tertiary structure. The only substantial difference is the absence in the yeast molecule of residues corresponding to residues 119–139 in the horse liver molecule, which form a loop that Jörnvall suggests might interfere with the formation of a tetrameric molecule.

The similarities in location of the active site and the sequence that in the horse liver enzyme acts as a second zinc-binding site (Brändén et al., 1975) are also strong indications that the two tertiary structures are related. We therefore felt justified in locating the amino acid substitutions in relationship to the horse liver enzyme, with the understanding that elucidation of the precise mechanism of these substitutions awaits both the solution of the yeast ADH tertiary and quaternary structures and a more precise understanding of the kinetics of the enzyme in relation to its structure.

The effect of the different substitutions on kinetics of the mutant enzymes is varied. One mutant, S-AA-5, has the most easily understood alterations in terms of its effect on the molecule. It is a substitution of an Arg for a His at position 44. This residue almost certainly corresponds to the arginine at position 47 in the horse liver enzyme, and it can be seen from Fig. 2 that this residue bonds to the pyrophosphate moiety of the cofactor. Substitution of a more strongly for a more weakly ionized residue at physiological pH should result in stronger binding to the cofactor, and this is what is seen in the kinetics (Table I). The $K_m$ for both NAD and NADH has been reduced half in the mutant enzyme, while $K_m$ values for the substrates are unchanged.

In the other mutants, the $K_m$ values for substrate and cofactor are either unchanged or increased, in some cases markedly. In the case of mutant N-15, for example, the $K_m$ for ethanol is increased tenfold, while that for acetaldehyde is essentially unchanged. The substitution in this mutant, an arginine for a tryptophane, has occurred in the pocket through which the substrate molecule must pass in order to reach the active site. Whatever the alteration in the properties of this pocket (and it is apparently not simply a change in the overall charge), it has the effect of excluding ethanol to a greater extent than acetaldehyde. Further, the charge is close to the nicotinamide ring of the cofactor. This charge difference may interfere with the binding, and with the formation of the substrate–cofactor complex.

Other substitutions are much less certain in their effects. The substitution for C-40, which has the effect of increasing substrate $K_m$ valves, is located at some distance from the active site. It may be that this substitution, of an arginine for a proline, may distort the molecule in such a way as to close the hydrophobic cleft (at the bottom of the molecule in the figure) through which the substrate approaches the active site. The substitutions in mutants S-AA-1 and C-40-80 are less precisely known, and are some distance from the active site, yet the S-AA-1 substitution has a large effect on substrate kinetics. This may be because it affects interactions at the quaternary level of organization.

It can be seen that both the kinetic effects and the amino acid substitutions can vary widely, yet all yield roughly the same phenotype. To turn to the question posed at the beginning of this chapter, it is apparent that a particular phenotypic alteration can be caused by alterations in many parts of the molecule. There are many different ways to skin this particular molecular cat.

## 5. Altered Kinetics of the Mutants

It is possible to construct an explanation for the connection between the altered kinetics of the mutant enzymes and the change in the NADH–NAD ratio in the cell. We do this with some hesitation, since any explanations offered at our current level of understanding of the ADH molecule are decidedly *post hoc*. All we are certain of is that the kinetic changes (including the $V_{max}$ changes, which without exception are in the direction of a lower $V_{max}$ for the mutants) result in altered ratios of the reduced to the oxidized cofactor.

It is immediately apparent that the mutant enzymes that have been investigated in detail all exhibit markedly altered kinetics, and that the kinetics have in many cases been altered in strikingly different ways. Just as there are many different possible amino acid substitutions that can produce these kinetic changes (see below), there are a surprising number of kinetic alterations that can produce a shift away from NAD toward NADH in the regularly metabolizing cell.

We know that the equilibrium concentrations of NAD and NADH in the wild-type cell grown under normal conditions on complete medium are approximately 1:1. The ratio can be changed 2- to 3-fold by growth in the presence of high concentrations of ethanol or acetaldehyde, provided that only one of the two cytoplasmic enzymes is present. The equilibrium concentrations of ethanol and acetaldehyde have not been measured, but the concentrations in the medium at log phase of growth have been (Wills, 1976a), and may reflect the intracellular concentrations. The ethanol concentration is approximately 96 mM and the acetaldehyde concentration approximately 1.2 mM. These can be compared with the

$K_m$ for ethanol and acetaldehyde for ADH-I, which are 24 and 3.4 mM, respectively. Thus, the concentration of acetaldehyde in the cell is likely to be substantially below the $K_m$, and that of ethanol to be substantially above it. Over a moderate range, changes in the $K_m$ for ethanol are less likely to affect the rate of reaction of the enzyme than are equivalent changes in the $K_m$ for acetaldehyde.

Bearing these factors in mind, the variety of changes seen in the mutant enzymes become more comprehensible. Not all possible patterns of change are seen, and those that are seen are explicable on the basis of the dynamics of the system.

Table I lists the mutant enzymes for which detailed kinetics have been obtained and the alterations in their kinetic parameters relative to wild type along with the approximate magnitude of the alterations. The mutants were obtained with and without mutagenesis, and under a variety of selective conditions. Nearly every mutant has a strikingly different pattern.

*S-AA-1*, a spontaneous mutant. In this mutant, the $K_m$ values for ethanol and acetaldehyde have been increased twofold and decreased by half, respectively. This would be expected to have a greater effect on the "back" reaction from acetaldehyde to ethanol than on the "forward" reaction from ethanol to acetaldehyde. However, in addition to these changes, the relative turnover number of the back reaction has been greatly decreased relative to the forward reaction. In a nonequilibrium situation, the ratio of the forward to the back reaction should be skewed in the direction of the forward reaction relative to the wild-type enzyme, and the NADH to NAD ratio should increase.

*S-AA-5*, spontaneous. Here, the substrate $K_m$ values and the ratio of the forward to the back reaction rates are unchanged, but the cofactor $K_m$ have each been decreased by half. This should only have the expected effect on the relative reaction rates if the intracellular concentration of NAD were below the $K_m$ for the wild-type enzyme, while that for NADH was above it. It is not possible to make such measurements with precision for the intracellular compartment containing ADH. But if this were the case, then a drop in the $K_m$ for NAD would permit the forward reaction to increase, while a similar drop in the $K_m$ for NADH would have little effect.

*C-40*, selected in a turbidostat under conditions of continuous logarithmic growth in the presence of increasing concentrations of allyl alcohol. In this mutant, the substrate $K_m$ valves have been increased markedly, while the $Kzm$ for the cofactors and the relative reaction rates are substantially unchanged. Because the concentration of ethanol is high, an increase in the $K_m$ will have little effect, while an equivalent increase in

## Table I
### Kinetic Information on Mutant ADH-I Enzymes Conferring Resistance to Allyl Alcohol, Relative to the Wild-Type Enzyme[a]

| Mutant | Position of substitution | $K_m$ For substrates | $K_m$ For cofactors | $V_f/V_b$ |
|---|---|---|---|---|
| S-AA-1 | Distant | EtOH up 2×, Acet down 1/2 | Unchanged | Increased 2.5× |
| S-AA-5 | Near cofactor site | Unchanged | Both down 1/2 | Unchanged |
| C-40 | Distant | EtOH up 4×, Acet up 2× | Unchanged | Unchanged |
| P5-55-19-7 | n.d. | EtOH up 10×, Acet up 8× | NAD up 3×, NADH up 6× | Unchanged |
| $D_B$-AA3-N15 | Near substrate and cofactor sites | EtOH up 10×, Acet unchanged | NAD up 10×, NADH up 30× | Increased 2× |
| $D_B$-AA3-N10 | n.d. | EtOH up 2×, Acet up 2× | NAD up 10×, NADH up 15× | Slight decrease |
| $D_B$-AA5-E1 | n.d. | Unchanged | NAD up 3×, NADH up 6× | Increased 1.5× |

[a]Each mutant is discussed more fully in the text. n.d., Not determined. $V_f$, the $V_{max}$ in the direction ethanol → acetaldehyde; $V_b$, the $V_{max}$ in the reverse direction.

the $K_m$ for acetaldehyde should have the effect of slowing the back reaction markedly.

*P5-S5-19-7*, spontaneous. The $K_m$ valves for both substrates and both cofactors in this mutant enzyme have been increased markedly. With regard to the substrate $K_m$, the same argument can be made for this enzyme as for C-40. Working against this, however, is the increase in $K_m$ for NAD, which should slow the forward rate if the NAD concentration is low. This increase is only about one-third that of the increases in the substrate $K_m$, however, and this difference may be enough to tilt the balance in the direction of the forward reaction.

$D_B$-*AA3-N15*, selected after nitrosoguanidine mutagenesis. Large kinetic changes are exhibited by this mutant. The $K_m$ for ethanol is increased tenfold, while that for acetaldehyde is unchanged. The $K_m$ for NAD is also increased tenfold, while that for NADH is increased 30-fold, the largest change found in any of the mutant enzymes so far. Finally, the relative rates of the forward to back reactions have been doubled. The enormous increase in the NADH $K_m$ ensures that the back reaction will be markedly slowed, permitting NADH to build up.

$D_A$-*AA3-N10*, isolated after nitrosoguanidine treatment. Substrate $K_m$ values are doubled. While the cofactor $K_m$ values are increased markedly, tenfold for NAD and 15 fold for NADH. There is a slight decrease in the ratio of the forward to back velocities. Again, the NADH $K_m$ is more markedly increased than that for NAD, and this, coupled with the increase in ethanol $K_m$, may be enough to tilt the nonequilibrium reaction in the direction of accumulating NADH.

$D_B$-*AA5-E1*, from ethyl methane sulfonate mutagenesis. This is the mutant with the smallest effect on mobility of those investigated. The $K_m$ values for substrate are unchanged, but those for cofactor are markedly increased, threefold for NAD and sixfold for NADH. There is a slight increase in the forward to back velocity ratio. Again the situation is a difficult one to analyze, since the change in NAD $K$zm should have a greater effect than that for NADH. Yet if the change in NADH $K_m$ is large enough to bring it above the concentration of NADH in the cytoplasm, this may be enough to tilt the balance, particularly in view of the increased relative forward velocity. Of all the mutants so far investigated, this one has the most puzzling kinetics.

## 6. Evolutionary Implications

The field of population genetics has been racked for two decades by the question of what proportion of the amino acid substitutions seen in the course of evolution or evidenced as polymorphisms is subject to meaningful amounts of selection. It has proved painfully apparent that there is no mathematical way to distinguish between the effects of selection and drift (Lewontin, 1974). I have pointed out elsewhere (Wills, 1981) that the only approach that will eventually resolve this conflict is to proceed on a case-by-case basis. If enough cases are found, both in nature and in the laboratory, where selective differentials result from amino acid substitutions, then the argument for selection as a primary determiner of such changes will be progressively strengthened. In addition to the considerable light cast on the physiology of yeast by the current investigations, I feel that we have moved in the direction of strengthening the selectionist point of view.

The story that is emerging is similar to that seen for the hemoglobin molecule. A large fraction of the ADH molecule appears to be potentially involved in alterations that affect its kinetics, as evidenced by the fact that we have not yet found two identical allyl alcohol resistance mutants. This is all the more remarkable because of the specificity of the selection pressure applied; what is required for survival is that the mutant molecule have altered kinetics that shift the redox balance inside the cell. Further, our investigations have shown that the ADH isozyme system has a least one additional function, that of helping to stabilize the ratio of reduced to oxidized pyridine cofactors in the cell.

ACKNOWLEDGMENTS. These investigations have been or are being supported by the National Institutes of Health, the Guggenheim Foundation, the National Science Foundation, and the Department of Energy. I wish to thank many co-workers, students, and assistants who helped to make this system such a productive one, particularly Hans Jörnvall, Paul Kratofil, David Londo, Tracy Martin, and Julie Phelps.

## References

Beier, D. R., and Young, E. T., 1982, Characterization of a regulatory region upstream from the *ADR2* locus of *S. cerevisiae, Nature* **300**:724–728.
Berger, E. M., 1977, Are synonymous mutations adoptively neutral? *Am. Nat.* **111**:606–607.
Brändén, C., Jörnvall, H., Eklund, H., and Furugren, B., 1975, Alcohol dehydrogenases, in: *The Enzymes,* 3rd ed. (P. Boyer, ed.), Academic Press, New York, pp. 103–190.
Ciriacy, M., 1975a, Genetics of alcohol dehydrogenase in *Saccharomyces cerevisiae.* I. Isolation and analysis of *adh* mutants, *Mut. Res.* **29**:315–326.
Ciriacy, M., 1975b, Genetics of alcohol dehydrogenase in *Saccharomyces cerevisiae.* II. Two loci controlling synthesis of the glucose-repressible ADH-II, *Mol. Gen. Genet.* **138**:157–164.
Ciriacy, M., *Cis*-dominant regulatory mutations affecting the formation of glucose-repres-

sible alcohol dehydrogenase (ADH-II) in *Saccharomyces cerevisiae, Mol. Gen. Gent.* **145**:327–333.

Ciriacy, M., 1979, Isolation and characterization of cis- and trans-acting regulatory elements involved in the synthesis of glucose-repressible alcohol dehydrogenase (ADH II) in *Saccharomyces cerevisiae, Mol. Gen. Genet.* **176**:427–431.

Denis, C., Ciriacy, M., and Young, E. T., 1981. A positive regulatory gene is required for accumulation of the functional messenger RNA for the glucose-repressible alcohol dehydrogenase from *Saccaromyces cerevisiae, J. Mol. Biol.* **148**:355–368.

Englesberg, E., and Wilcox, G., 1974, Regulation: Positive control, *Annu. Rev. Genet.* **8**:219–242.

Goodman, M. G., Moore, G. W., and Matsuda, G., 1975, Darwinian evolution in the genealogy of hemoglobin, *Nature* **253**:603–607.

Harris, H., 1966, Enzyme polymorphisms in man, *Proc. R. Soc. B* **164**:298–310.

Jörnvall, H., 1977a, The primary structure of yeast alcohol dehydrogenase, *Eur. J. Biochem.* **72**:425–442.

Jörnvall, H., 1977b, Differences between alcohol dehydrogenases: Structural properties and evolutionary aspects, *Eur. J. Biochem.* **72**:443–452.

Jörnvall, H., Fairwell, T., Kratofil, P., and Wills, C., 1980, Differences in α-amino acetylation of isozymes of yeast alcohol dehydrogenase, *FEBS Lett.* **111**:214–218.

Kato, T., Berger, S. J., Carter J. A., and Lowry, O., 1973, An enzymatic cycling method for nicotinamide-adenine-dinucleotide with malic and alcohol dehydrogenases, *Anal. Biochem.* **53**:86–97.

Kimura, M., 1968, Evolutionary rate at the molecular level, *Nature* **217**:624–626.

King, J. L., and Jukes, T. H., 1969. Non-Darwinian evolution: Random fixation of selectively neutral mutations, *Science* **164**:788–798.

Lewontin, R. C., 1974, *The Genetic Basis of Evolutionary Change*, Columbia University Press, New York.

Lewontin, R. C., and Hubby, J. L., 1966, A molecular approach to the study of genetic heterozygosity in natural populations. II. Amount of variation and degree of heterozygosity in natural populations of *Drosophila pseudoobscura, Genetics* **54**:595–626.

Lipman, D. J., Smith T. F., Beckman R. J., and Waterman, M. S., 1982, Hierarchical analysis of influenza ⁻A hemagglutinin gene sequences, *Nucleic Acid Res.* **10**:5375–5389.

Lutstorf, U., and Megret, R., 1968, Multiple forms of alcohol dehydrogenase in *Saccharomyces cerevisiae, Arch. Biochem. Biophys.* **126**:933–944.

Megnet, R., 1967, Mutants partially deficient in alcohol dehydrogenase in *Schizosaccharomyces pombe, Arch Biochem. Biophys.* **121**:194–210.

Neuberg, C., and Reinfurth, E., 1918, Die Festlegung der Aldehydstufe bei der alkoholischen Gärung, *Biochem. Z.* **89**:365–414.

Northrop, J. H., 1954, Apparatus for maintaining bacterial cultures in steady state, *J. Gen. Physiol.* **38**:105–115.

Ohta, T., 1973, Slightly deleterious mutant substitutions in evolution, *Nature* **246**:96–97.

Rando, R. R., 1974, Allyl alcohol induced irreversible inhibition of yeast alcohol dehydrogenase, *Biochem. Pharmacol* **23**:2328–2331.

Sanger, F., Air, G. M., Barrell, B. G., Brown, N. L., Coulson, A. R., Fiddes, J. C., Hutchison, C. A., Slocombe, P. M., and Smith, M., 1977, Nucleotide-sequence of bacteriophage φX174 DNA, *Nature* **265**:687–695.

Sofer, W. H., and Hatkoff, M. A., 1972, Chemical selection of alcohol dehydrogenase negative mutants in *Drosophila, Genetics* **72**:545–549.

Williamson, V. M., Young, E. T., and Ciriacy, M., 1981, Transposable elements associated with constitutive expression of yeast alcohol dehydrogenase II, *Cell* **23**:605–614.

Wills, C., 1973, In defense of naive pan-selectionism, *Am. Nat.* **107**:23–34.
Wills, C., 1976a, Production of yeast alcohol dehydrogenase isozymes by selection, *Nature* **261**:26–29.
Wills, C., 1976b, Controlling protein evolution, *Fed. Proc.* **35**:2098–2101.
Wills, C., 1978, Rank-order selection is capable of maintaining all genetic polymorphisms, *Genetics* **89**:403–417.
Wills, C., 1981, *Genetic Variability*, Oxford University Press, Oxford.
Wills, C., and Jörnvall, H., 1979, The two major isozymes of yeast alcohol dehydrogenase, *Eur. J. Biochem.* **99**:323–331.
Wills, C., and Martin, T., 1980, Alteration in the redox balance of yeast leads to allyl alcohol resistance, *FEBS Lett.* **119**:105–108.
Wills, C., and Phelps, J., 1975, A technique for the isolation of yeast alcohol dehydrogenase mutants with altered substrate specificity, *Arch. Biochem. Biophys.* **167**:627–637.
Wills, C., and Phelps, J., 1978, Functional mutants of yeast alcohol dehydrogenase affecting kinetics, cellular redox balance, and electrophoretic mobility, *Biochem. Genet.* **16**:415–432.
Wills, C., Kratofil, P., and Martin, T., 1982, Functional mutants of yeast alcohol dehydrogenase, in: *Genetic Engineering of Microorganisms for Chemicals* (A. Hollaender, ed.), Plenum Press, New York, pp. 305–329.
Wright, S., 1931, Evolution in Mendelian populations, *Genetics* **16**:97–159.

*CHAPTER 9*

# Gene Recruitment for a Subunit of Isopropylmalate Isomerase

## JOST KEMPER†

### 1. The Leucine Operon in *Salmonella typhimurium* Wild-Type Strains

The leucine operon in *Salmonella typhimurium* consists of four structural genes, *leuABCD* (Burns *et al.*, 1966; Margolin, 1963), and a control region located adjacent to the *leuA* gene. The major regulation of the expression of the leucine operon occurs via modulation of the transcription attenuation of a leader region that responds to the intracellular level of leucyl-tRNA$^{leu}$ (Gemmill *et al.*, 1979; Wessler and Calvo, 1981). In addition, mutations at a locus *flr* unlinked to the leucine operon result in constitutive expression of the leucine genes as well the isoleucine/valine gene (Friedberg *et al.*, 1974). The relationship of the leucine genes to the enzymes in the leucine biosynthetic pathway is shown in Fig. 1. The enzymes α-isopropylmalate synthetase (*leuA* gene product) and β-isopropylmalate dehydrogenase (*leuB* gene product) have been purified from *S. typhimurium* and characterized (Kohlhaw *et al.*, 1969; Parson and Burns, 1970). The work described here is particularly concerned with the second leucine biosynthetic enzyme, isopropylmalate isomerase.

### 2. The Wild-Type Isopropylmalate Isomerase

The isopropylmalate isomerase interconverts α-isopropylmalate and β-isopropylmalate via temporary formation of the intermediate isopro-

---

*JOST KEMPER†* • Institute of Molecular Biology, University of Texas at Dallas, Richardson, Texas 75080.

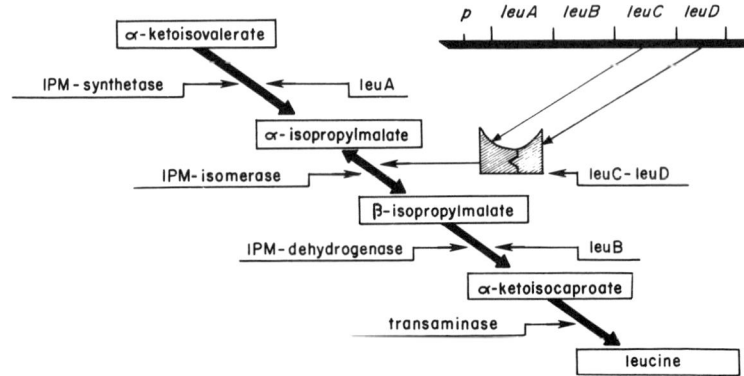

Figure 1. Biosynthetic pathway for leucine and the relationship of the leucine-specific enzymes to the genes of the leucine operon. $p$, Promoter for the leucine operon; IPM, isopropylmalate isomerase.

pylmaleate (also called dimethylcitraconate); the reaction equilibrates at a ratio of α-isopropylmalate–isopropylmaleate–β-isopropylmalate of 8.7:1:3.4 (Gross et al., 1963). The native enzyme has a $K_m$ for α-isopropylmalate of $3 \times 10^{-4}$ M (Fultz and Kemper, 1981). Genetic analyses of leucine auxotrophs of both *Escherichia coli* and *S. typhimurium* suggested that the isopropylmalate isomerase is a complex enzyme with two different subunits coded for by two separate genes *leuC* and *leuD* (Margolin; 1963; Somers et al., 1973; Yang and Kessler, 1974). This was verified by the partial purification of the isopropylmalate from *S. typhimurium* strains. Two polypeptides were shown to copurify through successive ammonium sulfate fractionation and were identified on sodium dodecyl sulfate–polyacrylamide gels as having a molecular weights of 51,000 (*leuC* gene product) and 23,500 (*leuD* gene product) (Fultz and Kemper, 1981). No activity of the individual isomerase subunits for either α- or β-isopropylmalate was detectable. The isopropylmalate isomerase enzyme is relatively unstable; all activity is lost within minutes in dilutions and at 32°C, while at 4°C in concentrated crude extracts 75% of the specific activity remains after 2 days. Using crude extracts from *leuC* and *leuD* mutant strains, it was possible to demonstrate *in vitro* complementation. It is important to note that the uncomplexed polypeptide components of the isopropylmalate isomerase were fairly stable, showing only a loss of approximately 40% after 16 hr at 5°C (Fultz and Kemper, 1981).

Purification of the isopropylmalate isomerase from yeast (Bigelis and Umbarger, 1975) and from *Neurospora crassa* (Reichenbecher and Gross, 1978) has shown that in both these species the isopropylmalate isomerase is composed of only one polypeptide, with a molecular weight of 90,000. The finding that this polypeptide is nonrandomly cleaved into two polypeptide fragments of 56,000 and 37,000 dalton (Reichenbecher and Gross, 1978) suggests that the gene coding for this enzyme might be the result of a fusion of two genes during evolution of this species. Thus, the native form of the *N. crassa* isopropylmalate isomerase would contain two distinct domains in its tertiary structure, which represent the two ancestral polypeptides that are now joined in a region that is a preferential site for peptide cleavage. Examples of multifunctional enzymes that consist of a single polypeptide in one species and of two polypeptides in different species have been reported (Crawford, 1975; Grieshaber and Bauerle, 1972; Miozzari and Yanofsky, 1979). The underlying principle of gene fusion could be visualized as a mechanism in the development of more efficient enzymes.

## 3. Strains Carrying *leuD* Mutations Revert to Leucine Prototrophy

Originally discovered in an attempt to explain rare occurrences in crosses between *leu* mutants of *S. typhimurium*, it was found that certain *leuD* mutant strains, which included those carrying *leuD* deletions, reverted with a low frequency to leucine prototrophy. Using a Leu$^+$ derivative of the deletion strain $\Delta(leuD-ara)700$, which extends from *leuD* into or through the arabinose operon, it was shown by phage P22-mediated transduction that this Leu$^+$ strain still carried the original *leuD* deletion [$\Delta(leuD-ara)700$] and that the strain had acquired an additional mutation, which was then termed *supQ* (R-16).

## 4. Model for Leucine Biosynthesis in *leuD–supQ* Mutant Strains

The basic principle for the mechanism of suppression of *leuD* mutations by *supQ* mutations is outlined in Fig. 2. The genes *supQ* and *newD*, located substantially distant from the leucine region of the chromosome, code for two polypeptides that form a polypeptide complex, whose normal function in the cell remains to be determined. The *newD* polypeptide is capable of forming a complex with the *leuC* polypeptide, resulting in a functional isopropylmalate isomerase. The hybrid *leuC–newD* complex, however, can only be formed in the absence of both the *leuD* and *supQ* polypeptides, since these have a higher affinity for their "normal part-

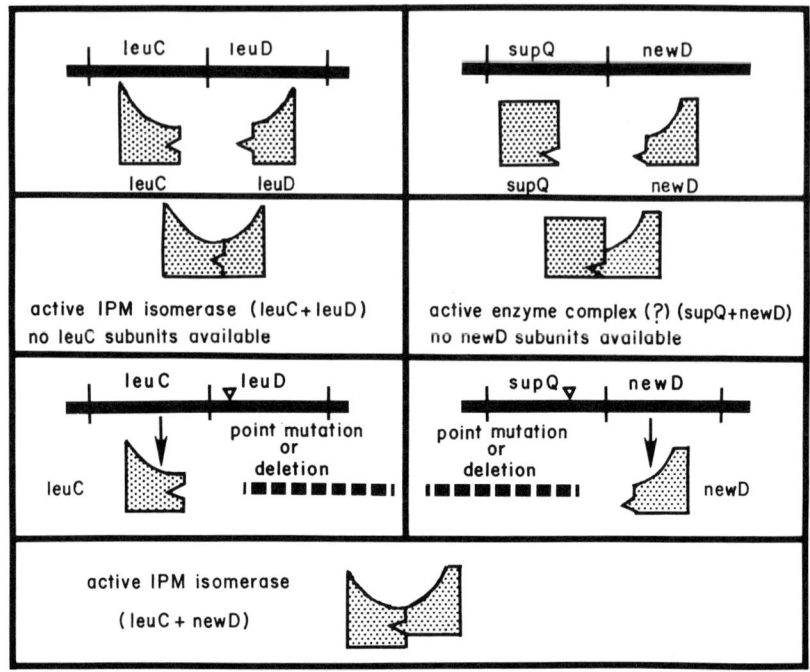

**Figure 2.** Model for *supQ* suppression of *leuD* mutant strains.

ners" *leuC* and *newD*, respectively (Fultz *et al.*, 1979). In the following sections data will be presented to support the above model as well as to rule out several alternative explanations.

## 5. Leucine Biosynthesis in *leuD–supQ* Mutant Strains

In several cases where a deleted, mutated, or absent function was restored through additional mutations, it was found that new pathways were established that circumvented the block in the original reaction chain. This was especially true in those cases where the utilization of different carbon sources was studied (Hegeman and Rosenberg, 1970; Wu *et al.*, 1968). It was therefore important to show that the pathway for leucine biosynthesis in the *leuD–supQ* mutant strains was not altered. This was accomplished by constructing strains that combined the *leuD–supQ* mutations with additional mutations in the *leuA, leuB,* or *leuC* gene. In all cases, the resulting strains were leucine auxotrophic (Kemper and Margolin, 1969). This then verified that the overall leucine biosynthetic reactions were unaltered in *leuD–supQ* mutant strains, and in particular that the *leuC* polypeptide was required and that the *supQ* mutation must have resulted in restoration or generation of an isomerase activity that could convert α-isopropylmalate into β-isopropylmalate.

## 6. Genetic Characterization of the *leuC–newD* Isopropylmalate Isomerase

### 6.1. A Free *leuC* Polypeptide Is Needed

As described above, the *leuC* polypeptide was required for leucine prototrophy. A systematic analysis of different *leuD* mutations showed that these fell into two classes, those that were suppressible by the originally isolated *supQ* mutation and those that were not. Most mutants of the first group were able to give rise to independent Leu⁺ revertants of the *supQ* type; since the spontaneous mutation frequency for *supQ* mutations is rather low ($10^{-8}$–$10^{-10}$), this test is only possible for those mutants that either do not yield revertants of different types or only yield them with a very low frequency. No Leu⁺ revertant of the second group of *leuD* mutants was ever found to be due to a *supQ* mutation.

It was then shown that the first group of *leuD* mutants (those suppressible by *supQ*) consisted of deletions and chain-terminating and frameshift mutations, while all likely missense mutations belonged to the second group. This strongly suggested that the common and functionally pertinent characteristic of the first group of mutants was the production of either no or only a relatively short amino-terminal *leuD* protein fragment, which probably was incapable of complex formation with the *leuC* polypeptide, resulting in "free," i.e., uncomplexed *leuC* polypeptides. On the other hand, *leuD* missense mutants would produce an altered *leuD* polypeptide, which would be likely to have retained the ability to form an, albeit nonfunctional, complex with the *leuC* polypeptide, unless the missense mutation had affected the particular region of the polypeptide essential for the complex formation. From the relative position within the *leuD* gene of all *leuD* mutations tested, we have reasons to believe that mainly the segment toward the carboxy-terminal end is essential for complex formation. (No *leuD* mutations mapping in the promoter-distal region of the *leuD* gene were ever found that were not suppressible by *supQ*.)

Additional support for the notation that a free *leuC* polypeptide is required for suppression by *supQ* (i.e., for complex formation with the *newD* polypeptide) comes from an analysis of the following strains: Strain JK11 (*leu*D657 *supQ*1 *trp*50$_{amber}$) is Leu⁻ Trp⁻, because the small deletion *leuD*657 results in formation of a *leuD* polypeptide of substantial size that still has the ability to complex with the *leuC* polypeptide. Strain JK47 (*leuD*657 *supQ*1 *trp*50$_{amber}$ *leuD*804$_{amber}$) is Leu$^+$ Trp$^-$, since the additional *leuD*$_{amber}$ mutation prevents production of a full-sized polypeptide. Strain JK51 (*leuD*657 *supQ*1 *trp*50$_{amber}$ *leuD*804$_{amber}$ *sup*$_{amber}$) is Leu$^-$ Trp$^+$ because the amber suppressor restoring the Trp$^+$ phenotype also results in the production of an approximately full-sized *leuD* polypeptide, which again can complex with the *leuC* polypeptide (Kemper and Margolin, 1969).

From these data it was concluded that a free, uncomplexed *leuC* protein is required and that this *leuC* protein must form a complex with a different polypeptide (the *newD* gene product) that is not capable of competing successfully with *leuD* missense polypeptides for the *leuC* binding sites. In agreement (and support) of the above conclusion is the observation that *leuD*–*supQ* mutant strains of the second group (e.g., *leuD* missense mutants) showed slow growth in the absence of exogenous leucine supply upon extended incubation.

### 6.2. *SupQ* Mutations Result in Availability of the *newD* Gene Product

6.2.1. Localization of *supQ* Mutations

Since it was observed that seemingly spontaneous *supQ* mutations occurred more frequently in presence of P22 phage, it was suspected that the *supQ* gene might be localized near the phage P22 attachment site *ataA*. Figure 3 shows this region of the *S. typhimurium* chromosome. Using P22-mediated transduction, it was shown that 32 independently isolated *supQ* mutations were localized near *proAB*, showing cotrans-

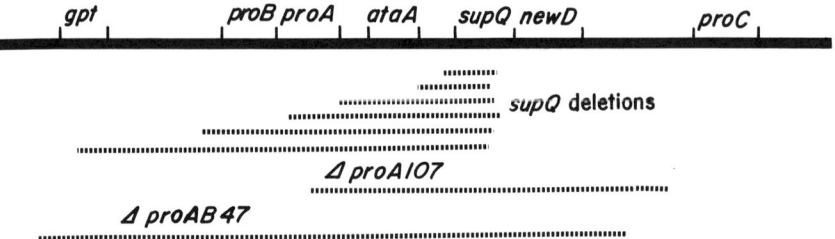

**Figure 3.** Region of the *Salmonella typhimurium* chromosome containing the *supQ newD* genes; distances are not to scale (see Fig. 5). *ataA*, Primary attachment site for phage P22; *gpt*, guanosine phosphoribosyltransferase; the endpoints of all deletions are not accurately determined.

duction frequencies ranging from 99.9% to as low as 10.7 or 5.4%. This would indicate that some *supQ* mutations are located very distant to the *proA* gene. Comparing these cotransduction frequencies with values that have been reported for the *leu–ara–pyrA* (Kemper, 1974a; Margolin, 1963) or *trp–cysB* regions (Sanderson and Hall, 1970) and considering a model for generalized, random transduction (Kemper, 1974a), one can estimate that a cotransduction frequency of 10% corresponds to a distance of approximately 20 genes (60% of the length of a transducing genome). Regardless of the accuracy of these estimates, it seems untenable that all *supQ* mutations are mutations within one gene. Furthermore, it seems unlikely that the nucleotide sequence of one gene can be altered by a large number of point mutations and/or deletions and still provide the genetic information for the same or similar proteins. The large difference in cotransduction frequencies is therefore best explained by assuming that some of the *supQ* mutations are deletions that have effectively brought the relatively distant *supQ* region into the direct vicinity of the *proA* gene (Kemper, 1974b).

When independent Leu$^+$ revertants of the *supQ* type were isolated in *leuD* deletion strains in the presence of proline supplementation, a significant number of the *supQ* mutants had simultaneously acquired a proline requirement. Mapping by P22-mediated transductions showed that the *supQ* (Leu$^+$) Pro$^-$ character of these strains was due to single deletions that extended from the *supQ* gene into the *proAB* operon, some extending even further into or through the *gpt* gene (guanosine phosphoribosyl transferase) (see Fig. 3) (Kemper, 1974b). While many *supQ* mutations (if not most of those primarily isolated) were deletions, it is important to note that point mutations also can give rise to the *supQ* (Leu$^+$) phenotype. This could only be shown indirectly by the fact that 2-aminopurine mutagenesis, which predominatly causes transitions, resulted in a 100- to 1000-fold increase in the spontaneous *supQ* mutation frequency. Most of the 2-aminopurine-induced *supQ* mutant strains are temperature-sensitive. The possible significance of this will be discussed in Section 6.3.4.

It should be noted that the *supQ* mutations themselves do not result in an identifiable phenotype; strains carrying a wild-type leucine operon and a *supQ* mutation are indistinguishable from wild-type strains under all conditions so far tested.

6.2.2 Localization of the newD Gene

The existence of a large number of different *supQ* mutations and the fact that many of them are deletions made it extremely unlikely that all of these *supQ* mutations produce an altered *supQ* product that is more or less equally functional and required to restore leucine prototrophy by complexing with the *leuC* polypeptide. According to the model, the necessary polypeptide to complex with the *leuC* polypeptide is the product of the *newD* gene, which is identical and unaltered in all *supQ* mutant strains. Since the *supQ* and the *newD* polypeptides normally form a high-affinity complex, the common and essential property of all *supQ* mutations is only to make free *newD* polypeptides available. While the data presented so far indicate neither the existence nor the location of the *newD* gene, the following experiments show that *newD* is required, and is located adjacent to the *supQ* gene. The leucine prototrophic strain Δ(*leuD–ara*)798 *supQ*394 was mutagenized (nitrosoguanidine treatment) and leucine auxotrophic mutants were collected after penicillin selection. Mutations causing leucine auxotrophy in the Δ(*leuD–ara*)798 *supQ*394 strain were either mutations in the *leuA, leuB*, or *leuC* genes of the leucine operon, since these genes are required for *supQ* function, or they were mutations in the proposed *newD* gene. The possibility that the *supQ*394 mutation had reverted to its original status could be excluded, since *supQ*394 was a deletion of substantial size. Seven of the *newD* mutations thus obtained were used for further mapping, to identify the location and extent of the *newD* gene. Recombination frequencies between individual *newD* mutations ranged from $>10^{-7}$ to $<10^{-5}$, typical for intragenic recombination. Furthermore, cotransduction frequencies with *pro* mutations allowed us to establish the gene order *proBA ataA supQ newD* (see Fig. 3) (Kemper, 1974c).

### 6.2.3. Deletions of supQ and newD

The proline deletions Δ(*proAB*)47 and Δ(proA)107 in strains isolated by Itikawa and Demeric (1968) could be shown to extend through the *supQ newD* region; these strains thus have lost the ability to give rise to mutations (*supQ newD*$^+$) that restores leucine prototrophy to *leuD* mutant strains (Kemper, 1974c). The fact that these extensive proline mutations cause only a proline requirement indicates that the normal functions(s) of *supQ newD* is inessential under all growth conditions so far tested, making any identification of this function considerably more difficult.

### 6.2.4. Specialized P22-Transducing Particles Carrying newD

Because of the location of the P22 attachment site *ataA* in close vicinity to the *supQ* gene (see Fig. 3), it was possible to demonstrate the production of a very high frequency of *newD*-carrying specialized transducing particles in *supQ* mutant strains. The frequency of *newD* specialized transducing particles is several orders of magnitude higher than that of the production of *proAB* specialized transducing particles because of the directionality in the phage P22 DNA-packaging process (Kwoh and Kemper, 1978a). A substantial number of these *newD*-carrying specialized transducing particles were capable of transducing by integration into the P22 attachment site *ataA* (Fultz *et al.*, 1979). This provided a valuable tool for transferring the *newD* gene into different strains and for constructing merodiploid strains.

### 6.2.5. F' Factors Carrying the newD Gene

The production of F' factors from *S. typhimurium* Hfr strains is very inefficient (Sanderson *et al.*, 1972); however, an existing *E. coli* F' factor was easily utilized. This F' factor carries the *E. coli proAB* gene as well as the P22 attachment site, which exists in *E. coli* (Hoppe and Roth, 1974). Integration of a *newD*-carrying specialized P22 transducing genome into the attachment site of this F' factor yielded the F'*prolac*P22d*supQ*394*newD*$^+$ (Fultz *et al.*, 1979), which can be transferred into *S. typhimurium* or *E. coli* strains. This then represents an easy method to study gene dosage effects for the *newD* gene, as well as to test the dominance relationship (see Section 6.3.2).

## 6.3. Nature of *supQ* Mutations

While *supQ* mutations seem to represent a very inhomogeneous group, and certainly include point mutations and deletions of substantial size, their one common characteristic is that they lead to the availability of the *newD* polypeptide; the following sections will analyze how this is accomplished.

### 6.3.1. SupQ Mutations Are Not Promoter Mutations

Although it has been shown that point mutations can generate a promoter (Wuestoff and Bauerle, 1970), it seems unlikely that this can occur with the very high frequency observed and at apparently so wide a variety of different sites; furthermore, it is highly unusual that these promoters would all be temperature-sensitive. *SupQ* mutations that are deletions could generate promoters by fusing DNA sequences such that RNA polymerase recognition sites are generated; again it seems highly unlikely that this would occur with such a high efficiency (*supQ* mutations of the deletion type are generally very effectively expressing *newD*) and at such a variety of different sites, since the *supQ* deletion strains vary widely in the extent of their deletions. Finally, *supQ* mutations could lead to the expression of the *newD* gene by fusing the *newD* gene to a different "operon." Although again the large spectrum of *supQ* deletions is difficult to explain, and although no transcriptional units (operons?) are known to be located between *newD* and *proBA*, this possibility seems theoretically to provide a viable explanation. However, in those cases in which the *supQ* deletions extend into the proline operon, it would be obvious that the *newD* gene would be fused to the *proAB* operon. Since the direction of the transcription of the *proAB* genes is not known, one cannot predict whether such a fusion could produce functional, active *newD* polypeptides; if so, however, it would certainly place the *newD* gene under proline regulation. This is not the case; the *supQ pro* deletion strains obviously require proline supplementation for growth, yet express the *newD* gene effectively under these conditions.

### 6.3.2. SupQ Mutations Are Not Regulatory Mutations

It could be considered that the *newD* gene in the wild-type strain is repressed under all growth conditions so far tested by a system that functionally involves the *supQ* gene or region, and that the *supQ* mutations are lifting this repression, leading to constitutive expression of the *newD* gene. Unless this regulation is only *cis*-active, e.g., activation of a promoter (see Section 6.3.1.), one would expect that the wild-type $supQ^+$ would be dominant over mutant *supQ*. Utilizing the F'*prolac*-P22d*supQ*394*newD*$^+$ factor transferred into a *leuD* mutant strain and carrying a wild-type $supQ^+$ $newD^+$ region, it was found that the mutant *supQ* was dominant, i.e., the merodiploid strain was Leu$^+$ (Fultz *et al.*, 1979). This would eliminate the possibility that *supQ* represents a repressor gene for expression of the *newD* gene. It should be noted that the *supQ* point mutations could be of the $O^c$ (operator-constitutive) type and thus be dominant; however, the same could not be true for the extensive *supQ* deletion strains. Finally, although the primary effect of *supQ* is not the regulation of *newD* gene expression, it can, of course, not be excluded that large *supQ* deletions might also delete a putative regulatory gene that might regulate the expression of *newD* (and *supQ*).

### 6.3.3. SupQ Codes for a Protein Subunit of the *supQ newD* Protein Complex

The remaining possibility, which is thus an integral part of the proposed model, is that the *supQ* gene codes for a protein that prevents the availability of the *newD* polypeptide; either the *supQ* protein leads to a stoichiometric inactivation of the *newD* protein, or the *supQ* polypeptide forms a more or less tight complex with the *newD* polypeptide. The data from the studies of merodiploid $supQ^-$ $newD^+/supQ^+$ $newD^+$ strains would favor the latter explanation.

### 6.3.4. Different supQ Mutations Vary in Their Efficiency to Restore Leucine Prototrophy

The strength or efficiency of a *supQ* mutation is measured by the growth rate of the strain in minimal medium compared to the growth rate in leucine-supplemented medium, in which the growth rate is in all cases identical to that of the wild-type strain. Most *leuD–supQ* are leucine-bradyotrophic and temperature-sensitive, with a consistent pattern such that in all cases of a growth limitation a better growth is obtained at lower temperatures. In particular, most point mutation *supQ* strains completely fail to grow at 37°C, and even at 24°C show a lower growth rate than the wild-type strain. According to the proposed model, all *supQ* mutant strains would result in the availability of the same identical *newD* gene product. It is therefore suspected that the difference in growth rate reflects a difference in the amount of the available *newD* gene product and not in any qualitative difference. This explanation agrees with the fact that the fastest growing *supQ* mutant strains are those carrying a deletion. However, even the "best" *supQ* deletion strains will display a certain temperature sensitivity at 42°C. It remains possible that this level of temperature sensitivity represents a basic temperature sensitivity of the *newD* product or of the *leuC–newD* polypeptide complex.

The temperature sensitivity of most *supQ* mutant strains is believed to result from different amounts of free *newD* polypeptides, e.g., at higher temperatures residual *supQ* products or fragments might be capable of complex formation with the *newD* polypeptide, thus reducing the amount of free available *newD*. As stated above, the growth limitations and temperature sensitivity cannot be due to the effect of the *supQ* mutations on functions in the cell other than leucine biosynthesis, since leucine supplementation completely compensates for all limitations.

Proof for the model that different *supQ* mutant strains produce different quantities of the identical *newD* polypeptide would require biochemical identification and quantitation of the *newD* polypeptide. While this is not possible until a method to identify the *newD* polypeptide has been developed, strong support for the model is obtained from the following data.

### 6.3.5. Effects of NewD Gene Dosage

Utilizing the F' factor F'prolacP22dsupQ394newD$^+$ described above, which carries a functional newD gene, strains were constructed that contained two copies of the newD gene. The extent of the growth rate limitation due to suboptimal, endogenous leucine production is measured by the $L/M$ value, which represents the ratio of the doubling time in leucine-supplemented medium to that in minimal medium. Table I shows that the small growth limitation in strain JK1394 ($L/M = 0.9$) is completely lifted upon addition of a second copy of the newD gene, strain JK472 ($L/M = 1.0$). The second pair of strains carry the leuD657 mutation, which is a small internal deletion in the leuD gene. It was previously concluded that the leuD mutant polypeptide produced as a result of the leuD657 mutation is able to form an (inactive) complex with the wild-type leuC polypeptide, rendering most of the leuC polypeptides unable to form a complex with the newD polypeptide, thus in effect lowering the concentration of free leuC polypeptides (Kemper, 1974b). The resulting growth limitation in strain JK513 ($L/M = 0.3$) is very significantly lowered by the addition of a second copy of the newD gene, strain JK533 ($L/M = 0.7$). These data strongly imply that the insufficient concentration of the newD gene product is at least partially responsible for the slow growth of leuD–supQ mutant strains in minimal medium. It further suggests that the expression of the newD gene in supQ mutant strains is not regulated so as to maintain a constant concentration of the newD gene product per cell.

Since leucine biosynthesis requires the formation of a complex of the leuC with the newD polypeptide, it was important to show that the amount of leuC polypeptide produced (in absence of a leuD polypeptide) is not

Table I.
Growth-Rate Dependence of supQ newD Mutant Strains with One or Two Copies of the newD Gene

| Strain | Relevant genotype | Number of newD gene copies | Doubling time[a] | | L/M |
|---|---|---|---|---|---|
| | | | L | M | |
| JK1394 | (leuD–ara)798 supQ394 newD$^+$ | 1 | 52 | 58 | 0.9 |
| JK472 | (leuD–ara)798 supQ394 newD/F'supQ394 newD$^+$ | 2 | 68 | 68 | 1.0 |
| JK513 | leuD657 supQ394 newD$^+$ | 1 | 50 | 168 | 0.3 |
| JK533 | leuD657 supQ394 newD$^+$ F'supQ394 newD$^+$ | 2 | 58 | 83 | 0.7 |

[a]Doubling time, in minutes, of bacterial culture in minimal medium ($M$) and in leucine-supplemented medium ($L$).

imposing any limitations. Similar to the experiments described above, two sets of strains were constructed carrying *leuD–supQ* mutations and an additional copy of the wild-type *leuC* gene by transferring an F' factor containing the *leuOABC* region of the *S. typhimurium* chromosome, F' *leuOABC*$^+$*leuD*54046*ara*$^+$. Since the leucine operon is maximally derepressed due to the inadequate endogenous supply of leucine (Fultz *et al.*, 1979), it can be expected that the presence of the F'(*leuC*) factor would approximately double the amount of *leuC* polypeptides produced. The *L/M* values for the strain pair carrying the *supQ*98 allele were 0.46 (single *leuC* copy) and 0.49 (two *leuC* copies), while the *L/M* values for the strain pair carrying the *supQ*198 alleles were 0.74 (single *leuC* copy) and 0.77 (two *leuC* copies. Taken together, the data strongly suggest that the main growth limitation of different *leuD–supQ* mutant strains is due to the limiting amount of the *newD* polypeptide that is produced or that becomes available. Strains are currently under construction that carry a functional *newD* gene on a high-copy-number plasmid; if such strains produce a significantly increased amount of the *newD* polypeptide, it can be expected that the amount of the *leuC* polypeptide might become the limiting factor.

### 6.4. Direction of Transcription in the *supQ newD* Region

For the following, the orientation usually presented of the *proB proA ataA supQ newD* region shall be called left to right in this order (see Fig. 3). It can be excluded that the *supQ newD* genes are transcribed as one transcriptional unit left to right, i.e., with a promoter located between *supQ* and *ataA*, since most *supQ* deletions would eliminate this promoter and since strong arguments have been presented that the *supQ* mutations themselves do not represent new promoters.

A potential transcription of the *newD* gene left to right would thus require existence of a separate promoter between *supQ* and *newD*. Although many cases exist where genes coding for enzymes of one specific biochemical pathway are located at different positions on the chromosomes and are transcribed from separate promoters (e.g., *arg* or *pyr* genes), genes located adjacent to each other and coding for subunits of a complex (enzyme) generally have been found to be transcribed as one transcriptional unit [e.g., *pyrBI* (Pauza *et al.*, 1982)]. Furthermore, the fact that the amount of free, uncomplexed *newD* polypeptide is strictly dependent on the nature of the *supQ* mutation (i.e., amount of *supQ* polypeptide) suggests that the expression of the *supQ* and *newD* genes is under a common control. If *supQ* and *newD* are not cotranscribed, it would require that both genes are under a common regulation, and that such a regulatory region is located between *supQ* and *newD*. The size of this putative promoter–regulatory region would have to be relatively small, since the mapping data indicate that *newD* and *supQ* mutations are very closely linked.

Although this latter possibility of a separate promoter for the *newD* gene and a left to right transcription cannot be excluded, the remaining possibility of a right to left transcription is favored. This case, in which *newD* and *supQ* would be most likely transcribed as one transcriptional unit, is compatible with all data so far obtained.

### 6.5. What Is the Original Function of *supQ newD*?

This question is not yet resolved and presents particular difficulties, since a complete deletion of the *supQ newD* genes (see above) does not result in any identifiable phenotypic difference from the wild-type strain. However, some important conclusions can be drawn that would eliminate certain possibilities for the functions of the *supQ newD* genes. First of all, *supQ* and *newD* are not normally involved in any essential step of leucine biosynthesis, since no leucine-deficient mutant has ever been isolated that was not located in the leucine operon itself.

#### 6.5.1. The supQ newD Genes Are Not Translocated (Duplicated) leuC leuD Genes

A number of arguments will be presented to support the above conclusion. If *supQ newD* were translocated, dormant *leuC leuD* genes, then obviously *supQ* would be the *leuC* gene and *newD* the *leuD* gene equivalent. Furthermore, since *leuC* and *leuD* mutations (in the leucine operon) are readily isolated, the *supQ* (= *leuC3*) *and newD* (= *leuD*) genes must not be functional or not be expressed.

In the latter case one would assume that the *supQ* (= *leuC3*) *and newD* (= *leuD*) genes are structurally unaltered, i.e., that the *supQ* and *newD* polypeptides are functionally fully equivalent to the *leuC* and *leuD* polypeptides, but that these genes are not normally transcribed. One would more or less expect this, since the translocated genes would be separated from their normal promoter. In this case, however, *supQ* mutations that bring about expression and/or availability of the *newD* (= *leuD*) polypeptide would have to be generating a new promoter, a possibility that was eliminated (see Section 6.3.1.) as extremely unlikely, if not impossible.

Assuming that the *supQ* (= *leuC*) and *newD* (= *leuD*) genes had been translocated into the vicinity of an active promoter from which they could be transcribed, one would need to assume further that the *supQ* (= *leuC3*) and *newD* (= *leuD*) polypeptides are not functional, either individually or as a complex. This is easily possible, even if not to be expected, since in the obvious absence of a selection pressure to maintain function, mutations could accumulate in the translocated genes. The number of mutations would be related to the time since the translocation occurred; the distribution of the mutations in the *supQ* (=*leuC*) and *newD* (=*leuD*) genes and the nature of the mutations, however, would have to follow a rather specific pattern as outlined below.

1. The *supQ* (=*leuC3*) gene would have accumulated a number of severe mutations, or a large number of mutations, such that the *supQ* (=*leuC*) polypeptide is no longer functional. Also, it is not possible that the changes in the *supQ* (=*leuC*) gene are reversible by a single-step mutation since no *leuC* (i.e., leucine operon) mutant has ever been found to revert by mutations mapping in the *supQ* region (in direct similarity to the *leuD* revertants that are the subject of this chapter). These conditions are easily possible.
2. The *newD* 3(=*leuD3*) gene would have accumulated only a few mutations, such that the *newD* (=*leuD*) polypeptide is still functional. However, it is necessary to assume that some alteration has occurred, since the *newD* polypeptide is not fully equivalent to the wild-type *leuD* polypeptide, as described in the sections on growth limitations. These conditions are also easily possible, especially since the *leuD* wild-type gene is approximately one-half the size of the *leuC* gene (based on a comparison of the molecular weights of the *leuD* and *leuC* polypeptides) and would thus represent a smaller target for mutations.
3. In particular, the *newD* (=*leuD*) gene would have to contain a mutation that renders the *newD* (=*leuD*) polypeptide significantly less active in complexing with the *leuC* (wild-type) polypeptide, since it fails to compete with *leuD* missense polypeptides produced in *leuD* missense mutants of the leucine operons. Again this condition is easily possible.
4. The following considerations are very important. The *newD* (=*leuD*) polypeptide, despite the mutational alteration that significantly lowered its ability to complex in the wild-type *leuC* polypeptide, is still capable of forming a high-affinity, tight complex with the *supQ* (=*leuC*) polypeptide. This is based on the fact that apparently the *supQ* (=*leuC*) polypeptide must be eliminated (in the *supQ* mutants) to free the *newD* (=*leuD*) polypeptide for complexing and restoration of a Leu$^+$ phenotype. It is therefore necessary to assume that both *supQ* (=*leuC*) and newD (=leuD) have accumulated complementary mutations that affected the ability to

form a complex, such that the mutationally altered *supQ* (=*leuC*) and *newD* (=*leuD*) polypeptides maintained the ability to form high-affinity, tight complexes. In contrast, at least the *newD* (=*leuD*) polypeptide lost the affinity to the original wild-type *leuC* polypeptide. The occurrence of such compensating mutations is only understandable if there exists a selection to maintain a function. However, the transposed, duplicated *supQ* (=*leuC*)–newD (=*leuD*) genes are, by definition, unnecessary and furthermore the *supQ* (=*leuC*) gene did accumulate mutations that apparently interfered with its function as a *leuC* gene. Unless these mutations occurred under conditions of selective pressure to maintain a certain (unknown) *supQ newD* expression that no longer exists, there seems to be overwhelming indirect evidence to exclude the possibility that *supQ newD* are translocated *leuC leuD* genes.

5. Finally, one more argument against the translocation theory can be made: Assuming that *supQ newD* represent the *leuC leuD* genes, the direction of transcription of *supQ newD* would need to be identical to that of the leucine operon, i.e., from the direction of the *leuC* to the *leuD* gene. This means a transcription of *supQ newD* from left to right, which was already deduced to be impossible (see above). The remaining possibility of a separate promoter between *supQ* and *newD* again disagrees with the similarity to the leucine operon, where no such promoter (even of low efficiency) has has been found.

It therefore seems reasonable to assume that the *supQ newD* genes are not translocated *leuC leuD* genes, in agreement with the failure to detect any recombination between *leuD* and *newD* despite very extensive tests.

6.5.2. Speculations and Wild Guesses

Although no definite answers are available at this time, it seems obvious that at least a functional similarity exists between the *leuD* and *newD* polypeptides. It is difficult to present more arguments, since no separate properties can be assigned to the two individual polypeptides (*leuC* and *leuD* gene products) of the wild-type isomerase. It is thus not necessary to assume that the *supQ newD* system has even a remote similarity in its reactions to those of the isopropylmalate isomerase. For instance, it is possible that the major enzymatic activitiy and/or substrate specificity is contained in the *leuC* and *supQ* polypeptides and that *leuD* and *newD* polypeptides have a similar activator property to activate the respective complexes.

It was noted already in 1961 (Gross et al., 1962; Jungwirth et al., 1961) that there is a close similarity in the biochemistry of the reaction catalyzed by the isopropylmalate isomerase to that of the isomerization of citrate and isocitrate (aconitase reaction) (see Fig. 4). The actual biochemical reactions are identical, with the substrates differing in only one respect: α-isopropylmalate contains two methyl groups in the position of one of the carboxyl groups in citrate. The only known difference in the reaction mechanisms of the two enzymes is that the aconitase requires a ferrous ($Fe^{2+}$) ion as cofactor (Morrison, 1954), whereas the isopropylmalate isomerase has no known cofactor (this work; Bigelis and Umbarger, 1976; Gross et al., 1963). It should be noted that all information concerning the aconitase reaction is based on studies of this enzyme from eukaryotic species, and that so far no aconitase mutants have been identified in *S. typhimurium* or *E. coli*. The *supQ newD* genes, however, certainly do not represent the (only) genes for the bacterial aconitase, since a strain with a deletion of this region [e.g., Δ(*gpt proBA supQ newD*)47] has no obvious requirements (except proline), and contains a functional aconitase, as verified by *in vitro* enzyme assays. It remains possible that the *supQ newD* genes are in fact related to the aconitase reaction and might perhaps represent duplicate genes or genes for isozyme species. The lack of identified aconitase mutants in *S. typhimurium* and

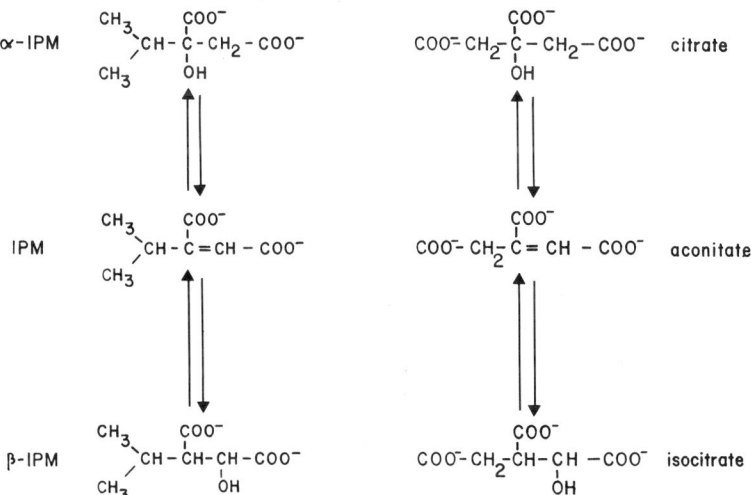

Figure 4. A comparison of the chemical reactions in the leucine biosynthetic pathway mediated by the isopropylmalate (IPM) isomerase with the reactions in the tricarboxylic acid cycle that are mediated by the aconitase; IPM*, isopropylmaleate or dimethylcitraconate.

E. coli, although mutants in genes for almost all other enzymes of the tricarboxylic acid cycle have been isolated, seems to support the above possibility. These questions will be readdressed and, one hopes, answered when aconitase mutants have been isolated and characterized in *S. typhimurium*

A frequently raised question in connection with a discussion of the *supQ newD* function is that of the properties of a putative *supQ–leuD* polypeptide complex. Although the answer is not known, this complex certainly would not have isopropylmalate isomerase activity. Since in order to have isopropylmalate isomerase activity the *supQ* polypeptide would need to be able to functionally replace the *leuC* polypeptide in complexing with the *leuD* polypeptides, and since *newD* was shown to be functionally similar to *leuD* and *supQ* and *newD* polypeptides form an efficient complex, this would mean that the wild-type *supQ newD* genes would need to code for a functional isopropylmalate isomerase, which is clearly not the case. If the *supQ–leuD* polypeptide complex is formed at all, one would suspect that it might have an activity that is similar to that of the wild-type *supQ–newD* polypeptide complex, whatever that may be. Any further analysis needs to be postponed until the wild-type function of *supQ newD* can be identified, especially since until that time no selections are possible to obtain or identify strains carrying a functional *supQ*, but inactive *newD* gene.

### 6.6. Are *supQ newD* Genes Existing or Functional in *Escherichia coli*?

While all strains so far discussed were *S. typhimurium* LT2 derivatives, it was shown that the *supQ newD* genes also exist in *S. typhimurium* LT7 strain by using *leuD* deletion strains with the genetic background of LT7 and selecting for Leu$^+$ revertants. However, when similar experiments were performed with *leuD* deletion strains of *E. coli* B/r, using 2-aminopurine, nitrosoguanidine, and nitrous acid mutagenesis, no Leu$^+$ revertants could be obtained. Based on the extensiveness of these experiments with *E. coli* and in comparison to parallel experiments with *S. typhimurium*, one can conclude that the *supQ* mutation frequency in *E. coli* (if not zero) is at least $5 \times 10^3$-fold lower than that in *S. typhimurium* (Fultz *et al.*, 1979). Therefore, either *E. coli* does not have genes similar to *supQ* and *newD*, or, if the potential for *leuD* suppression does exist, its activation is a very rare event. It might require either multiple mutations or perhaps the inactivation of a gene essential for growth. If multiple mutations are required in genes that are contiguous, then mutagenesis with nitrous acid or nitrosoguanidine could have resulted in Leu$^+$ colonies, since nitrous acid can cause deletions (Schwartz and Beckwith, 1969) and nitrosoguanidine is known to cause multiple neighboring mutations (Guerola et al., 1971).

If the process of generating an *E. coli newD* product inactivates an essential gene, then selection for a mutation analogous to *supQ* might be possible in an *S. typhimurium* background. Because of considerable homology between *S. typhimurium* and *E. coli* (Middleton, 1971) with respect to the map position of genes, one would expect that putative *supQ newD* genes in *E. coli* would map in the region between *proAB* and *proC*. Therefore, two F' factors were utilized, containing a segment of the *E. coli* genome including *proAB* an *proC*, the region where *supQ* and *newD* map in *S. typhimurium*. The F'*proABlac* and F'*lacproC* were transferred into *S. typhimurium* strains carrying the $\Delta(leuD\ ara)798$ and $\Delta(gpt\ proAB\ newD)47$ deletions and both of the resulting hybrids were mutagenized with nitrosoguanidine. No Leu$^+$ colonies arose (Kohlhaw et al., 1969), indicating that *E. coli* does not have the potential for *supQ* mutations in these two regions of the genome, or that both regions are required simultaneously.

A remaining possibility, that the expression of the *newD* gene in *E. coli* is prevented due to the presence of other genes, was eliminated by the following experiments. The *newD* gene-carrying F' factor F'*prolac*P22d*supQ*394*newD*$^+$ was transferred into *E. coli leuD* deletion strains, resulting in Leu$^+$ hybrid strains. This proved that the *newD* gene was expressed in *E. coli* and that the *S. typhimurium newD* polypeptide can complex with the *E. coli leuC* polypeptide. While *S. typhimurium* strains carrying a *leuD* deletion and the *supQ*394 allele show some growth limitations ($L/M = 0.90$), two *E. coli leuD* deletion mutants containing the F'*prolac*P22d*supQ*394*newD*$^+$ have the same doubling times in minimal and leucine-supplemented media. It is therefore concluded that the *newD* gene product is not limiting the growth of these strains. This could be due to a higher level of expression of the *newD* gene in *E. coli* than in *S. typhimurium*, a greater affinity of the *E. coli leuC* than the *S. typhimurium leuC* polypeptide for the *newD* gene product, or a greater affinity of the *E. coli leuC–newD* complex for the enzyme substrates $\alpha$- and $\beta$-isopropylmalate.

At first glance, the apparent absence of genes in *E. coli* equivalent to *supQ newD* might support the notion that these genes are actually dormant genes in *S. typhimurium* and perhaps, despite all previous counterarguments, represent translocated leucine genes. However, a comparison of the corresponding genome regions of *E. coli* and *S. typhimurium* (see Fig. 5) shows that this particular region exhibits major differences between *E. coli* and *S. typhimurium*, including the well-known absence of the *lac* operon in *S. typhimurium*. It is thus quite possible that genes analogous to *supQ* and *newD* are absent in *E. coli*. It should be noted

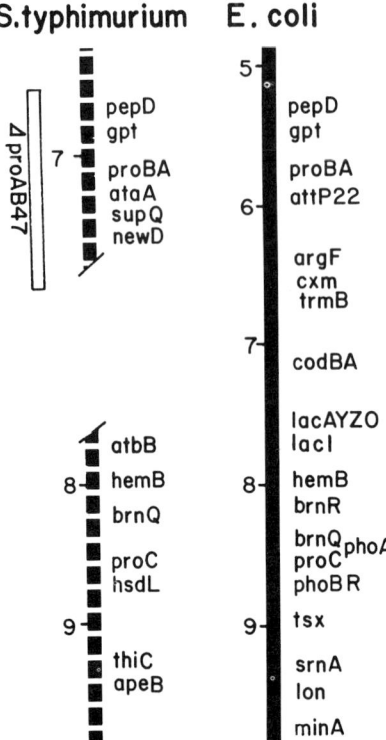

Figure 5. Comparison of a section of the *S. typhimurium* and *E. coli* linkage maps. The continuity of the *S. typhimurium* chromosome is interrupted for better alignment with the *E. coli* chromosome.

that the *leuD* deletion strains used in these studies are *E. coli* B/r derivatives, whereas the F′*prolac* is derived from *E. coli* K12 and that B/r strains, as well as *S. typhimurium* strains, lack the *argF* gene (Riley and Anilionis, 1978), which lies adjacent to *att* P22 between the *proAB* and *proC* genes on the K12 chromosome. The *argF* gene represents a duplicate gene for the ornithine carbamoyltransferase.

## 7. Biochemical Characterization of the *leuC–newD* Isopropylmalate Isomerase

### 7.1. *In Vitro* Specific Activity of the Hybrid *leuC–newD* Isopropylmalate Isomerase

*In vitro* specific activity of the mutant isopropylmalate isomerase was expected to be significantly lower than the specific activity of the wild-type isopropylmalate isomerase, since the mutant strains show a growth limitation caused by the inefficiency of the mutant isomerase (either in amount or in catalytic efficiency). Also, the mutant isopropylmalate isomerase, a complex of the *leuC* protein and the newly recruited *newD* protein, was expected to have a lower affinity than the wild-type enzyme for its substrates. Both assumptions were shown to be true, for only very small amounts of isomerase activity were detected when the mutant enzymes was assayed under the same conditions used for the wild-type enzyme. However, when increased levels of α-isopropylmalate were used in assaying crude extracts of *leuD–supQ* mutants, isopropylmalate isomerase activity was easily demonstrated. The specific activity levels obtained were approximately the same as those of the wild-type strain; this raises the question of why the mutant strains show a growth limitation (see Section 7.2). Assays using crude extracts of strains with a wild-type isopropylamalate isomerase contain α-isopropylmalate with a concentration of $5 \times 10^{-3}$ M, sufficient for substrate-excess conditions, since the $K_m$ for α-isopropylmalate of the wild-type enzyme is $3 \times 10^{-4}$ M. Preliminary data indicate that the *S. typhimurium leuC–newD* mutant isomerase has an approximately 100-fold higher $K_m$ ($\sim 3 \times 10^{-2}$ M); assays using extracts of these mutant strains are run with an α-isopropylmalate concentration of $1.5 \times 10^{-1}$ M, which should represent a condition of substrate excess. The enzyme activity observed when using the high concentration of α-isopropylmalate was shown to be specific for the isopropylmalate isomerase because no activity was detected when a crude extract of a *leuD* mutant strain was assayed with α-isopropylmalate at high concentration.

## 7.2. Mutant Isopropylmalate Isomerase Activity Is Limited by the Endogenous Concentration of α-Isopropylmalate

Based on the above data, it is concluded that the growth limitations in *leuD–supQ* mutant strains are not primarily due to a low amount of active (*leuC–newD*) isopropylmalate isomerase but are due to the low substrate affinity of the mutant isomerase to α-isopropylmalate and the fact that the intracellular concentration of α-isopropylmalate cannot be increased to the required high level of approximately 100 mM. The reason for the production of only low concentrations of α-isopropylmalate despite a limitation in leucine biosynthesis lies in the properties of the α-isopropylmalate synthetase. This enzyme, condensing α-ketoisovalerate and acetyl-CoA to α-isopropylmalate, is the first specific enzyme in the leucine biosynthesis pathway and as such it is feedback-inhibited by leucine (Calvo and Calvo, 1967). Although the growth rate of the *leuD–supQ* mutant strains is limited by the endogenously generated level of leucine, a certain internal concentration of leucine is maintained that is intimately related to the $K_m$ of the leucyl-tRNA synthetase for leucine. This $K_m$ value is in the range of $10^{-6}$ M, the same order of magnitude as the $K_i$ value of the α-isopropylmalate synthetase enzyme for leucine, thus preventing any increase in production of α-isopropylmalate. J. Calvo had previously isolated and characterized a mutation, *leuA*2010, in the *leuA* gene that renders the α-isopropylmalate synthetase (*leuA* gene product) insensitive to feedback inhibition by leucine without affecting its enzymatic activity (Calvo and Calvo, 1967). When a strain was constructed combining this mutation, *leuA*2010, with the mutation *flr*19 described earlier, which causes constitutive expression of the leucine operon, the unbridled leucine biosynthesis results in conversion of more than 50% of the glucose into leucine, which is excreted into the medium (Calvo and Calvo, 1967). Strains were constructed that combine the *leuA*2010 and *flr*19 mutations with a mutation in the *leuCD* genes or in the *leuB* gene, respectively. These strains accumulate and excrete large amounts of α-isopropylmalate, and α- and β-isopropylmalate, respectively, and were used routinely for isolation and purification of these leucine precursors, which are not commercially available (Fultz *et al.*, 1980). Strains with an internal overproduction of α-isopropylmalate provide a possibility to test the notion that the major limiting factor in *leuD–supQ* mutant strains is the higher $K_m$ value of the hybrid mutant enzyme for its substrate α-isopropylmalate. Two strains were constructed carrying the *flr*19, *leuA*2010, and Δ(*leuD–ara*)798 mutations, and in addition a *supQ* mutation, the relatively efficient *supQ*394 allele, or the inefficient *supQ*98 allele. The data of Table II, showing an L/M value of 1.0 for both strains carrying the *flr*19 and *leuA*2010 mutations, strongly support the above notion. Additional support comes from the observation that the strain JK552 (see Table II) excretes leucine, as dem-

Table II
Growth-Rate Dependence of *supQ newD* Mutant Strains in the Presence or Absence of the *flr*19 and *leuA*2010 Mutations, Which Cause Overproduction of α-Isopropylmalate

| Strain | Relevant genotype | Doubling time[a] | | L/M |
|---|---|---|---|---|
| | | L | M | |
| JK1394 | Δ(*leuD–ara*)798 *supQ*394 | 52 | 58 | 0.9 |
| JK539 | Δ(*leuD–ara*)798 *supQ*394 *flr*19 *leuA*2010 | 95 | 92 | 1.0 |
| JK18 | Δ(*leuD–ara*)798 *supQ*98 | 72 | 160 | 0.45 |
| JK552 | Δ(*leuD–ara*)798 *supQ*98 *flr*19 *leuA*2010 | 85 | 83 | 1.0 |

[a] See footnote to Table I.

onstrated by the growth of a *leuB* mutant strain on solid minimal medium in the proximity of growing JK552 bacteria.

### 7.3. Growth Limitations in Strains with an Unbridled Leucine Biosynthesis Pathway

The data of Table II indicate that strains carrying the *flr*19 and *leuA*2010 mutations, although showing an *L/M* value of 1.0, have significantly increased doubling times compared to those of the isogenic strains that do not carry the two regulatory mutations. This longer doubling time is characteristic of all strains that carry both the *flr*19 and *leuA*2010 mutations (Fultz *et al.*, 1980) and is due to the completely unbridled action of one biosynthetic pathway that utilizes a major part of the provided carbon source. An additional explanation is given by the fact that α-ketoisovalerate is not only the substrate for α-isopropylmalate synthetase, but is also the immediate precursor of valine, as well as a precursor of pantothenic acid. The $K_m$ of ketopantoate hydroxymethyltransferase to α-ketoisovalerate is approximately $1.1 \times 10^{-3}$ M (Powers and Snell, 1976), the $K_m$ of transaminase B for α-ketoisovalerate is $2 \times 10^{-3}$ M (F. B. Armstrong, personal communication), and the $K_m$ of α-isopropylmalate synthetase for α-ketoisovalerate is $6 \times 10^{-5}$ M (Kohlhaw *et al.*, 1969). Therefore, in these strains, in which the leucine operon is not regulated, α-ketoisovalerate is primarily used in leucine biosynthesis, and as a result primarily a pantothenic acid and partly a valine limitation are generated. Thus, a significantly shorter doubling time is obtained when isoleucine, valine, and pantothenic acid are added to the media. The apparent requirement for isoleucine is probably not directly related to the utilization of α-ketoisovalerate, but might be an outcome of feedback inhibition of the first enzyme common to the isoleucine and valine biosynthetic pathways by the large amounts of valine (Calhoun and Hatfield, 1975; O'Neill and Freundlich, 1972). It should be pointed out that the *L/M* value is strictly a characteristic of the strain's capacity to synthesize leucine and was found to be unaltered when the overall doubling time was increased by use of a poor carbon source or decreased by supplementation with amino acids (except leucine), vitamins, and nucleotide precursors.

Although *in vitro* activity of the hybrid *leuC–newD* mutant isopropylmalate isomerase is clearly detectable, it is relatively low and unstable, and lost in the first steps of an ammonium sulfate purification procedure. Purification of this enzyme is therefore postponed until methods can be devised to increase the initial amount or activity of this enzyme. The primary approach to this goal is the cloning of the necessary genes on amplifiable plasmids (see Section 9). When this is accomplished, the question can also be readdressed of whether different *supQ* alleles of different efficiency do de facto result in the availability of the identical *newD* polypeptide, as the proposed model suggests.

## 8. Theoretical Steps in the Evolution of a Complex Enzyme

The described case of generating an active enzyme using an existing polypeptide (*leuC*) by recruitment of a new subunit (*newD*) from a different system can be viewed as perhaps similar to a primary step in the evolution of a complex enzyme (with altered or improved activity). A number of secondary and further mutational changes can be visualized that would lead in steps to the further improvement and development of an enzyme species with very high specific activity.

A first step would increase the expression of the *newD* gene without a change in the *newD* polypeptide itself. It is not known if the *newD* (and *supQ*) gene is under a specific regulation in *S. typhimurium* wild-type strains, what this regulation is, and whether it is still active on the *newD* gene in *supQ* mutant strains. The isolation of *supQ* mutants as Leu$^+$ revertants of *leuD* mutants in a single step, as well as the finding of a gene dosage effect in a merodiploid strain, suggest that the *newD* (and *supQ* gene) might be constitutively expressed, at least under those growth conditions so far tested, although the possibility cannot be excluded that some regulatory (repressing) elements are localized in the region between *supQ* and *ataA* and are deleted in the deletion *supQ* strains. This would certainly agree with the fact that such deletion *supQ* strains are significantly more efficient than point mutations in *supQ*, although there are other possibilities that would explain the inefficiency of *supQ* point mutations. If the *supQ newD* genes represent (duplicate) genes for an essential enzyme, such as, for instance, the aconitase, which is needed under all growth conditions, it would of course be expected that these genes are expressed at all times. While this does not imply that these genes are not regulated in some fashion, it is probably impossible to detect this regulation without prior identification of the actual nature of the *supQ newD* genes.

Constitutive expression of a particular gene has been frequently identified as the first step in the activation of a new or altered biochemical pathway, e.g., for the utilization of an alternate or novel carbon source (Hegeman and Rosenberg, 1970; Jensen, 1976; Rigby *et al.*, 1974; Riley and Anilionis, 1978; Wilson *et al.*, 1977).

Another step in the development of a more efficient system might be a change that would bring the *newD* gene under joint regulation with the leucine operon *leuDABC* either at the original location of *newD*, at a new, different site, or perhaps as a part of the leucine operon. In this context it is interesting to note that a translocation of a mutated *supQ* allele from the proline region to the leucine region of the chromosome has been observed. The *supQ*227 allele, originally isolated in a Δ(*leuD–ara*)798 strain, was mapped in the "normal" *supQ newD* location, showing approximately 45% cotransduction with the *proA* gene; this indicates that the *supQ*227 mutation was either a relatively small deletion or possibly a point mutation. This strain was then used as donor in a P22-mediated transduction to select Leu$^+$ derivatives of the Δ(*leuD–ara*)700 strain. Only one such Leu$^+$ transductant was analyzed in detail: the strain still contained the original deletion Δ(*leuD–ara*)700; its Leu$^+$ phenotype was due to a segment inserted near the leucine operon (showing 79% cotransduction) supposedly carrying the functional *newD* gene and perhaps the mutated *supQ*227 allele from the parental strain. The strain would mostly likely be merodiploid for the *newD* gene, carrying a wild-type *supQ*$^+$ *newD*$^+$ region in its normal location on the chromosome. It is not known whether the translocation of the *newD* gene also encompasses its "normal" promoter, or whether the *newD* gene is now possibly transcribed from the leucine operon promoter. It would be interesting to see whether, through additional deletions, this *newD* gene could be brought into direct contact with the leucine operon, essentially becoming a part of this operon. An additional important observation should be noted: the original Δ(*leuD–ara*)798 *supQ*227 strain displayed a characteristic temperature sensitivity, showing no growth at 40°C in minimal medium and significantly slower growth at 37°C in minimal than in leucine-supplemented medium; the Δ(*leuD–ara*)700 *supQ*227 derivative, however, had completely lost all temperature sensitivity. Since only one such clone was analyzed, one cannot deduce whether these changes are due to the new location of the translocated *newD* gene, to a possibly different expression, or to the possible occurrence of an additional mutation in the isolated transductant.

Further, additional mutations could occur and be selected, which, cumulatively, would increase the specific activity of the *leuC–newD* isopropylmalate isomerase. One could anticipate that mutations in *leuC* as well as in *newD* could increase the affinity of these two subunits for complex formation. Other mutations in *leuC*, *newD*, or both genes could result in an increase in the affinity of the enzyme to the substrate α-isopropylmalate; a lowered $K_m$ value would be a major step toward the development of a more efficient system. It might be possible to set up a special selection system for this latter type of mutation: the parental strain should carry a *leuD* deletion, a relatively efficient *supQ* mutation, and one additional mutation "*leu*$^{br}$" that lowers the intracellular concentration of α-isopropylmalate to such an extent that the conversion by the *leuC–newD* isopropylmalate isomerase is too inefficient to result in a Leu$^+$ phenotype (or at least under only extremely slow growth). The required mutation *leu*$^{br}$ would be a mutation of the *leuA* gene that significantly lowers but does not eliminate the α-isopropylmalate synthetase activity. This *leuA*$^{br}$ mutant would need to be isolated first, separate from the other required mutations, and would be characterized as a leucine-bradyotrophic strain, ideally with extremely slow growth in minimal medium. Fast-growing Leu$^+$ revertants of the *leuA*$^{br}$ *leuD supQ* triple mutant strain would be either revertants of the *leuA*$^{br}$ mutation or, hopefully, the type of *leuC* or *newD* mutant that would improve the hybrid isopropylmalate activity. An alternate approach to obtain secondary *leuC* and *newD* mutants would be to grow a *leuD supQ* parental strain for extended periods of time under chemostat conditions, thus selecting for faster growing variants.

Finally, a different method to isolate secondary mutations that improve the efficiency of the *leuC–newD* isopropylmalate isomerase is based on the temperature-sensitive nature of all *supQ* mutants so far isolated. One of the most efficient, the primary *supQ* strain Δ(*leuD–ara*)798 *supQ*394, fails to grow at 42°C and does not seem to acquire spontaneous mutations that would allow growth at 42°C. Following nitrosoguanidine mutagenesis, however, three derivatives of Δ(*leuD–ara*)798 *supQ*394 were isolated that did grow at 42°C, although with lower growth rates than the wild-type strain. The secondard mutations that resulted in the ability to grow at 42°C were shown to be cotransducible with *proAB*; i.e., they mapped in the *supQ newD* region, but it is not known whether they are actually mutations of the *newD* gene itself.

### 9. Characterization of the *newD* (and *supQ*) Gene(s)

Since the limited amounts of the *leuC–newD* isopropylmalate isomerase that were found in *leuD supQ* mutant strains were insufficient for purification and further analysis, it was decided to clone the necessary genes. Using DNA from the *newD*-carrying F′ factor F′ *prolac*P22*supQ*394*newD*$^+$, cleavage with the restriction enzyme *Bam*HI, and the plasmid pBR322 as host vector, the recombinant plasmid pJKS100 was constructed. This plasmid confers only ampicillin resistance; it contains two *Bam*HI restriction fragments (1400 and 2600 daltons) inserted in the *Bam*HI restriction site, located internally in the tetracycline resistance gene. The plasmid pJKS100 transformed into *S. typhimurium* or *E. coli leuD* mutant strains confers ampicillin resistance and leucine prototrophy. The Leu$^+$ phenotype of these transformants includes ''*supQ-*

strain" characteristics, in that these strains are temperature-sensitive on minimal medium, failing to grow or growing significantly slower at 42°C than in the presence of leucine. *In vitro* isopropylmalate isomerase activity was demonstrated in crude extracts of such Leu$^+$ transformant strains; the amount, however, was only slightly higher than that of strains carrying a functional *newD* gene in the genome itself. Using restriction enzyme *Pvu*II cleavage, a reduced size plasmid pJKS103 was constructed that confers the same Leu$^+$ characteristics onto *leuD* transformant strains. This plasmid, pJKS103, contains the 1.4-kb *Bam*HI–*Bam*HI restriction fragment, as well as a 0.8-kb *Bam*HI–*Pvu*II restriction fragment. It could be shown that both fragments are essential for the Leu$^+$ phenotype used; therefore it can be concluded that the *newD* gene is partly in one and partly in the other fragment, or that the *newD* gene and its functional promoter are on different fragments.

It is not known at this time whether the cloned *newD* gene is expressed from its own promoter or whether it is fused to a plasmid promoter. In this latter case, the promoter would need to be the *tet* gene promoter; although this promoter is known to be inefficient, the relatively high copy number of pJKS103 (approximately 35–40 per genome) could possibly compensate and allow production of sufficient amounts of the *newD* gene product.

A detailed restriction analysis of the *newD* gene-carrying fragments is underway and the *newD* gene and its responsible promoter are to be sequenced. The data should allow us to identify the *newD* gene and the size of its protein product, as well as compare it to the *leuD* gene of the leucine operon.

Determinations of the growth rates of *leuD* strains carrying pJKS100 and pJKS103 consistently showed a slightly better growth in the absence of leucine than in its presence. This is a unique observation, which deserves further analysis; it might suggest that the overproduction of the *newD* gene polypeptide is detrimental to the cell. To approach this question, which might provide a valuable step toward the identification of the wild-type function of *newD* (and *supQ*), the *newD* gene should be fused to a system that can be regulated. For instance, fusion of the *newD* gene to the *lac*UV5 promoter on the pMB9 derivative pKB252 or to the *lac* wild-type promoter on pBH20 would allow induction of *newD* protein production with *T*PTG (isopropyl thio-β-D-galactoside).

This fusing of the *newD* gene to an inducible control region would also ensure production of significant amounts of the *newD* protein. To ensure a concomitant increase in the amount of a functional *leuC–newD* isopropylmalate isomerase complex, additional *leuC* gene copies will be provided from a plasmid carrying the *leuABC* genes or through construction of one plasmid in which the *leuC* and *newD* genes are jointly fused to an inducible promoter/operator system.

Finally, for identification of the *supQ* gene, the corresponding region of the chromosome will be cloned on a high-copy-number, amplifiable plasmid. This would not only provide an important step toward identification of the *supQ newD* gene function(s) but also allow a comparison of the *supQ* gene to the *leuC* gene of the leucine operon.

## References

Bigelis, R., and Umbarger, H. E., 1975. Purification of yeast isopropylmalate isomerase. High ionic strength hydrophobic chromatography, *J. Biol. Chem.* **250**:4315–4321.

Bigelis, R., and Umbarger, H. E., 1976, Yeast α-isopropylmalate isomerase. Factors affecting stability and enzyme activity, *J. Biol. Chem.* **251**:3545–3552.

Burns, R. O., Calvo, J., Margolin, P., and Umbarger, H. E., 1966, Expression of the leucine operon, *J. Bacteriol.* **91**:1570–1576.

Calhoun, D. H., and Hatfield, G. W., 1975, Autoregulation of gene expression, *Annu. Rev. Microbiol.* **29**:275–299.

Calvo, R. A., and Calvo, J. M., 1967, Lack of end-product inhibition and repression of leucine synthesis in a strain of *Salmonella typhimurium*, Science **156**:1107–1109.

Crawford, I.P., 1975, Gene rearrangements in the evolution of the tryptophan synthetic pathway, *Bacteriol. Rev.* **39**:87–120.

Friedberg, D., Mikulka T. W., Jones, J., and Calvo, J.M., 1974, *flrB*, a regulatory locus controlling branched-chain amino acid biosynthesis in *Salmonella typhimurium*, *J. Bacteriol.* **118**:942–951.

Fultz, P. N., and Kemper J., 1981, Wild-type isopropylmalate isomerase in *Salmonella typhimurium* is composed of two different subunits, *J. Bacteriol.* **148**:210–219.

Fultz, P. N., Kwoh, D. Y. and Kemper, J., 1979, *Salmonella typhimurium new D* and *Escherichi coli leuC* genes code for a functional isopropylmalate isomerase in *Salmonella typhimurium–Escherichia coli* hybrids, *J. Bacteriol.* **137**:1253–1262.

Fultz, P. N., Choung, K. K. L., and Kemper, J., 1980, Construction and characterization of *Salmonella typhimurium* strains that accumulate and excrete α- and β-isopropylmalate, *J. Bacteriol.* **142**:513–520.

Gemmill, R. M., Wessler S. R., Keller, E. D., and Calvo, J. M., 1979, *leu* operon of *Salmonella typhimurium* is controlled by an attenuation mechanism, *Proc. Natl. Acad. Sci. USA* **76**:4941–4945.

Grieshaber, M., and Bauerle, R., 1972, Structure and evolution of a bifunction enzyme of the tryptophan operon, *Nature New Biol.* **236**:232–235.

Gross, S. R., Jungwirth, C., and Umbarger, E., 1962, Another intermediate in leucine biosynthesis, *Biochem. Biophys. Res. Commun.* **7**:5.

Gross, S. R., Burns, R. O., and Umbarger, H. E., 1963, The biosynthesis of leucine. II. The enzymic isomerization of β-carboxy-β-hydroxyisocaproate and β-hydroxy-β-carboxyisocaproate, *Biochemistry* **2**:1046–1052.

Guerola, N., Imgraham, J. L., and Cerda-Olmeda, E., 1971, Induction of closely linked multiple mutations by nitrosoguanidine, *Nature New Biol.* **230**:122–125.

Hegeman, G. D., and Rosenberg, S. L., 1970, The evolution of bacterial enzyme systems, *Annu. Rev. Microbiol.* **24**:429–462.

Hoppe, L., and Roth, J., 1974, Specialized transducing phages derived from *Salmonella* phage P22, *Genetics* **76**:633–654.

Itikawa, H., and Demerec, M., 1968, *Salmonella typhimurium* proline mutants, *J. Bacteriol.* **95**:1189–1190.

Jensen, R. A., 1976, Enzyme recruitment in evolution of new functions, *Annu. Rev. Microbiol.* **30**:409–425.

Jungwirth, C., Margolin, P., Umbarger E., and Gross S. R., 1961, The initial step in leucine biosynthesis, *Biochem. Biophys. Res. Commun.* **5**:435.

Kemper, J., 1974a, Gene order and co-transduction in the *leu-ara-fol-pyrA* region of the *Salmonella typhimurium* linkage map, *J. Bacteriol.* **117**:94–99.

Kemper, J., 1974b, Evolution of a new gene substituting for the *leuD* gene of *Salmonella typhimurium*: Characterization of *supQ* mutations, *J. Bacteriol.* **119**:937–951.

Kemper, J., 1974c, Evolution of a new gene substituting for the *leuD* gene of *Salmonella typhimurium*: Origin and nature of *supQ* and *newD* mutations, *J. Bacteriol.* **120**:1176–1185.

Kemper, J., and Margolin, P., 1969, Suppression by gene substitution for the *leuD* gene of *Salmonella typhimurium*, *Genetics* **63**:263–279.

Kohlhaw, G., Leary, T. R., and Umbarger, H. E., 1969, α-Isopropylmalate synthetase from *Salmonella typhimurium*: Purification and properties, *J. Biol. Chem.* **244**:2218–2225.

Kwoh, D. Y., and Kemper, J., 1978a, Bacteriophage P22-mediated specialized transduction in *Salmonella typhimurium:* High frequency of aberrant prophage excision, *J. Virol.* **27**:519–534.

Margolin, P., 1963, Genetic fine structure of the leucine operon in *Salmonella*, *Genetics* **48**:441–457.

Middleton, R. D., 1971, The genetic homology of *Salmonella typhimurium* and *Escherichia coli*, *Genetics* **69**:303–315.

Miozzari, G. F., and Yanofsky, C., 1979, Gene fusion during the evolution of the tryptophan operon in Enterobacteriaceae, *Nature* **277**:486–489.

Morrison, J. F., 1954, The activation of aconitase by ferrous ions and reducing agents, *Biochem. J.* **58**:685–692.

O'Neill, P., and Freundlich, M., 1972, Two forms of biosynthetic acetohydroxy acid synthetase in *Salmonella typhimurium*, *Biophys. Res. Commun.* **48**:437–443.

Parson, S. J., and Burns, R. O., 1970, β-Isopropylmalate dehydrogenase (*Salmonella typhimurium*), *Meth. Enzymol.* **17A**:793–799.

Pauza, C. D., Karels, M. J., Navre, M., and Schachman, H. K., 1982, Genes encoding *Escherichia coli* aspartate transcarbamoylase: The *pyrBI* operon. *Proc. Natl. Acad. Sci. USA* **79**:4020–4024.

Powers S. G., and Snell, E. E., 1976, Ketopantoate hydroxymethyltransferase. II. Physical, catalytic, and regulatory properties, *J. Biol. Chem.* **251**:3786–3793.

Reichenbecher, V. E., Jr., and Gross, S. R., 1978, Structural features of normal and complemented forms of the *Neurospora* isopropylmalate isomerase, *J. Bacteriol.* **133**:802–810.

Rigby, P. W. J., Burleigh, B. D., Jr., and Hartley, B S., 1974, Gene duplication in experimental enzyme evolution, *Nature* **251**:200–204.

Riley, M., and Anilionis, A., 1978, Evolution of the bacterial genome, *Annu Rev. Microbiol.* **32**:519–560.

Sanderson, K. E., and Hall, C. A., 1970, F-prime factors of *Salmonella typhimurium* and an inversion between *S. typhimurium* and *Escherichia coli*, *Genetics* **64**:215–228.

Sanderson, K. E., Ross, H., Ziegler L., and Makela, H. P., 1972. $F^+$, Hfr, and F' strains of *Salmonella typhimurium* and *Salmonella abony*, *Bacteriol. Rev.* **36**:608–637.

Schwartz, D. O., and Beckwith, J. R., 1969, Mutagens which cause deletions in *Escherichia coli*, *Genetics* **61**:371–376.

Somers, J. M., Amzallag, A., and Middleton, R. B., 1973, Genetic fine structure of the leucine operon of *Escherichia coli* K-12, *J. Bacteriol.* **113**:1268–1272.

Wessler, S. R., and Calvo, J. M., 1981, Control of *leu* operon expression in *Escherichia coli* by a transcription attenuation mechanism, *J. Mol. Biol.* **149**:579–597.

Wilson, A. C., Carlson, S. S., and White, T. J., 1977, Biochemical evolution, *Annu. Rev. Biochem.* **46**:573–639.

Wu, T. T., Lin, C. C., and Tanaka, S., 1968, Mutants of *Aerobacter aerogenes* capable of utilizing xylitol as a novel carbon, *J. Bacteriol.* **96**:447–456.

Wuesthoff, O. G., and Bauerle, R. H., 1970, Mutations creating internal promoter elements in the tryptophan operon of *Salmonella typhimurium*, *J. Mol. Biol.* **48**:171–196.

Yang, H.-L., and Kessler, D. P., 1974, Genetic analysis of the leucine region in *Escherichia coli* B/r: Gene–enzyme assignments, *J. Bacteriol.* **117**:63–72.

*CHAPTER 10*

# Arrangement and Rearrangement of Bacterial Genomes

## MONICA RILEY

## 1. Introduction

In the preceding chapters, experimental systems have been described in which mechanisms of evolution of bacterial genes by point mutation have been fruitfully studied. In this chapter, the focus will be on rearrangements, some on a small scale, occurring within genes, and others on a large scale, causing gross chromosomal rearrangements of the whole bacterial genome. On the whole, less is known about the molecular events of chromosome rearrangement than is known about point mutations, and even less is known about the pressures for stabilization and destabilization of genome structure. Occurrences of large-scale changes in the genome have no doubt played important roles in evolution. In this chapter progress in understanding the molecular mechanisms of rearrangement and the mechanisms of resistance to change will be summarized with emphasis on experiments that have been done with *Escherichia coli* and its close relatives.

## 2. Chromosomal Rearrangements: Mechanisms of Change

### 2.1. Duplications

An important source of genetic diversity is believed to be gene duplication and subsequent divergence of the replica genes. In a sense,

---

*MONICA RILEY* • Department of Biochemistry, State University of New York at Stony Brook, Stony Brook, New York 11794.

duplication is a type of chromosomal rearrangement: it is a change that affects an entire group of nucleotide pairs in a single event. By comparative study of pairs or groups of genes that are present in the contemporary bacterial genome—genes that appear to have descended from a common ancestor—we can learn some things about mechanisms of divergence. Also, taking another tack, by selecting mutants that undergo gene duplication in the laboratory, we can learn other types of things about mechanisms of gene duplication. Work along both of these lines will be summarized.

Although most of the chromosomal DNA of haploid *E. coli* is unique, present in a single copy, there are quite a few examples in the wild-type genome of pairs or groups of duplicate or highly similar genes. Some of these seem to reflect a need for increased gene dosage, such as the seven sets of genes for ribosomal RNAs, while others either have already diverged in function or are in a position to do so. Many pairs or sets of isozyme genes are probably evolutionarily related, but the genes for many of these enzymes have not yet been isolated and compared in detail.

The two closely similar enzyme complexes chorismate mutase-prephenate dehydrogenase and chorismate mutase-prephenate dehydratase are encoded by adjacent genes *pheA* and *tyrA*, suggesting an ancient tandem duplication event (see Fig. 1). Three aspartokinase isozymes show immunological relatedness (Zakin *et al.*, 1978), but if the three genes *tyrA*, *metL,* and *lysC* arose by tandem duplication of a common ancestor, the duplicate genes were subsequently separated on the *E. coli* genome (Fig. 1). Future studies on such sets of genes could provide information on their evolutionary histories.

In a few cases one of a pair of closely related genes is active and the other is silent, but can be reactivated by one or a few mutations. In these cases, we might be watching evolution in progress. A duplicate gene would be expected to lose its original function as it proceeded toward its new role in the cell. In *E. coli* B/r, there is an active, constitutive *rbs* operon at map position 2, and a silent, inducible operon at map position 83 (Abou-Sabé *et al.*, 1982). In *E. coli* K12, the *rbs* operon at position 83 is active. There is another example: in *E. coli* K12 and W, an active gene for acetylornithine transaminase *argD* at map position 73 is repressed by arginine, while an inactive arginine-inducible gene *argM* is located near map position 88 (Vogel, 1963; Vogel and Bacon, 1966; Riley and Glansdorff, 1983). Perhaps the silent gene in each case is now free to change in the direction of assuming a new function.

More molecular detail is available for other evolutionarily related gene pairs. Nucleotide sequences have been determined for some related genes or parts of genes. Some of this information is gathered and analyzed

ARRANGEMENT AND REARRANGEMENT OF BACTERIAL GENOMES 287

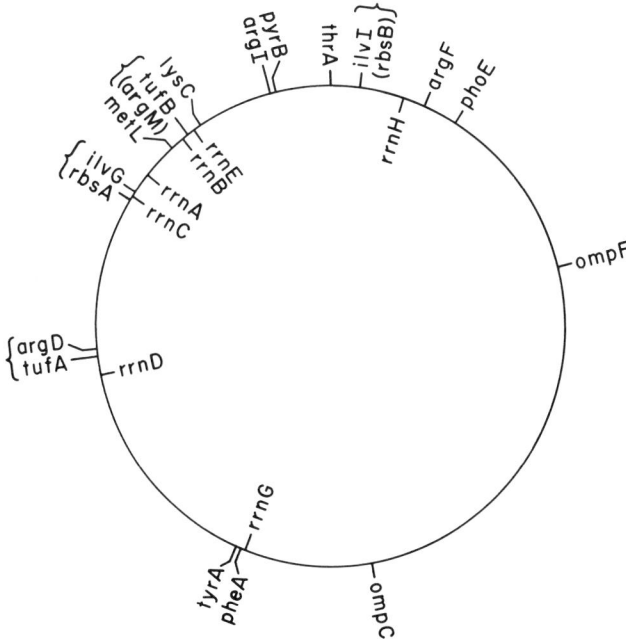

**Figure 1.** Map positions of selected pairs or groups of closely related genes in *E. coli*. The relationships between the genes are discussed in the text. [Map positions are taken from Bachmann and Low (1980), Abou-Sabé *et al.* (1982), and Vogel and Bacon (1966).]

in Table I. Of the pairs or groups of genes contained in Table I, some code for the same function, such as *tufA* and *tufB*, both for the protein elongation factor EF-Tu; *ilvG* and *ilvI*, both for acetohydroxy acid synthases; and *argI* and *argF*, both for ornithine transcarbamylases. The other genes shown in Table I have diverged in function. The *pyrB* gene is for aspartate transcarbamylase and maps adjacent to and is related to the *argI* gene. The *hisJ* gene is for a transport protein that binds histidine; *argT* is for a transport protein that binds arginine, lysine, and ornithine. The *pabA* gene and the first part of the *trpD* gene, called *trp(G)D*, both encode glutamine amido transferases, but with different specificities. The *ompC* and *ompF* genes encode separate outer-membrane passive-diffusion pore proteins. The *phoE* gene is closely related to the *omp* genes, but codes for a porin that has a different specificity and regulation system. In Table I, only data for the *ompC* and *ompF* sequences have been presented, since all three pairwise comparisons give similar results (see Fig. 2 and discussion below). All of the gene pairs presented in Table I are from *E. coli*, except for the *hisJ–argT* pair, which is from *S. typhimurium*.

## Table I
### Comparison of Coding Regions of Pairs of Genes Presumed to Be Related by Duplication[a]

| Genes | Rearrangement | | | Degree of similarity of paired regions[b] | | | |
|---|---|---|---|---|---|---|---|
| | Number of discontinuities | Type | Nucleotide[d] position at start of discontinuity | Nucleotide positions[d] | Percent[c] similarity of nucleotides | Percent similarity of paired regions[c] | | |
| | | | | | | All codons | Identical codons | Synonymous codons |
| tufA–tufB | 0 | — | — | 1–1182 | 99 | 99.7 | 97 | 3 |
| argI–argF | 0 | — | — | 36–110 | 73 | 76 | 28 | 48 |
| argI–pyrB[e] | 8 | 5-bp gap[f] | 39 | 36–110 | 69 | 45 | 25 | 20 |
| | | 2-bp gap | 43 | | | | | |
| | | 5-bp gap | 65 | | | | | |
| | | 4-bp gap | 68 | | | | | |
| | | 1-bp gap | 71 | | | | | |
| | | 2-bp gap | 74 | | | | | |
| | | 2-bp gap | 77 | | | | | |
| | | 10-bp gap | 82 | | | | | |
| hisJ–argT | 0 | — | — | 67–222 | 65 | 59 | 21 | 38 |
| | | | | 223–348 | 83 | 92 | 51 | 41 |
| | | | | 349–501 | 65 | 55 | 33 | 22 |
| | | | | 502–594 | 77 | 90 | 42 | 48 |
| | | | | 595–783 | 70 | 65 | 35 | 30 |
| ompC–ompF | 9 | 30-bp substitution/gap | 539 | 467–538 | 81 | 83 | 54 | 29 |
| | | 45-bp gap | 929 | 554–928 | 72 | 70 | 46 | 24 |
| | | 3-bp gap | 980 | 974–1087 | 69 | 66 | 26 | 40 |
| | | 15-bp substitution/gap | 1088 | 1103–1210 | 64 | 64 | 31 | 33 |
| | | 21-bp substitution/gap | 1211 | — | — | — | — | — |

|  |  |  |  |  |  |  |  |
|---|---|---|---|---|---|---|---|
|  | 9-bp substitution/gap | 1319 | 1217–1318 | 72 | 65 | 47 | 22 |
|  | 3-bp gap | 1434 | 1329–1439 | 77 | 73 | 49 | 24 |
|  | 30-bp substitution/gap | 1440 | 1469–1507 | 82 | 67 | 50 | 17 |
| pabA–trp(G)D | 9-bp gap | 223 | 86–238 | 49 | 35 | 18 | 17 |
|  | 36-bp duplication | 266 | 239–271 | 72 | 73 | 45 | 28 |
|  | — | — | 334–346 | 80 | 75 | 50 | 25 |
|  | 24-bp inversion/gap | 484 | 347–484 | 59 | 54 | 19 | 35 |
|  | 3-bp gap | 539 | 502–550 | 41 | 25 | 13 | 13 |
|  | 37-bp inversion | 551 | 587–652 | 59 | 59 | 36 | 23 |
| ilvI–ilvG | 3-bp gap | 715 | 613–1098 | 58 | 58 | 21 | 37 |
|  | 18-bp gap | 1179 | 1099–1221 | 27 | 17 | 0 | 17 |
|  |  |  | 1222–1522 | 58 | 47 | 26 | 21 |
|  | 30-bp gap | 1606 | 1523–1605 | 32 | 20 | 4 | 16 |
|  |  |  | 1636–2058 | 53 | 45 | 24 | 21 |
|  |  |  | 2059–2301 | 25 | 10 | 5 | 5 |

[a] References to nucleotide sequence data are as follows: *tufA*, Yokota *et al.* (1980); *tufB*, An and Frisen (1980); *argI*, Piette *et al.* (1982); *argF*, Piette *et al.* (1981, 1982), Moore *et al.* (1981); *hisJ* and *argT*, Higgins and Ames (1981); *ilvG*, Lawther *et al.* (1981); *ilvI*, Squires *et al.* (1983); *pyrB*, Roof *et al.* (1982), Turnbough *et al.* (1983); *ompC*, Mizuno *et al.* (1983); *ompF*, Inokuchi *et al.* (1982); *pabA* and *trp(G)D*, Kaplan and Nichols (1983).

[b] The word "paired" in this context means that regions of discontinuity are excluded from the comparison. The identifications of the discontinuities and the paired regions are taken as suggested by the cited authors, with only minor adjustments to improve comparability of data for the various gene pairs.

[c] In all but one case, rounded to the nearest percentage point.

[d] Where the numbering system for the nucleotides differs for the pair of genes being compared, the numbers used in the table are those assigned to the first gene named in the first column.

[e] Incomplete gene sequence beginning at the amino-terminal end.

[f] A gap is either the insertion or deletion of one or more bases.

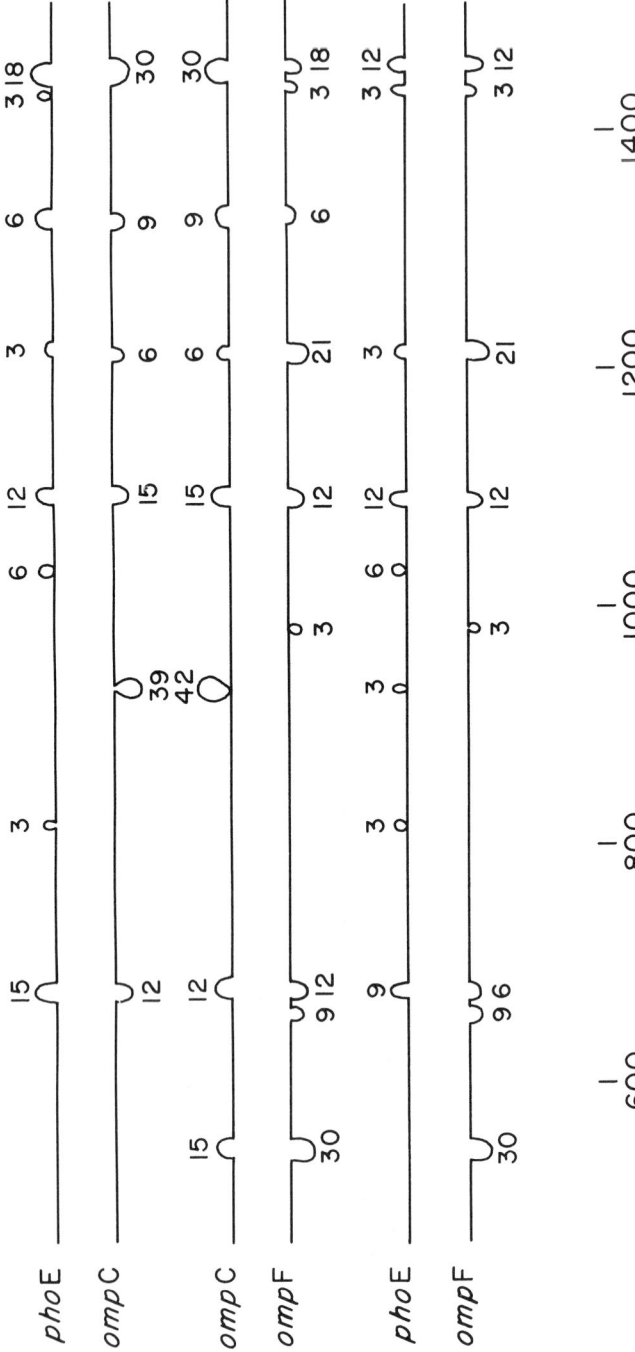

**Figure 2.** Topography of pairwise relationships between the *ompC*, *ompF*, and *phoE* genes. The nucleotide sequences of the coding regions of the genes as determined by Inokuchi *et al.* (1982), Overbeeke *et al.* (1983), and Mizuno *et al.* (1983) were compared. The sequences are indicated as parallel lines in regions that align properly base for base. Excess nucleotides in one gene relative to another, or a group of nucleotides that appear completely unrelated to their counterparts, are shown as loops. The number beside each loop is the number of bases in that loop. The scale at the bottom is the numbering of nucleotides in the *ompC* gene.

Examination of Table I shows that some gene pairs have diverged to a greater extent than others. A gene pair that has a high degree of similarity might have retained similarity because of shared fastidious functional constraints dictated by the nature of the gene products, so that the genes were unable to accommodate mutational change. On the other hand, the similarity of the genes might simply reflect a more recent time of duplication compared to a pair of genes that is less similar. The very nearly identical *tufA–tufB* pair might have duplicated more recently than, say, the less similar *ilvG–ilvI* gene pair.

The mode of divergence has not been the same for all structural genes. In some cases, such as the *tufA–tufB*, *argI–argF*, and *hisJ–argT* gene pairs, the two nucleotide sequences match base for base, differing only by base pair substitutions. In other cases, the genes match base for base in some places, but in other places one gene has more DNA than the other, or there is discontinuity in homology, suggesting that small oligonucleotide rearrangements have occurred in the course of divergence of the gene pair. The small rearrangements have taken the form of insertions or deletions, collectively referred to as gaps, as well as duplications and inversions.

A comparison of the structural genes of *ompC, ompF,* and *phoE* has been made (Mizuno *et al.*, 1983) and is shown diagrammatically in Fig. 2. The relationships between different gene pairs are not identical, but are of the same kind. The data for one pair, *ompC–ompF*, has been chosen as illustrative and is given in greater detail in Table I. Not only do the two genes *ompC* and *ompF* differ by simple base pair substitutions, but there are also nine places within the two structural genes where there seem to have been insertions, deletions, and substitutions of oligonucleotides ranging in size from three to 45 bases. Each rearranged segment is a multiple of three bases and thus the reading frames of neighboring nucleotides were not disturbed by these alterations.

An even more striking example of the effect of DNA rearrangement emerges from analysis of the data for the *argF*, *argI*, and *pyrB* structural genes. The nucleotide sequences of the $NH_2$-terminal portions of these three structural genes have been determined. The *argI* and *argF* genes are both genes for ornithine transcarbamylases. The sequences of the parts of the *argF* and *argI* structural genes that have been determined were aligned and were found to match base for base and to be 73% similar in nucleotide sequence (Piette *et al.*, 1982; and Table I). No rearrangements seem to have taken place during the divergence of these two closely related genes. In sharp contrast, when one aligns and analyzes the sequences of the *pyrB* and *argI* genes, one sees that insertion or deletion of small oligonucleotides has played an important part in the evolutionary divergence of these two genes (Table I and Fig. 3). On average there has been a small rearrangement event about every ten nucleotides. The *pyrB* gene for aspartate transcarbamylase is more distantly related to *argI* than is *argF*. This greater distance is embodied exclusively in the rearrangements that have occurred. If one sets aside the additions and deletions, the remaining common stem, or paired portions, of the *pyrB* and *argI* genes are 69% similar, almost the same degree of similarity found for the *argI* and *argF* genes (Table I). Thus, insertion/deletion, that is, the introduction of gaps, has been the overriding factor in determining the degree of separation in relatedness. In contrast to the loops in the *ompC-ompF-phoE* genes, which are comprised of multiples of three nucleotides, the loops in the *argI/F-pyrB* genes are comprised of odd numbers of nucleotides that have the effect of causing frameshifts in parts of the genes. In this way, a single insertion/deletion event has the ability to bring about major change, even though most of the bases in the genes retain a high degree of homology.

Other types of rearrangements have been observed by analyzing the nucleotide sequences of the *pabA* and *trp(G)D* genes (Kaplan and Nichols, 1983). As summarized in Table I, inversions and duplications were found as well as gaps. The loci of change in both the *pabA-trp(G)D* genes and the *ompC-ompF-phoE* genes were often observed to be bracketed by short repeated sequences (Kaplan and Nichols, 1983; Mizuno *et al.*, 1983).

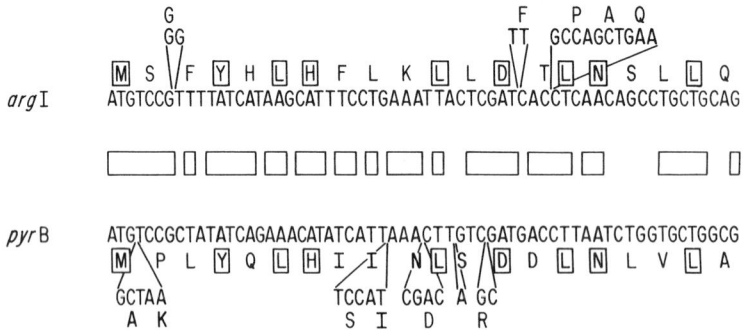

**Figure 3.** Topography of relationships between the *argI* and *pyrB* genes. Amino-proximal amino acid and nucleotide sequences as determined by Gigot *et al.* (1977, 1978), Piette *et al.* (1982), Roof *et al.* (1982), and Turnbough *et al.* (1982) have been analyzed. The nucleotide sequences were aligned so as to maximize match. Open boxes signify identical nucleotides in the aligned portion. Excess nucleotides in one sequence relative to another are shown as loops. Single letter amino acid designations are placed at the center base of a codon when that codon is divided between a loop and the aligned portion. Amino acid designations are boxed when they are identical in both *argI* and *pyrB*.

It seems likely that these repeated sequences were involved in some way in the molecular events that caused the small DNA rearrangements.

Turning away from the rearrangement events, one can look at other aspects of the nucleotide sequences summarized in Table I. The percent similarity of nucleotide sequences is very similar to the percent similarity of amino acid sequences for all of the gene pairs except for the *argI–pyrB* pair, which has been discussed above. This relationship contrasts with the findings for divergence relationships between homogous genes in different enteric bacteria. The *trpA* genes of *E. coli* and *S. typhimurium* differ in 25% of the nucleotides and 15% of amino acids (Nichols and Yanofsky, 1979). The corresponding differences for the *trpB* genes are 15% of the nucleotides and 4% of the amino acids (Crawford *et al.*, 1980). Similar figures have been obtained for variation in nucleotide sequences of the *lpp* genes (for lipoprotein) of *E. coli, Erwinia amylovora,* and *Seratia marcescens* (Nakamura and Inouye, 1980; Yamagata *et al.*, 1981). These contrasting results show that identical codons, as opposed to synonymous codons, are used more often in evolutionarily related genes in the same genome than they are in homologous genes in the genomes of diverged bacterial genera. Presumably duplication and divergence within a genome occurred more recently than the separation of bacterial genera, and thus the equilibrium value for the use of synonymous codons for the same amino acid has not yet been achieved for duplicated genes in the same genome.

One last comment will be made about the data summarized in Table I. The degree of similarity of sequences is not uniform throughout the gene pair. For instance, the *ilvI–ilvG* gene pair shows clearly three regions of relatively high conservation interspersed with three nonconserved regions. For other gene pairs, differences in the degree of conservation are not as dramatic when intervals on the order of 50–150 nucleotides are examined. However, if smaller intervals are compared, heterogeneity in degree of conservation can be observed. Clear examples of such microheterogeneity of conservation have been selected from the sequences of the *ompC–ompF* gene pair and the *pabA–trp(G)D* gene pair, presented in Table II.

In summary, comparison of nucleotide and amino acid sequences of evolutionarily related genes is providing information on mechanisms of evolutionary divergence. Base substitutions have accumulated to different extents in various genes and in various subsections within the genes, presumably reflecting both the severity of functional constraints of the gene product and the amount of time left to go before achieving mutational equilibrium. Other important mechanisms operate in addition to base substitutions. Small rearrangements of oligonucleotides within genes have

## Table II
### Examples of Heterogeneity in Base Pair Similarities within Genes

| | Degree of similarity of paired regions | | |
|---|---|---|---|
| Genes | Nucleotide positions | Percent similarity of nucleotides | Percent similarity of amino acids |
| ompC–ompF | 614–637 | 92 | 100 |
| | 638–682 | 56 | 27 |
| | 683–805 | 76 | 80 |
| | 806–841 | 31 | 24 |
| | 842–928 | 86 | 90 |
| pabA–trp(G)D | 86–127 | 71 | 79 |
| | 128–157 | 30 | 10 |
| | 158–178 | 76 | 43 |
| | 179–238 | 33 | 15 |

played an important role in the process of divergence of duplicate genes. Rearrangements observed in genes include gaps (insertions or deletions), substitutions, inversions, and duplications. Changes of this kind, especially when the oligonucleotide involved is not a multiple of three, can cause widespread changes in the coding characteristics of a gene as a result of a single evolutionary event. It appears that base substitutions can bring about finely tuned adjustments, whereas DNA rearrangements can cause major changes in single leaps.

All of the duplicate genes discussed above occur naturally in the *E. coli* genome. Comparative studies are telling us about the kinds of changes that have occurred during the divergence of these genes in the past, but they do not throw light on the mechanisms of the processes by which the original gene duplications themselves were generated. In order to learn more about these mechanisms, studies have been carried out in several laboratories in which duplications were selected and the mechanisms of their formation were delineated. Duplications arise spontaneously at relatively high rates ($\sim 10^{-4}$) in enterobacterial genomes, but they also revert to wild type at high frequency, presumably by homologous recombination that excises redundant sequences (Anderson and Roth, 1977). Selection either for increased gene dosage or for maintenance of a heterogenetic state stabilizes a duplication.

Some of the duplications that have been isolated after imposing a selective pressure were found to be bordered by repeated sequences that existed in the bacterial genome before the selection was imposed. In other words, existing duplications give rise to other duplications. C. Hill and co-workers found that mutants selected for duplication and heterozygosity of the *glyT* gene fell into several classes that were duplicated for different lengths. The duplications did not have randomly located endpoints, but instead each duplicated segment was found to be bounded by a pair of *rrn* gene sets in direct repeat orientation (Hill *et al.*, 1977). (Duplicate *rrn* gene sets contain genes for ribosomal RNAs and transfer RNAs.) As predicted, the duplication was not stable. When selective pressure was removed, the duplicated region was excised by homologous recombination to generate cytoplasmic nonreplicating circles. When analyzed, the contour length and the position of *rrn* genes in these circles corresponded to the lengths of sections of chromosomal DNA that lay between two *rrn* loci. A mechanism for the formation of duplications of this type has been proposed (Hill *et al.*, 1977) that entails unequal crossover between replicate genes, either between two recently replicated arms of DNA, "sister chromatids," or between the DNAs of two different chromosomes in the same bacterium. Recombination of this type yields one product bearing a deletion between the two replicate genes, and one product with a tandem duplication of the segment that lies between the replicate genes (Fig. 4). *Escherichia coli* is not unique; the *rrn* gene sets have also been shown to be hot spots for unequal recombination in *S. typhimurium* (Lehner and Hill, 1980).

Other repeated sequences besides *rrn* loci have been found to generate tandem duplications of the genes that lie between them. Duplications of 140 kb of *E. coli* DNA occur spontaneously in the *glyS–xyl* region (Capage and Hill, 1979). The recombination has been found to occur by unequal crossover between two duplicate genetic elements that border the 140-kb region. The bordering repeated elements are 3.8kb long, are called *rhs*, for "recombination hot spot," and there are three copies in the *E. coli* K12 genome (Lin *et al.*, 1983).

Using this principle of bordering repeated elements, a system has been engineered in the laboratory in order to allow directed rearrangement in *S. typhimurium*. Copies of the transposon Tn10 were inserted in two

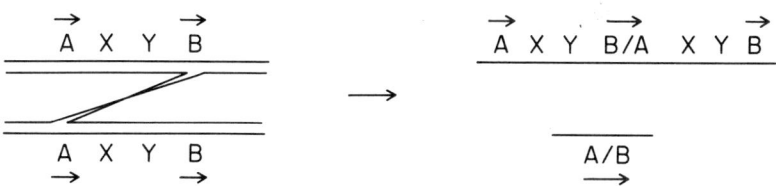

**Figure 4.** Diagram of unequal recombination between multicopy genes, generating both a duplication and a deletion. Here *A* and *B* signify multicopy genes, *X* and *Y* unique genes. The arrows show the relative orientation of the duplicate genes. The two recombining DNA duplexes could be either the arms of a replicating fork from one chromosome or genetically homologous parts of two chromosomes.

locations in the *S. typhimurium* genome, flanking a region to be duplicated (Chumley and Roth, 1980). When the Tn10 elements were in direct repeat configuration and the appropriate selection for gene duplication was applied, tandem duplications were isolated that had two copies of Tn10 bordering the duplication, just as in the *rrn* system.

Although examples in this chapter are being drawn almost exclusively from experiments with enterobacteria, it is impossible not to refer to the interesting mutants of *Bacillus subtilis* that have undergone multiple rearrangement events entailing duplication, inversion, and transposition (Anagnostopolous, 1976). One of these rearranged, partially diploid mutants, *trpE*30, undergoes spontaneous rearrangement between two alternate genetic configurations. It also undergoes additional duplication and rearrangement when it is transformed with wild-type DNA (Schneider and Anagnostopoulos, 1983). Plausible models that explain these events invoke homologous recombination between repeated genetic elements at the borders of the rearranging segments. One of these repeated sequences seems likely to be the DNA of prophage SP, since a strain that was deleted in the region of the prophage in the chromosome did not undergo the duplicative rearrangement after transformation with wild-type DNA that had been seen in the nondeleted strain.

The mechanism of duplication in all of the above examples is most easily understood as unequal recombination between directly repeated genetic elements on either side of the segment that is duplicated. At least one other mechanism seems to be capable of generating duplications. Tandem duplications have been obtained that do not appear to have repeated sequences at the borders of the duplicated segment. Short duplications of the *argE* gene, selected for relief of polarity in an *arg* operon, have been shown to be strictly tandem at the level of resolution of the electron microscope (~200bp) (Charlier *et al.*, 1979). The mutations arose with equal frequencies in *recA* and Rec$^+$ backgrounds, indicating that homologous recombination was not involved in their formation. Perhaps the mechanism of formation of this type of duplication is a type of unequal crossover by illegitimate recombination that requires only a short recognition sequence. Or perhaps the defect leading to duplication is in replication, some sort of reiterative copying. However, the size of the duplicated DNA is large relative to the size of errors known to be made during either replication or repair.

## 2.2. Transpositions

Most duplications that arise by selection in the laboratory seem to be in tandem with the parental gene, although there are examples of separation between the original and the new copy (Jackson and Yanofsky, 1973). However, the duplicate genes found in nature often are separated from each other in the genome, presumed having been transported away from their original, tandem locations to their present, separate locations (see Fig. 1). In this section, instances and mechanisms of transposition will be discussed.

Rearrangement by homologous recombination between repeated genetic elements acounts for one class of transposition events. When duplicated genes are flanked by directly repeated sequences, homologous recombination can cause excision of a circle containing the repeated sequence. Reintegration of the nonreplicating circle into the genome can occur anywhere that the same repeated sequence resides in the genome (Fig. 5).

Transposition by this mechanism, using *rrn* genes as the foci of recombination, has been documented. The *E. coli* genes that lie between the *rrnB* and *rrnE* loci were first duplicated as described above and then mutants were selected in which one set of genes was transposed, making use of the property of the transposed mutants of being genetically more stable, that is, being no longer susceptible to loss by simple excision (Hill and Harnish, 1982). In different instances, transposition was observed to occur by insertion of the *rrnB–rrnE* segment into either the *rrnC, rrnD, rrnG,* or *rrnH* genes, giving mutants with two copies of the duplicated segment, one at the original location, one at the other *rrn* locus. Haploid segregants were isolated that had lost the original gene segment between *rrnB* and *rrnE,* retaining only the transposed genes.

Similar manipulations were carried out using Tn10 as the repeated sequence. The Tn10 was inserted on both sides of the *his* operon in *S. typhimurium.* This strain was used as a donor, transferring the Tn10–*his*–Tn10 DNA to another strain that carried a Tn10 insertion elsewhere, e.g., in the *pyrB* gene (Chumley and Roth, 1980; Schmid and Roth, 1980). Circular intermediates were identified, showing that tranposition resulted from the same mechanism of excision and reinsertion as was observed in the *rrn* system, both excision and reintegration steps taking place by homologous recombination.

It is probable that F' episomes are formed in Hfr bacteria by excision of a circular self-replicating segment of DNA following recombination between two copies of an IS element (Ohtsubo and Hsu, 1978; Timmons *et al.,* 1983), and, in retrospect, homologous recombination between F factor and chromosomal IS elements probably accounts for the transpositions of chromosomal genes observed in Hfr strains many years ago (Berg and Curtiss, 1967). Some transposition events resist analysis in these terms. One particular site-specific integration of an F' into the chromo-

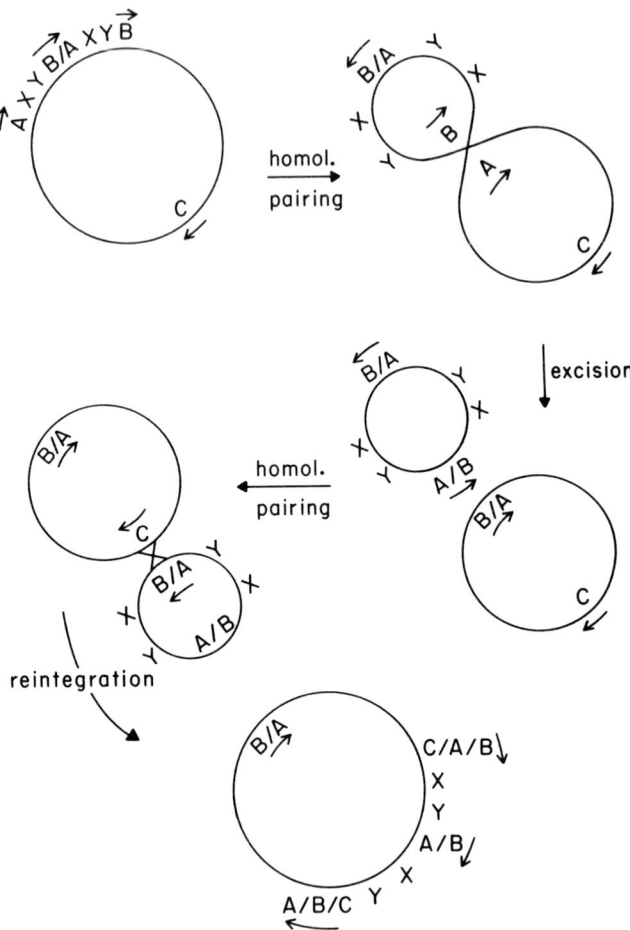

**Figure 5.** Diagram of transposition of duplicated genes. Here $A$, $B$, and $C$ signify multicopy genes, and $X$ and $Y$ are unique genes that were duplicated as in Fig. 2. The arrows show relative orientation of duplicate genes. Hybrid genes are indicated with a slash. In the first two steps, homologous recombination leads to excision of one copy of the duplicated genes $X$ and $Y$ together with one of the bordering multicopy genes. In the next two steps, the circle undergoes homologous recombination with another member of the multicopy family $C$ that is distant from the original genetic location. Recombination leads to integration into the chromosome and consequently transposition to the new location.

some at an abnormal genetic location was shown to be *recA*-dependent and therefore presumably took place by homologous recombination, but no region of homology between the episome and the chromsomal DNA at this location could be detected by Southern hybridization of the appropriate DNA fragments (LaVerne and Ray, 1980).

One of the naturally occurring duplicate genes in *E. coli, argF,* resides in a transposonlike structure in the genome. It is near one end of a 10-kb segment of chromosomal DNA that is flanked by duplicate copies of the IS1 element in direct repeat orientation (Hu and Deonier, 1981; York and Stodolsky, 1981). One is tempted to think of this 10-kb genetic unit as a transposon, one that arose by some kind of duplication of the *argI* region, which was then transposed to the present location by an illegitimate insertion mechanism. However, there is no detectable homology between sequences immediately flanking the *argI* and *argF* genes as judged by heteroduplex analysis (Kikuchi and Gorini, 1975), showing that the 10 kb of DNA between the two IS1 sequences could not have arisen by simple duplication of 10 kb of the DNA from the *argI* region of the chromosome. The duplication and transposition of the *argF* gene seem to have occurred by a complex multistep process.

There are faint clues to the types of events that may have taken place for the transposition of other duplicate genes in *E. coli.* It seems likely that the duplicate genes *ilvG* and *ilvI* at map positions 83 and 2 were separated by the same event that separated the duplicate *rbs* operons, which also map at positions 83 and 2. These genes are shown in brackets in Fig. 1. Likewise, the duplicate *tufA* and *tufB* genes at map positions 73 and 89 are near the closely related gene pair *argD* and *argM,* also bracketed in Fig. 1. Close to both of these map locations are some of the genes for ribosomal RNA, transfer RNA, ribosomal proteins, and subunits of RNA polymerase. Smith *et al.* (1979) suggested that in *E. coli* a duplication, transposition, and divergence process took place during establishment of more than one group of protein synthesis genes in *E. coli,* whereas in *B. subtilis,* which has only one cluster of protein synthesis genes (and only one copy of a *tuf* gene), no such duplication and transposition took place. In support of this idea, at least 16 kb of DNA from these two locations in *E. coli* including the *argD* and *argM* genes seem to be highly homologous as judged by Southern hybridization and positions of restriction sites (Riley and Glansdorff, 1983).

In summary, the mechanism of transposition that occurs by homologous recombination between repeated elements that border the transposed segment has been well documented, but other types of genetic dispersal mechanisms that do not entail homologous recombination require more study for better understanding.

## 2.3. Inversions

Rearrangement by inversion occurs in nature. In most respects, the arrangements of genes in the maps of *E. coli* and *S. typhimurium* are very much alike, but there is a relatively large inversion of about 10% of the genome in *E. coli* relative to *S. typhimurium* (Casse *et al.*, 1973). There are also many examples of small inversions between *E. coli* and *S. typhimurium* (Riley and Anilionis, 1978). Even *E. coli* strains differ among themselves. One laboratory strain of *E. coli* carries an inversion relative to other laboratory strains: genes between the *rrnD* and *rrnE* gene sets are inverted in *E. coli* K12 strain 3110 relative to the *E. coli* K12 strains from which it descended (Hill and Harnish, 1981).

Inversions bordered by *rrn* loci in inverted orientation were isolated in the laboratory following P1 transduction of strains that contained an *rrn*-mediated transposed duplication (Hill and Harnish, 1981). Some of the segregants from the transductants were found to have undergone both gene conversion and inversion between *rrnD* and either *rrnB* or *rrnE*, presumably by homologous recombination between the bordering inverted *rrn* loci (Fig. 6).

Putting selective pressure on the wild-type genome, inversions were selected in *S. typhimurium* by selecting for mutants in which a promoter of the *his* operon was either restored or replaced (Roth and Schmid, 1981). Most His$^+$ mutants that were obtained arose by point mutation, duplication, or transposition. Only a very few were inversions. Since inverted repeat sequences seem unlikely to have been present in the genome at appropriate locations to fulfill the requirements of this mutant selection, probably the inversions that were isolated were formed by a type of rare illegitimate recombination event that does not involve recombination be-

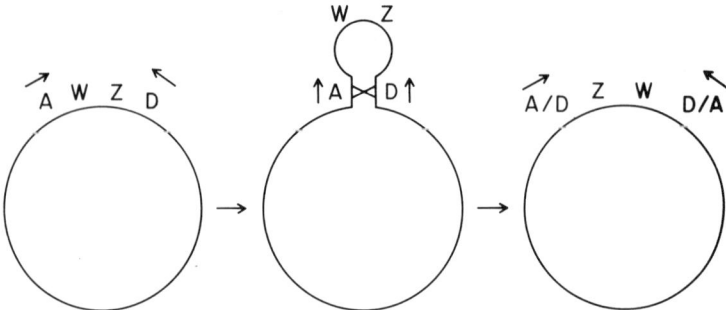

**Figure 6.** Diagram of inversion of genes that are bracketed by duplicate genes. Here *A* and *B* signify multicopy genes, *W* and *Z* are unique genes, and arrows show relative orientation. Homologous recombination between two inverted sequences leads to inversion of the unique genes that lie between them.

tween like sequences. In another experimental system it was also observed that inversions were rare. Inversions of *lac* were rarely found even in a system that provided a sensitive assay for their detection (Konrad, 1977).

There is a type of *recA*-independent, site-specific recombination that causes very specific inversions to occur in the *S. typhimurium* genome and in certain phages. Phase variation in *S. typhimurium* entails inversion of a 995-bp segment of the bacterial genome, an event that changes the regulation of genes specifying two types of flagellar antigens, H1 and H2 (Silverman *et al.*, 1980; Scott and Simon, 1982). The 995-bp segment is bordered by two 14-bp inverted repeat sequences. A protein called Hin is encoded within the 995-bp unique segment. The Hin enzyme recognizes the 14-bp repeated sequence and, to a lesser extent, the surrounding nucleotide sequences. The Hin enzyme causes inversion by catalyzing *recA*-independent recombination between the inverted 14-bp sequences, leading to loss of expression of one flagellar gene and gain of expression of the other. The Hin protein catalyzes related reactions as well: deletion when the 14-bp repeats are in directly repeated configuration and cointegration of two circular DNAs when each circle contains one of the repeated sequences (Scott and Simon, 1982). This site-specific recombination system from *S. typhimurium* bears close resemblance to two phage inversion systems, phage Mu G-inversion, governing tail fiber genes, and the closely similar phage P1 C-inversion (Kamp *et al.*, 1978; Kamp and Kahmann, 1981; Kennedy *et al.*, 1983). The Gin and Cin proteins carry out similar reactions and fulfill each other's functions in complementation tests (Iino and Kutsukake, 1980). These inversion-catalyzing proteins are also similar to the resolvases of the closely related transposon Tn3 (Kostriken *et al.*, 1981). Resolvase, the product of the *tnpR* gene, shares 33% homology in amino acid sequence with the Hin protein (Simon *et al.*, 1980), but the respective target nucleotide sequences are less alike (Kostriken *et al.*, 1981).

A phenomenon phenotypically similar to phase variation occurs in cultures of *Neisseria gonorrhoeae* in that expression of pilus genes is highly variable and colonial morphology oscillates between opaque and translucent forms at high frequency. The basis for the variability is a form of genetic rearrangement. The type of rearrangement and the agent of change has not yet been established (Meyer *et al.*, 1982).

## 2.4. Additions and Deletions

Additions and deletions of all sizes seem to have occurred during the evolution of bacterial DNAs. Massive addition of DNA might have taken place by entire genome doublings (Riley *et al.*, 1978), although there is no experimental evidence to show that this was the case. Massive deletions appear to have occurred in the past, generating mycoplasmas from various bacteria. *Acholeplasma* species have evolved from the genomes of streptococci by a set of large deletions (Neimark and London, 1982).

Such major changes involve many hundreds of kilobases. On a smaller scale, segments of DNA in the range of tens of kilobases can be acquired by the bacterial genome, thereby adding new metabolic functions. Acquisition can either be by duplicating existing DNA, as discussed in Section 2.1, or by physically acquiring new DNA from a source outside of the cell. Plasmids that carry genes for the catabolism of many substances have been identified (Bennett and Richmond, 1978). Bacteria that do not have chromosomal genes for these functions can acquire the function by acquiring the plasmid DNA as a cytoplasmic factor. Another important step may take place. New DNA from a plasmid or some other source may be incorporated into the chromosome, making the new DNA a permanent part of the cell's genome. Both homologous recombination as in sexual exchange, and nonhomologous recombination as in transposon hopping, can mediate such events. There is circumstantial evidence that one of the mechanisms of divergence of enterobacterial genomes has been by the uptake of new DNA into some genomes and not into others. Examples of this kind of history will be considered in turn in this section.

The genes for nitrogen fixation may be such a case. *Klebsiella* species are able to fix nitrogen, whereas other enteric bacteria, such as *E. coli*, are not. *Klebsiella* bacteria may have acquired the *nif* genes as an addition to the chromosome from an outside source at a particular time during evolution of the genome, whereas *E. Coli* did not (Ruvkin and Ausubel, 1980). Likewise, some enteric bacteria have the *rtl* and *atl* operons that enable growth on ribitol and D-arabitol, respectively, whereas others do not. *Klebsiella pneumoniae* and *E. coli* C do have these genes, while *E coli* K12 does not. Instead, *E. coli* K12 has the *gat* operon for catabolism of galactitol, mapping in the same genetic location. It has been shown by transductional exchange and by restriction endonuclease mapping that the two sets of genes, *rtl atl* on the one hand and *gat* on the other, are mutually exclusive and constitute chromosomal alternatives at that site: adjacent genetic material on either side is the same in both *E. coli* K12 and C; only the alternative alleles differ (Link and Reiner, 1983; Woodward and Charles, 1983).

The *rtl atl* region has a transposonlike structure. A 7.2-kb segment including the *rtl* and *atl* genes is flanked by two 1.4-kb inverted repeats of imperfect homology, and thus the *rtl atl* region seems to constitute a vestigial transposable element that was acquired in the evolutionary past by *K. pneumoniae* and *E. coli* C, but not by *E. coli K12* (Link and Reiner,

1982). The 1.4-kb segment is not present in *E. coli* K12. Link and Reiner (1983) propose that ancestral *E. coli* bacteria contained *gat*. After acquisition of the *rtl atl* transposon by the insertion of the transposon near the *gat* gene, deletions occurred starting at one end of the transposon and extending through *gat*, removing *gat* and surrounding genetic material, thus establishing the chromosomal structure of *E. coli* C and *K. pneumoniae* as it appears today. A survey of natural *E. coli* isolates showed that most strains were either Rtl$^+$ or GAT$^+$, but not both (Link and Reiner, 1983; Woodward and Charles, 1983). Where both gene sets were present, they were adjacent, as expected from the model.

There is a suggestion that other addition/deletion events have taken place in enteric genomes. A comparison of the genetic maps of *E. coli* K12 and *S. typhimurium* LT2 suggests that one of the ways in which these bacterial genomes have diverged besides point mutation was by the addition or deletion of a relatively small number of segments of DNA on the order of 25–100 kb in size (Riley and Anilionis, 1978). Certain metabolic capacities that are present in one bacterium but not in the other are derived from genes that map in regions where there appears to be excess genetic material in one chromosome relative to the other. The *S. typhimurium* genes for inositol utilization, for instance, and for tricarboxyllic acid transport, not known in *E. coli*, are located on "loops" of apparent excess DNA in the *S. typhimurium* genome. Likewise, the *Lac* operon and the duplicate *argF* gene of *E. coli*, not known in *S. typhimurium*, are located on an apparent loop that lies between the *proA* and *proC* genes.

One of the apparent loops of excess DNA that was defined by genetic linkage data was shown to have physical reality and did not simply reflect abberations of recombination frequencies. The contour lengths of DNAs from *S. typhimurium* and *E. coli* between the *rrnA* and *rrnB* loci were measured (Lehner and Hill, 1980). As predicted by the genetic maps, the distance between *rrnA* and *rrnB* was longer in *S. typhimurium* (155 kb) than in *E. coli* (126 kb). It seems clear that either the *S. typhimurium* genome has gained DNA in this region or the *E. coli* genome has lost DNA here in the course of evolution.

In this laboratory, some work has been done on delineating the molecular basis of the "*lac* loop" that seems to be present in *E. coli* and absent in *S. typhimurium* (Riley and Anilionis, 1978). Southern hybridizations between *S. typhimurium* chromosomal DNA and 21 kb of *E. coli* DNA including the *lac* operon showed that there is 12–13 kb of *E. coli* DNA that has no detectable homology to *S. typhimurium* DNA, whereas there is demonstrable partial homology between the two bacterial DNAs on either side of the 12- to 13-kb region (Lampel and Riley, 1982).

Based on the information derived from comparing the two genetic maps, one might have expected a large loop to exist in *E. coli* between the *proA* and *proC* genes, extending for about 70 kb, with no homology to *S. typhimurium* DNA. However, the hybridization experiment (Lampel and Riley, 1982) showed that the region is more complex than this. Not just one loop, but at least three loops seem to lie between *proA* and *proC*. The *lac* loop, 12–13 kb, lies near *proC*. A putative transposon including the *argF* gene, discussed above, lies near *proA*, composed of about 10 kb of bacterial DNA flanked by two IS1 sequences of 800 bp each. Since there is a total of about 70 kb of excess DNA between *proA* and *proC*, at least one other loop must exist. On genetic grounds, one might expect the *E. coli codA, B* genes to lie on another loop, since the analogous genes are located elsewhere in the *S. typhimurium* genome. There may also be a loop signifying added DNA in the *S. typhimurium* genome. The suppressor genes *supQ* and *newD* of *S. typhimurium* are located in this region and are not believed to be present in *E. coli* (Kemper, 1974; Fultz *et al.*, 1979). Figure 7 diagrams the present state of knowledge of the loop structure in the *proA–proC* region of *S. typhimurium* and *E. coli*.

What is the source of the *lac* DNA that is present in some enterobacterial genomes, but not in others? Plasmids carrying *lac* determinants have been found in many Gram-negative bacteria. One of these, from a strain of *Yersinia enterocolitica*, was found to carry its *lac* genes within a transposon, Tn951 (Cornelis *et al.*, 1978). The transposon was analyzed in some detail. It was found to have short inverted repeat sequences on either side of a 16.6-kb segment of DNA within which lay 5.6 kb of *lac* genes. The *lac* genes of Tn951 were shown to have homology to the *E.*

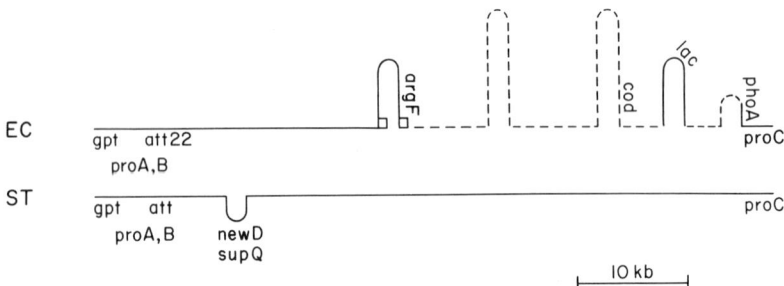

**Figure 7.** Comparison of the *proA–proC* sections of the *E. coli* and *S. typhimurium* genomes. A loop represents a part of one genome that does not appear to have a counterpart in the other genome (see text). Boxes represent IS1 sequences. Dashed lines represent the approximate length of *E. coli* DNA, but indicate lack of knowledge concerning the topography of the relationship to *S. typhimurium* DNA in that region.

*coli lacI, Z,* and *Y* genes but not to the *lacA* gene. The short inverted repeat sequences were determined to be the same as the sequences at the ends of the transposon Tn3 (Cornelis *et al.*, 1981). Part of a transposase gene that was analogous to the *tnpA* gene of Tn3 was found to be present in Tn951, but the gene was defective. Transposition of the Tn951 *lac* transposon could be mediated by complementaion with Tn3, showing that Tn951 is a defective transposon of the Tn3 family.

The *lac* determinants carried on many enterobacterial plasmids are similar to the determinants of the *E. coli lac* operon (Reeve and Braithwaite, 1974; Guiso and Ullman, 1976; Cornelis *et al.*, 1981), except that the lactose acetyltransferase *lacA* gene is not present. What is the mechanism of transmission of plasmid-borne *E. coli*-like *lac* genes? Is a transposon like Tn951 to be found as part of most natural *lac* plasmids? Between its two inverted repeat sequences, Tn951 has 7 kb of DNA on one side of the *lac* genes and 3 kb on the other side. Hybridization experiments showed that these flanking sequences of Tn951 are not present in the several *lac* plasmids that were tested (Cornelis, 1981). Therefore this particular transposon, Tn951, is not present in many *lac* plasmids. Whether *lac* is present as a transposon in other plasmids, and the nature of the source and the relatedness of the *lac* regions in wild-type chromosomes, are yet to be determined.

The generalization that emerges from these studies is that chromosomal genes can be acquired by bacteria by transposition from a visiting plasmid. Both the *rtl atl* and the *lac* genes of *E. coli* seem likely to have been acquired this way, although in both cases more needs to be known before the source of the genes and the precise mechanism of transmission are fully understood.

Clearly, transmissable plasmids and transposons have played a role in the distribution of bacterial genes during evolution. What is uncertain is the relative importance of this mechanism compared to other mechanisms such as point mutation (Reanny, 1976; Bennett and Richmond, 1978). Addition or deletion events may have been less frequent than point mutations, but each event may have brought about a larger change. If the large loops of excess *S. typhimurium* and *E. coli* genes as defined genetically (Riley and Anilionis, 1978) are taken as representing individual addition or deletion events, then there were relatively few (12–13) large-scale addition or deletion events in each genome. However, each loop contains many genes for many functions, so that even if the events were infrequent, the addition of a large segment of DNA, on the order of tens of kilobases, has the potential of adding many new functions in one genetic event. In this way, one giant step can make a set of changes that would otherwise require many base pair changes to achieve. More experimental analysis is needed to find out whether the 12–13 apparent large loops in the *E. coli* and the *S. typhimurium* genomes are indeed single loops or whether closer analysis will show that, as in the case of the *lac* loop, they are composed of a set of smaller loops. Even if the numbers of single addition/deletion events during divergence of *E. coli* and *S. typhimurium* turns out to be as high as 100, the number of such events is far lower than the number of base pair subsitiution events.

On an entirely different scale, there appear to be very small addition, deletion, or substitution events within closely related structural genes. As discussed above, there are loops of only 3–45 bp that have occurred about every 100–200 bp in the *ompC/F* and *phoE* genes, and loops of 2–10 bp in the *argI/F* and *pyrB* genes that have occurred very closely together (Figs. 2 and 3). The impact of the changes in the *ompC/F–phoE* case is relatively small, since the reading frames were not disturbed. The impact of the changes in the *argI/F–pyrB* case was larger because reading frames were altered in some regions.

In summary, additions and deletions of all sizes seem to have occurred during evolution of bacterial DNA: (1) events on the order of the size of a genome, (2) events in the size range 10–100 kb involving scores of genes, and (3) events of only 1–50 bp taking place within a single gene. Significant evolutionary changes seem to have taken place by this mechanism of addition or subtraction.

## 3. Conservation of Global Gene Order: Mechanisms of Stability

### 3.1. Integrity of Taxonomic Groupings

Horizontal transmission of genes from one genetic background to another by, for instance, conjugal plasmids carrying transposons is obviously a powerful tool for effecting dramatic evolutionary changes. But how frequently has this mechanism actually applied in the course of evolution of bacterial DNAs? As information on the degree of relatedness of specific parts of bacterial genomes accululates, we will learn more about how often such events have occurred. Some *E. coli* genes, such as the *trp* operon and the *thyA* and the *tnaA* genes, are related to their counterparts in other enteric genomes to about the same extent as the average of all the DNA of those organisms (Riley and Anilionis, 1980. In that sense the *trp*, *thyA*, and *tnaA* genes are typical genes with respect to the degree to which they have diverged. Other genes, such as the genes of the *nif* operon (Ruvkin and Ausubel, 1980; Mazur *et al.*, 1980) and the *tuf* genes (An and Frisen, 1980; Yokota *et al.*, 1980), are much more highly conserved than are average genes. The question arises whether such genes appear conserved because they have arisen in evolutionarily recent times, or whether they are conserved because functional requirements of the gene product cannot accommodate change. A history of horizontal transmission is suspected if genes are present in some but absent in other members of a closely related taxonomic group of bacteria, such as the *nif*, *rtl atl*, and *lac* genes in enterobacteria. When nucleotide sequences immediately flanking a conserved gene are not conserved to the same extent, it might be that the absence of conservation is a consequence of the absence of functional constraint, or it might be that gene movement by horizontal transmission has created junctions of discontinuity in homology.

Although horizontal transmission may play an important role in effecting relatively large changes instantaneously, there seem to be constraints that have limited the exercise of this mechanism during the evolution of the bacterial genome. If the horizontal exchange of plasmids were unlimited across taxonomic boundaries and if genes were frequently transposed from chromosomes to plasmids and back to chromosomes by illegitimate recombination, then the defined boundaries between distinctive groups of bacteria would not exist. The mechanisms of constraint on the *ad libitum* horizontal exchange of genetic material by conjugal plasmid transmission seem to be twofold: the phenomena of plasmid immunity and of restriction/modification protection. Many bacteria carry resident plasmids, and these residents confer immunity to a set of related plasmids, preventing their establishment. Other plasmids are degraded by restriction endonucleases. If the DNA of the foreign plasmid does not have the appropriate pattern of modified bases, then that DNA will be susceptible to degradation by the host cell's restriction enzymes. These two mechanisms may be sufficient to account for the limitations of horizontal exchange of genetic material between distantly related bacteria and the preservation of taxonomically distinct groups.

## 3.2. Genetic Maps

Although the order of genes on the maps of distantly related bacteria is entirely different, the order of genes on the maps of related bacteria is very much the same. The similarity of map position of most of the genes of *E. coli* K12 and *S. typhimurium* LT2 has been mentioned above. Among the enteric bacteria, the similarity extends to *K. pneumoniae*, and at least part of the *Serratia marcescens* maps, but not to the more distantly related *P. mirabilis*. The genetic maps of members of other groups of bacteria are unlike enterobacteria, but like one another. Among the Actinomycetes, the maps of *Streptomyces coelicolor*, *S. rimosus*, and *Nocardia mediterrani* are similar (Schupp *et al.*, 1975). The maps of the relatives *Rhizobium meliloti*, *R. leguminosarum*, and *Agrobacterium tumefaciens* are similar (Hooykaas *et al.*, 1982). Similarly, the general configuration of the maps of *Bacillus* species are alike and the maps of *Pseudomonas* species are alike. However, comparing the maps of enteric bacteria, actinomycetes, rhizobia, bacilli, and pseudomonads, it is evident that no one generic map coincides with another. In other words, the genes in the chromosomes of members of any particular group of bacteria tend to stay in the same arrangement, or, if there are changes, the changes appear to take place by rearrangement of large clusters of genes. Rapid random reshuffling of gene order does not occur, even though members of groups of bacteria that have highly similar genetic maps diverged from one another many millions of years ago (Fox *et al.*, 1980) and the presence of repeated sequences and active genetic sequences like insertion elements provide the opportunity for rearrangements. What factors can account for this persistent stability?

### 3.3. Possible Stabilizing Factors

What factors mitigate against the generation of a multitude of chromosomally rearranged variants? Is there a growth disadvantage in rearranged strains? The evidence so far is not clearcut. Major alterations of chromosomal structure, such as duplications that constitute as much as 22% of the genome, apparently have no effect on viability or vigor (Anderson and Roth, 1977). Some transpositions have no effect on gene function. The *lac* operon, for instance, functions in many different chromosomal locations (Casadaban, 1976). At least some inversions are not disadvantages, since they are found in nature (Casse *et al.*, 1973; Hill and Harnish, 1981).

To look at this questions more closely, mutants with chromosomal rearrangements have been examined to see whether the change in gene order imposed a disadvantage on the organism that was reflected in its growth rate relative to the wild type. Some transposition mutants that were created by recombination between *rrn* genes showed a 4–5% reduction in growth rate relative to the wild type, while others showed a smaller or no effect (Hill and Harnish, 1982). Various *rrn*-mediated inversion mutants grew 1–6% slower then wild type (Hill and Harnish, 1981). Roth and Schmid (1981) have reported a more severe effect of inversion in *S. typhimurium*. A particular category of mutant was selected that would bring a promoter to the *his* operon. The frequency of occurrence of inversion mutants was low compared to other kinds of mutants, and the growth of those inversion mutants that did arise was poor. In experiments that analyzed different mutant rearrangements of the *B. subtilis* chromosome, some of the mutants were found to be genetically stable, while others with a different order of genes were not stable, as if some classes of gene arrangements are permissible, while others are at a disadvantage and give way easily to the wild-type revertants (Anagnostopolous, 1976).

The specific factors that influence viability or vigor as a function of gene location have not yet been identified. Various suggestions have been advanced. Bachmann and Low (1976) pointed out that there is a gene dosage effect in relation to distance from the origin of replication, an effect that would be most pronounced in fast-growing cells that contain multiple growing forks. Perhaps there is a delicate balance between the regulation of the rate of gene expression and the average numbers of copies of the gene, a balance that would be disturbed by some classes of genome rearrangement but not by others.

Another factor that might tend to preserve the same gene order in related organisms is facilitation of genetic recombination between related chromosomes. Rearrangement would decrease opportunities for genetic exchange, thus shutting off a source of genetic diversity.

Another feature of the chromosome that may need to be preserved is the configuration of the folded chromosome. Is it possible that the observed regular spacing of most of the genes for glucose catabolism (Riley *et al.*, 1978) reflects a higher level of cellular organization, a coordination of function with gene location that would be disturbed by chromosome rearrangement? Is it possible that the grouping of genes in the various negatively supercoiled domains of chromosomal DNA must be maintained in order to retain optimal gene expression? Features of the tertiary structure of the *E. coli* nucleoid have been summarized recently by Pettijohn (1982). There are about 50 negatively supercoiled loops of DNA in the *E. coli* nucleoid and approximately one DNA-gyrase enzyme molecule per loop. The state of torsional tension of each loop or domain of nuclear DNA can be maintained separately and independently of the others. Perhaps the expression of genes or groups of genes is regulated in part by the amount of superhelicity in each domain. If this is the case, some chromosomal rearrangements might interrupt and reconnect certain domains in such a way that the levels of torsional tension become altered for some of the genes, leading to a deleterious effect on function. It is known that the expression of some genes is more sensitive than others to changes in the amount of superhelical coiling (Sanzey, 1979). Presence of a small number of essential genes in the chromosome that are sensitive to their location in a particular supercoiled domain could limit the amount of genome shuffling that could be tolerated. A mechanism of this kind could prevent mutants with new chromosomal rearrangements from overtaking and replacing the wild-type configuration. There is no direct experimental information that bears on this point at present; this line of thought remains completely speculative for the time being.

## 4. Conclusion

We have seen that chromosome rearrangements in bacteria are brought about in different instances by *recA*-dependent homologous recombination, *recA*-independent site-specific recombination, and perhaps nonspecific illegitimate recombination as well. There may also be a role for replication error in the case of small rearrangements. Not only base pair substitutions, but also both large- and small-scale chromosome rearrangements such as insertions and deletions, substitutions, inversions, and duplications have played important roles in bringing about evolutionary divergence.

The flexibility and versatility of bacterial recombination systems and the ubiquity of repeated sequences and active insertion elements in bacterial genomes set the stage for frequent and varied internal rearrangement of the bacterial chromosome. Also, the ubiquity of broad-host-range plasmids carrying transposons provides the opportunity for frequent horizontal transmission of genetic materials between relatively distantly related bacteria. Both of these mechanisms for change have doubtless played a role in creating genetic diversity during evolution, introducing relatively large changes with relatively few molecular events. However, constraints have clearly controlled these processes and have prevented unbridled shuffling and reshuffling of bacterial genes. Very little is known about the molecular nature of these constraints. Quite likely as we learn more about the influence of chromosome structure on the regulation of gene expression, the mechanisms responsible for genetic stability will become evident.

## References

Abou-Sabé, M., 1982, Evolution of the D-ribose operon of *Escherichia coli* B/r, *J. Bacteriol.* **150**:762–769.

An, G., and Frisen, J. D., 1980, The nucleotide sequence of *tufB* and four nearby tRNA structural genes of *Escherichia coli*, *Gene* **12**:33–39.

Anagnostopoulos, C., 1976, Genetic analysis of *Bacillus subtilis* strains carrying chromosomal rearrangements, in: *Modern Trends in Bacterial Transformation and Transfection* (A. Portolés, R. López, and M. Espinosa, eds.), Elsevier/North-Holland Biomedial Press, Amsterdam, pp. 211–230.

Anderson, R. P., and Roth, J. R., 1977, Tandem genetic duplications in phage and bacteria, *Annu. Rev. Microbiol.* **31**:473–505.

Bachmann, B. J., and Low, K. B., 1976, Recalibrated linkage map of *Escherichia coli* K-12, *Bacteriol. Rev.* **40**:116–167.

Bachmann, B. J., and Low K. B., 1980, Linkage map of *Escherichia coli* K-12, edition 6, *Microbiol. Rev.* **44**:1–56.

Bennett, P. M., and Richmond, M. H., 1978, Plasmids and their possible influence on bacterial evolution, in: *The Bacteria*, Vol. VI (L. N. Ornston and J. R. Sokatch, eds.), Academic Press, New York, pp. 1–69.

Berg, C. M., and Curtiss, R., III, 1967, Transposition derivatives of an Hfr strain of *Escherichia coli* K-12, *Genetics* **56**:503–525.

Capage, M., and Hill, C. W., 1979, Preferential unequal recombination in the *glyS* region of the *Escherichia coli* chromosome, *J. Mol. Biol.* **127**:73–87.

Casadaban, M. J., 1976, Transposition and fusion of the *lac* genes to selected promoters in *Escherichia coli* using bacteriophage lambda and mu, *J. Mol. Biol.* **104**:541–555.

Casse, F., Pascal, M.-C., and Chippaux, M., 1973, Comparison between the chromosomal maps of *E. coli* and *S. typhimurium*. Length of the inverted segment in the *trp* region, *Mol. Gen. Genet.* **124**:253–257.

Charlier, D., Crabeel, M., Cunin, R., and Glansdorff, N., 1979, Tandem and inverted repeats of arginine genes in *Escherichia coli*, *Mol. Gen. Genet.* **174**:75–88.

Chumley, F. G., and Roth, J. R., 1980, Rearrangement of the bacterial chromosome using Tn10 as a region of homology, *Genetics* **94**:1–14.

Cornelis, G., 1981, Sequence relationships between plasmids carrying genes for lactose utilization, *J. Gen. Microbiol.* **124**:91–97.

Cornelis, G., Ghosal, D., and Saedler, H., 1978, Tn951: A new transposon carrying a lactose operon, *Mol. Gen. Genet.* **160**:215–224.

Cornelis, G., Sommer, H., and Saedler, H., 1981, Transposon Tn951 is defective and related to Tn3, *Mol. Gen. Genet.* **184**:241–248.

Crawford, I. P., Nichols, B. P., and Yanofsky, C., 1980, Nucleotide sequence of the *trpB* gene in *Escherichia coli* and *Salmonella typhimurium*, *J. Mol. Biol.* **142**:489–502.

Fox, G. E., Stackebrandt, E., Hespell, R. B., Gibson, J., Maniloff, J., Dyer, T. A., Wolfe, R. S., Bolch, W. E., Tanner, R. S., Magrum, L. J., Zablen, L. B., Blakemore, R., Gupta, R., Bonen, L., Lewis, B. J., Stahl, D. A., Luehrsen, K. R., Chen, K. N., and Woese, C. R., 1980, The phylogeny of procaryotes, *Science* **209**:457–463.

Fultz, P.N., Kwoh, D. Y., and Kemper, J., 1979, *Salmonella typhimurium newD* and *Escherichia coli leuC* genes code for a functional isopropylamalate isomerase in *Salmonella typhimurium–Escherichia coli* hybrids, *J. Bacteriol.* **137**:1253–1262.

Gigot, D., Glansdorff, N., Legrain, C., Piérard, A., Stalon, V., Konigsberg, W., Caplier, I., Strosberg, A. D., and Herveé, G., 1977, Comparison of the N-terminal sequences of aspartate and ornithine carbamoyltransferases of *Escherichia coli*, *FEBS Lett.* **81**:28–32.

Gigot, D., Caplier, I., Strosberg, D., Piérard, A., and Glansdorff, N., 1978, Amino-proximal sequences of the *argF* and *argI* ornithine carbamoyltransferases from *Escherichia coli* K-12, *Arch. Int. Physiol. Biochim.* **86**:913–915.

Guiso, N., and Ullman, A., 1976, Expression and regulation of lactose genes carried by plasmids, *J. Bacteriol.* **127**:691–697.

Higgins, C. F., and Ames, G. F.-L., 1981, Two periplasmic transport proteins which interact with a common membrane receptor show extensive homology: Complete nucleotide sequences, *Proc. Natl. Acad. Sci. USA* **78**:6038–6042.

Hill, C. W., and Harnish, B. W., 1981, Inversions between ribosomal RNA genes of *Escherichia coli*, *Proc. Natl. Acad. Sci. USA* **78**:7069–7072.

Hill, C. W., and Harnish, B. W., 1982, Transposition of a chromosomal segment bounded by redundant rRNA genes into other rRNA genes in *Escherichia coli*, *J. Bacteriol.* **149**:449–457.

Hill C. W., Grafstrom, R. H., Harnish, B. W., and Hillman, B. S., 1977, Tandem duplications resulting from recombination between ribosomal RNA genes in *Escherichia coli*, *J. Mol. Biol.* **116**:407–428.

Hooykaas, P. J. J., Peerbolte, R., Regensburg-Tuink, A. J. G., de Vries, P., and Schilperoort, R. A., 1982, A chromosomal linkage map of *Agrobacterium tumefaciens* and a comparison with the maps of *Rhizobium* SPP, *Mol. Gen. Genet.* **188**:12–17.

Hu, M., and Deonier, R., 1981, Mapping of IS1 elements flanking the *argF* gene region on the *Escherichia coli* K-12 chromosome, *Mol. Gen. Genet.* **186**:82–86.

Iino, T., and Kutsukake, K., 1980, *Trans*-acting genes of bacteriophages P1 and mu mediate inversion of a specific DNA segment involved in flagellar phase variation of *Salmonella*, *Cold Spring Harbor Symp. Quant. Biol.* **45**:11–16.

Inokuchi, K., Mutoh, N., Matsuyama, S., and Mizushima, S., 1982, Primary structure of the *ompF* gene that codes for a major outer membrane protein of *Escherichia coli* K-12, *Nucleic Acids Res.* **10**:6957–6968.

Jackson, E. N., and Yanofsky, C., 1973, Duplication-translocations of tryptophan operon genes in *Escherichia coli*, *J. Bacteriol.* **116**:33–40.

Kamp, D., and Kahmann, R., 1981, The relationship of two invertible segments in bacteriophage mu and *Salmonella typhimurium* DNA, *Mol. Gen. Genet.* **184**:564–566.

Kamp, D., Chow, L. T., Broker, T. R., Kwoh, D., Zipser, D., and Kahmann, R., 1978, Site-specific recombination in phage mu, *Cold Spring Harbor Symp. Quant. Biol.* **43**:1159–1167.

Kaplan, J. B., and Nichols, B. P., 1983, Nucleotide sequence of *Escherichia coli pabA* and its evolutionary relationship to *trp(G)D*, *J. Mol. Biol.* **168**:451–468.

Kikuchi, A., and Gorini, L., 1975, Similarity of genes *argF* and *argI*, *Nature* **256**:621–624.

Kemper, J., 1974, Evolution of a new gene substituting for the *leuD* gene of *Salmonella typhimurium*: Characterization of *supQ* mutations, *J. Bacteriol.* **119**:937–951.

Kennedy, K. E., Iida, S., Meyer, J., Stalhammar-Carlemalm, M., Heistand-Nauer, R., and Arber, W., 1983, Genome fusion mediated by the site-specific DNA inversion system of bacteriophage P1, *Mol. Gen. Genet.* **189**:413–421.

Konrad, E. B., 1977, Method for the isolation of *Escherichia coli* mutants with enhanced recombination between chromosomal duplications, *J. Bacteriol.* **130**:167–172.

Kostriken, R., Morita, C., and Heffron, F., 1981, Transposon Tn3 encodes a site specific recombination system: Identification of essential sequences, genes and actual site of recombination, *Proc. Natl. Acad. Sci. USA* **78**:4041–4045.

Lampel, K. A., and Riley, M., 1982, Discontinuity of homology of *Escherichia coli* and *Salmonella typhimurium* DNA in the *lac* region, *Mol. Gen. Genet.* **186**:82–86.

LaVerne, L. S., and Ray, D. S., 1980, Site-specific integration of an F'; *lac pro* factor in the region of the replication origin (*oriC*) of *E. coli*, *Mol. Gen. Genet.* **179**:437–446.

Lawther, R. P., Calhoun, D. H., Adams, C. W., Hauser, C. A., Gray, J., and Hatfield, G. W., 1981, Molecular basis of valine resistance in *Escherichia coli* K-12, *Proc. Natl. Acad. Sci. USA* **78**:922–925.

Lehner, A. F., and Hill, C. W., 1980, Involvement of ribosomal ribonucleic acid operons in *Salmonella typhimurium* chromosomal rearrangements, *J. Bacteriol.* **143**:492–498.

Link, C. D., and Reiner, A. M., 1982, Inverted repeats surround the ribitol–arabitol genes of *E. coli* C, *Nature* **298**:94–96.

Link, C. D., and Reiner, A. M., 1983, Genotypic exclusion: A novel relationship between the ribitol–arabitol and galactitol genes of *E. coli*, *Mol. Gen. Genet.* **189**:337–339.

Liu, R.-J., Capage, M., and Hill, C. W., 1983, Characterization of two hot spots for the rearrangement of the *Escherichia coli* chromosome, *Abstr. Annu. Meet. Am. Soc. Microbiol.*, p. 129.

Mazur, B. J., Rice, D., and Hazelkorn, R., 1980, Identification of blue-green algal nitrogen fixation genes by using heterologous DNA hybridization probes, *Proc. Natl. Acad. Sci. USA* **77**:186–190.

Mizuno, T., Chou, M.-Y., and Inouye, M., 1983, A comparative study on the genes for three porins of the *Escherichia coli* outer membrane: DNA sequence of the osmoregulated *ompC* gene, *J. Biol. Chem.* 6932–6940.

Moore, S., Garvin, R., and James, E., 1981, Nucleotide sequence of the *argF* regulatory region of *Escherichia coli* K-12, *Gene* **16**:119–132.

Nakamura, K., and Inouye, M., 1980, DNA sequence of the *Serratia marcescens* lipoprotein gene, *Proc. Natl. Acad. Sci. USA* **77**:1369–1373.

Neimark, H., and London, J., 1982, Origins of the mycoplasmas: Sterol nonrequiring mycoplasmas evolved from streptococci, *J. Bacteriol.* **150**:1259–1265.

Nichols, B. P., and Yanofsky, C., 1979, Nucleotide sequences of *trpA* of *Salmonella typhimurium* and *Escherichia coli*: An evolutionary comparison, *Proc. Natl. Acad. Sci. USA* **76**:5244–5248.

Ohtsubo, E., and Hsu, M.-T., 1978, Electron microscope heteroduplex studies of sequence relations among plasmids of *Escherichia coli*: Structure of F100, F152, and F8 and mapping of the *Escherichia coli* chromosomal region *fep-supE-gal-att-uvrB*, *J. Bacteriol.* **134**:778–794.

Overbeeke, N., Bergmans, H., van Mansfeld, F., and Lugtenberg, B., 1983, Complete nucleotide sequence of *phoE*, the structural gene for the phosphate limitation inducible outer membrane pore protein of *Escherichia coli* K12, *J. Mol. Biol.* **163**:513–532.

Pettijohn, D., 1982, Structure and properties of the bacterial nucleoid, *Cell* **30**:667–669.

Piette, J., Cunin, R., Crabeel, M., and Glansdorff, N., 1981, Nucleotide sequences of the control region of the *argF* gene of *Escherichia coli*, *Arch. Int. Physiol. Biochim.* **89**:B127–128.

Piette, J., Cunin, R., Van Vliet, F., Charlier, D., Crabeel, M., Ota, Y., and Glansdorff, N., 1982, Homologous control sites and DNA transcription starts in the related *argF* and *argI* genes of *Escherichia coli* K12, *EMBO J.* **1**:853–857.

Reanny, D. C., 1976, Extra-chromosomal elements as possible agents of adaptation and development, *Bacteriol. Rev.* **40**:552–590.

Reeve, E. C. R., and Braithwaite, J. A., 1974, The lactose system in *Klebsiella aerogenes* V9A. 4. A comparison of the *lac* operons of *Klebsiella* and *Escherichia coli*, *Genet. Res.* **24**:323–331.

Riley, M., and Anilionis, A., 1978, Evolution of the bacterial genome, *Annu. Rev. Microbiol.* **32**:519–560.

Riley, M., an Anilionis, A., 1980, Conservation and variation of nucleotide sequences within related bacterial genomes: Enterobacteria, *J. Bacteriol.* **143**:366–376.

Riley, M., and Glansdorff, N., 1983, Cloning the *Escherichia coli* K12 *argD* gene specifying acetylornithine δ-transaminase, *Gene* **241**:335–339.

Riley, M., Solomon, L., and Zipkas, D., 1978, Relationship between gene function and gene location in *Escherichia coli*, *J. Mol. Evol.* **11**:47–56.
Roof, W. D., Foltermann, K. F., and Wild, J. R., 1982, The organization and regulation of the *pyrBI* operon in *E. coli* includes a rho-independent attenuator sequence, *Mol Gen. Genet.* **187**:391–400.
Roth, J. R., and Schmid, M., 1981, Arrangement and rearrangement of the bacterial chromosome, *Stadler Symp.* **13**:53–70.
Ruvkin, G. B., and Ausubel, F. M., 1980, Interspecies homology of nitrogenase genes, *Proc. Natl. Acad. Sci. USA* **77**:191–195.
Sanzey, B., 1979, Modulation of gene expression by drugs affecting deoxyribonucleic acid gyrase, *J. Bacteriol.* **138**: 40–47.
Schmid, M., and Roth, J. R., 1980, Circularization of transduced fragments: A mechanism for adding segments to the bacterial chromosome, *Genetics* **94**:15–29.
Schneider, A.-M. and Anagnostopoulos, C., 1983, *Bacillus subtilis* strains carrying two non-tandem duplications of the *trpE–ilvA* and the *purB–tre* regions of the chromosome, *J. Gen. Microbiol.* **129**:687–701.
Schupp, T., Hutter, R., and Hopwood, D. A., 1975, Genetic recombination in *Nocardia mediterranei*, *J. Bacteriol.* **121**:128–136.
Scott, T. N., and Simon, M., 1982, Genetic analysis of the mechanism of the *Salmonella* phase variation site specific recombination system, *Mol. Gen. Genet.* **188**:313–321.
Silverman, M., Zieg, J., Mandel, G., and Simon, M., 1980, Analysis of the functional components of the phase variation system, *Cold Spring Harbor Symp. Quant. Biol.* **45**:17–26.
Simon, M., Zeig, J., Silverman, M., Mandel, G., and Doolittle, R., 1980, Genes whose mission is to jump, *Science* **209**:1370–1374.
Smith I., Dubnau, E., Williams, G., Cabane, K., and Paress, P., 1979, Genetics of the translational apparatus in *Bacillus subtilis*, in: *Genetics and Evolution of RNA Polymerase, tRNA and Ribosomes* (S. Osawa, H. Ozeki, H. Uchida, and T. Yura, eds.), University of Tokyo Press, Tokyo, pp. 379–405.
Squires, C. H., Defelice, M., Devereux, J., and Calvo, J. M., 1983, Molecular structure of *ilvIH* and its evolutionary relationship to *ilvG* in *Escherichia coli* K12, *Nucleic Acids Res.* **11**:5299–5313.
Timmons, M. S., Bogardus, A. M., and Deonier, R. C., 1983, Mapping of chromosomal IS5 elements that mediate type II F-prime plasmid excision in *Escherichia coli* K-12, *J. Bacteriol.* **153**:395–407.
Turnbough, C. L., Jr., Hicks, K. L., and Donahue, J. P., 1983, Attentuation control of *pyrBI* operon expression in *Escherichia coli* K-12, *Proc. Natl. Acad. Sci. USA* **80**:368–372.
Vogel, H. J., 1963, Induction of acetylornithine δ-transaminase during pathway-wide repression, in: *Informational Macromolecules* (H. J. Vogel, V. Bryson, and J. O. Lampen, eds.), Academic Press, New York, pp. 293–300.
Vogel, H. J., and Bacon, D. F., 1966, Gene aggregation: Evidence for a coming together of functionally related, not closely linked genes, *Proc. Natl. Acad. Sci. USA* **55**:1456–1461.
Woodward, M. J., and Charles, H. P., 1983, Polymorphism in *Escherichia coli*: *rtl atl* and *gat* regions behave as chromosomal alternatives, *J. Gen. Microbiol.* **129**:75–84.
Yamagata, H., Nakamura, K., and Inouye, M., 1981, Comparison of the lipoprotein gene among the enterobacteriaceae. DNA sequence of *Erwinia amylovora* lipoprotein gene, *J. Biol. Chem.* **256**:2194–2198.
Yokota, T., Sugisaki, H., Takanami, M., and Kaziro, Y., 1980, The nucleotide sequence of the cloned *tufA* gene of *Escherichia coli*, *Gene* **12**:25–31.

York, M. K., and Stodolsky, M., 1981, Characterization of P1 *argF* derivatives from *Escherichia coli* K12 transduction, *Mol. Gen. Genet.* **181**:230–240.

Zakin, M. M., Garel, J.-R., Dautry-Varsat, A., Cohen, G. N., and Boulot, G., 1978, Detection of the homology among proteins by immunochemical cross-reactivity between denatured antigens. Application to the threonine and methionine regulated aspartokinases-homoserine-dehydrogenases from *Escherichia coli* K12, *Biochemistry* **17**:4318–4323

# *Glossary*

**Catabolism**—The degradation of a substrate, by which a cell obtains carbon and/or energy for growth.

**Catabolic pathway**—The biochemical pathway for the degradation of a substrate or "catabolite."

**Catabolite repression**—Preferential utilization of one substrate or catabolite over another when both are presented to a cell at the same time.

**Chemostat**—A device for growing cells continuously, at a constant growth rate, through control of the concentration of a limiting nutrient.

**Coliform**—The bacteria in the family Enterbacteriaciae; this term includes the genera *Escherichia*, *Klebsiella*, *Erwinia*, and *Salmonella*, among others.

**Constitutive**—Enzyme or enzymes synthesized continuously by the cell even in the absence of the substrate.

**Coordinate control**—Enzymes regulated by the same regulatory or control system.

**Cross-reacting proteins**—Proteins similar enough in structure to react with the same antibody.

**Direct cell count**—The microscopic determination of the total number of cells, both living and dead, in a culture.

**Doubling time**—The time required for a culture to double its population.

**Down-promoter mutant**—Mutation in the promoter region of an operon that decreases the rate of translation of the operon.

**Endogenous**—The basal level of enzyme synthesized in the absence of substrate or inducer; the enzyme can be induced to a higher level or a constitutive mutation can permit a higher amount to be made.

**Facilitated diffusion**—A transport system located in the cell membrane to aid or "facilitate" the transport of a substrate though the membrane and into the cell; metabolic energy is not used and the substrate cannot be accumulated against a concentration gradient.

**Genetic drift**—The concept that if a gene product is not required by a cell, mutations in the gene will not be selected against and will accumulate with time; the gene product will become altered and may lose its original function.

**Gratuitous inducer**—A compound that resembles the natural inducer of an enzyme enough to cause the induction of the enzyme even though it is not metabolized by the cell.

**Hyperconstitutive**—The production of high levels of enzyme without the presence of inducer; the amount of constitutive enzyme produced is higher than the normal induced level.

**Hypoconstitutive**—The production of a low level of enzyme without the presence of inducer; the amount of constitutive enzyme produced is higher than the endogenous or basal level but lower than the induced level.

**Inducible expression**—Regulation so that the amount of enzyme or enzymes synthesized by a cell is increased over the endogenous level if a specific chemical signal is present; this chemical may be the substrate or a product of the enzyme or enzyme pathway; the induction of an enzyme results in an increase in its rate of synthesis as compared to the rate of synthesis or total cellular protein.

$K_m$—The *Michaelis constant* of the substrate; the concentration of a substrate at which the enzyme catalyzes its reaction at one-half of the maximum rate.

**Magnoconstitutive**—See hyperconstitutive

**Negative control**—Regulation of the expression of an operon where the product of the regulator gene represses the transcription of the operon to mRNA and thus prevents the formation of the enzymes coded by the operon; the presence of the inducer prevents this repression and the enzymes of the operon are therefore *induced*.

**Operon**—Genes located adjacent to one another that are transcribed into a single, continuous strand of messenger RNA.

**Phosphoenolypyruvate:sugar phosphotransferase system**—An energy-requiring system, consisting of several enzymes, for the transport of substrates through the cell membrane; the substrate is phosphorylated as it passes through the cell membrane and enters the cell as a phosphate ester.

**Promoter**—The area of an operon where the RNA polymerase binds in order to begin the transcription of messenger RNA.

**Promoter-up mutation**—Same as up-promoter mutant.

**Positive control**—The regulation of an operon where the product of the regulator gene (the activator, usually in combination with the inducer) is required to permit the synthesis of the enzymes coded by the operon.

**Revertant**—A strain that had first lost a function because of one mutation and then regained that function (phenotype) due to another mutation; the second mutation may have reversed the actual genetic change caused by the first or it might have occurred at a different site; in the latter case it is termed a *suppressor* mutation.

**Regulator gene**—A gene controlling the expression of other genes.

**Regulon**—Those genes whose regulation is controlled by the same regulatory gene or by a common regulatory system.

**Spontaneous mutation**—A naturally occurring mutation.

**Structural gene**—A gene coding for the structure of an enzyme.

**Suppressor mutation**—A mutation which overcomes or partially overcomes the effect of a earlier mutation at a different site.

**Transcription**—The process by which the genes of an operon are copied into messenger RNA.

**Transketolase**-transaldolase rearrangements—A series of enzyme-catalyzed reactions that result in five-carbon sugars being converted into three- and six-carbon sugars.

**Up-promoter mutants**—Mutation at the promoter site of an operon resulting in more rapid expression of the genes of the operon and an increase in the rate of synthesis of the enzymes of the operon.

$V_{max}$—The maximum velocity of an enzymatic reaction; the maximum rate of activity of an enzyme in the presence of excess substrate.

# Index

Acetamide, 189–203
Acetanilide, 198–194
N-Acetylacetamide, 190–194, 197
Acetylornithine transcarbamylase, 286
*Acholeplasma*, 302
Acroline, 237, 243–244
Acrylamide, 190, 194
*Aeorbacter, see Klebsiella*
*Agrobacterium tumefaciens*, 308
Alcohol dehydrogenases
  biochemistry of, 236–242
  cytoplasmic enzymes
    ADH-I, 236–252
    ADH-II, 236–252
  kinetics of, 248, 250
  laboratory evolution of, 233–241
  mitochondrial enzyme, 237–241
  mutants of, 239–251
  regulation of, 236–242
  tertiary structure of, 246–247
Aldopentoses, 110–115
Allolactose, 177–179
Allyl alcohol, 237, 239–244, 250
Altered enzyme mutants, 205
D-Altrose, 159
Amidase
  altered enzyme mutants of, 205–221
  acetanilde metabolism by, 210–221
  butyramide metabolism by, 205–207, 214–221
  negative mutants of, 199–204
  phenylacetamide metabolism by, 207–210, 214–221
  regulatory mutants of, 191–199
  valeramide metabolism by, 207, 214–221

Amidase (*cont.*)
  wild type enzyme, 190–191, 214–221
Amidase genes, 222
Amidase inhibitors, 221–228
Amide analogue repression, 196–207
Amide structure, 189, 192
Amides, 187–228
Antimycin A, 239
D-Arabinose
  availability of, 110
  growth of *Escherichia* on, 127–131, 153–154, 159
  growth of *Klebsiella* on, 112, 120–131, 153–154
  growth of *Salmonella* on, 128–131, 153
  isomerization of, 111–115, 121–131, 153
  metabolism of, 113, 120–128
  mutants which grow on, 114–115, 120–131, 153, 159
  structure of, 109–110, 123
  transport of, 123
D-Arabinose isomerase, 111–115, 121–131, 153
L-Arabinose
  availability of, 110
  growth of *Klebsiella* on, 112
  metabolism of, 111
  structure of, 109–110
D-Arabitol
  availability of, 2–3
  growth of *Escherichia* on, 49, 154, 157, 305
  growth of *Klebsiella* on, 4–7, 28, 302–303
  metabolism of, 4–5, 136, 154–155

D-Arabitol (*cont.*)
  structure of, 2, 28, 67, 71
  transport of, 15–16, 154, 156
D-Arabitol dehydrogenase, 4–6, 9–11, 16–18, 28, 55–71, 82–86, 98–99, 117–118, 154
D-Arabitol operon (*dal*), 11, 15–19, 28, 55–63, 81–91, 96–103, 118
D-Arabitol operon promoter, 87
D-Arabitol operon repressor, 83, 87, 90–95
Aldopentoses
  availability of, 109–110
  metabolism of, 110–125
  mutants of, 114–129
  structure of, 109–110
L-Arabitol
  availability of, 2–3
  growth of *Klebsiella* on, 3–8, 13
  metabolism of, 4–9, 52
  mutants which grow on, 6–9, 13, 18
  structure of, 2, 70–71
L-Arabitol dehydrogenase, 4, 13
Arginine transport, 287
Aspartate transcarbamylase, 287, 292
Aspartokinase, 286

*Bacillus subtilis*, 296, 299, 309
Butyramide, 189–199, 201–208, 212
iso-Butyramide, 192

Catabolite-resistant mutants, 203–204, 209, 212–213
Chemostat studies, 14–15, 27–37, 50–51, 85
Chorismate mutase, 286
Chromosomal rearrangements, 285–299, 310
Constitutive mutants, 9–17, 28–52, 56–61, 72–73, 82, 98, 119–131, 137, 143, 145, 149–157, 166, 195–201, 205–206, 209, 212, 218, 225, 237–238, 286
Crotyl alcohol, 239

DNA additions, 301–306
DNA deletions, 25, 301–306
DNA inversions, 300–301
DNA sequencing
  of *dal* promoter, 87–91

DNA sequencing (*cont.*)
  of *dal* repressor protein, 87, 90–95
  of D-arabitol dehydrogenase, 85–90
  of pentitol operons, 80–103
  of *rbt* promoter, 91–93
  of *rbt* repressor protein, 82–86
  of ribitol dehydrogenase, 80–85
Down-promoter mutants, 202

Enzyme duplication, 160
Enzyme evolution, 27–52, 113–115, 157–160, 165–170, 184, 187–188, 226–228, 233–236, 251
Ethidium bromide, 239
*N*-Ethylacetamide, 192
*N*-Ethylformamide, 192
*Erwinia uredovora*, 8
*Escherichia coli*
  gene duplications in, 286, 293, 299–309
  growth on
    D-altrose, 159
    D-arabinose, 121, 127–131, 153–154, 159
    L-arabinose, 111
    D-arabitol, 17–18, 55–72, 81, 99, 154–155
    ethylene glycol, 157–158
    L-fucose, 123–128, 135–153
    L-galactose, 159
    galactosyl arabinose, 169–177
    lactobionate, 169–177
    lactose, 166–184
    lactulose, 169–174
    D-lyxose, 19–120, 131
    methylgalactoside, 171
    L-1,2-propanediol, 136–153
    xylitol, 17–18, 50–51, 155–156
    ribitol, 46–49, 55–73, 81, 99, 119–120, 127–128, 136
  leucine biosynthesis by, 271–274
  pentitol operons in, 17–18, 55–72, 81, 99, 302
Ethylene glycol
  growth of *Escherichia* on, 157–158
  metabolism of, 158
  mutants which grow on, 157
*Erwinia amylovora*, 293
Evolved β-galactosidase (*EBG*), 166–183, 224–227

Evolved β-galactosidase genes (*cont.*)
  *ebgA* gene, 176–177, 181
  *ebgR* gene, 176–177, 181
Evolved β-galactosidase operon (*ebg* operon), 168–181

Fluoracetamide, 191, 194, 200, 202–204
Formamide, 191–195, 199, 205, 212
D-Fructose, structure, 71
L-Fucose
  availability of, 123
  growth of *Escherichia* on, 123, 127–128
  growth of *Klebsiella* on, 135–153
  growth of *Salmonella* on, 128
  metabolism of, 123–130, 135, 139–145, 151–154
  structure of, 123
L-Fucose isomerase, 123–130, 140–149, 153–155
L-Fucose operon, 148
L-Fucose permease, 123
L-Fucose regulon, 148–151
L-Fuculokinase, 123–128, 140–149, 151, 153
L-Fuculose, 123–124
L-Fuculose 1-phosphate, 123–127
L-Fucolose 1-phosphate aldolase, 123–134, 127–128

Galactitol metabolism by *Escherichia*, 302–303
L-Galactose, 159, 159–160
D-Galactose dehydrogenase, 154–155
β-Galactosidase, 166–178, 200, 227
Galactosyl arabinoside, 169–177, 181
Galactosyl-β-D-fructose, *see* lactulose
Galactosyl-β-D-gluconate, *see* lactobionate
Gene divergence, 288–292
Gene duplication, 8, 24–26, 38–47, 50–52, 72–73, 97–99, 151, 159–160, 285–298
Gene inversion, 16, 96
Genetic doublings, 301
Genetic drift, 103, 183, 233, 252
Genetic maps, 307
Glutamic amido transferase, 287
Glycerophosphate–dihydroxyacetone phosphate shuttle, 240
Glycollamide, 190

Hemoglobin molecule, 235
Hexamide, 192
Histidine transport, 287
Hydroxyurea, 211, 213
Hyperconstitutive, 82, 98, 195

α-Isopropylmalate, 255–256, 273, 276
β-Isopropylmalate, 255–256, 271, 273, 276
α-Isopropylmalate dehydrogenase, 255–256
β-Isopropylmalate dehydrogenase, 255–256
Isopropylmalate isomerase, 255–256, 270–272, 275
Isopropylmalate isomerase mutants, 274–275, 277–279
α-Isopropylmalate synthetase, 255–256, 277

2-Ketopentoses, 2–3, 4, 111
*Klebsiella pneumoniae*
  gene duplication in, 307
  growth on
    aldopentoses, 111–131
    D-arabinose, 113–116, 120–131, 153–154
    L-arabinose, 111–112
    D-arabitol, 4–7, 11–17, 28–30, 55–64, 81–85, 98–99
    L-arabitol, 3–16
    L-fucose, 123–128, 148, 153–154
    D-lyxose, 112–120
    L-lyxose, 113–116, 129
    L-mannose, 160
    pentitols, 3–16, 55–64, 75–85, 96–99
    ribitol, 4–19, 28–35, 55–64, 72, 75–85, 98–99
    D-ribose, 111–112
    xylitol, 3–17, 28–47, 72, 111–132, 156, 302, 307, 153
    D-xylose, 111–112
    L-xylose, 113–116, 121, 125–129
  nitrogen fixation by, 302

L-Lactaldehyde, 137–140
L-Lactaldehyde dehydrogenase, 140, 142–145, 150, 157–158
L-Lactaldehyde metabolism, 137–140
Lactamide, 190–194, 198–199, 204, 207–209, 213
Lactase, 167–179
Lactobionate, 169–177
Lactobionate utilizing mutants, 169–174

Lactose, 166–184, 206
Lactose acetyltransferase, 305
Lactose operon, 37, 150, 167, 181–182, 197, 303–305
Lactose permease, 166–167, 178, 180–182
Lactose utilizing mutants, 166–169, 184
Lactulose, 169–177, 181
Lactulose utilizing mutants, 169–174
Leucine auxotrophic mutants, 257–258
Leucine biosynthetic pathway, 255–256, 266, 271, 275–277
Leucine constitutive mutants, 255, 264, 278
Leucine operon, 255–256, 268–269, 276–279
Leucine structural genes
  *leuA*, 255, 278
  *leuB*, 255, 276–278
  *leuC*, 256, 259–262, 265–279
  *leuD*, 256–276, 278–279
Leucine suppressor mutations
  *newD*, 258–275, 278–279, 304
  *supQ*, 257–276, 278–279, 304
Lysine transport, 287
D-Lyxose
  availability of, 110
  growth of *Escherichia* on, 119–120, 131
  growth of *Klebsiella* on, 112–119, 131
  growth of *Pseudomonas* on, 119
  isomerization of, 113–131
  metabolism of, 113–120
  mutants which grow on, 112–120, 131
  structure of, 109–110
  transport of, 119
D-Lyxose isomerase, 113–131
L-Lyxose
  availability of, 110
  growth of *Klebsiella* on, 112–115, 129, 131
  isomerization of, 113–115, 129–130
  metabolism of, 113–115
  mutants which grow on, 114, 129, 131
  structure of, 109–110
L-Lyxose isomerase, 113–115, 129–130

Magnoconstitutive, 82, 98, 195
D-Mannose
  isomerization of, 118–120
  metabolism of, 118–119, 131
  structure of, 71

D-Mannose isomerase, 118–120
L-Mannose, 160
Mannosephosphate isomerase, 120, 131
$N$-Methylacetamide, 190–192, 218
$N$-Methylformamide, 192
$\beta$-Methylgalactoside, 171–173
$\beta$-Methylgalactoside utilizing mutants, 171
$N$-Methylpropionamide, 218
Mycoplasmas, 302

*Neurospora crassa*, 256–257
Neutral mutations, 23, 45, 103
Nitrogen fixation, 302
*Nocardia mediterrani*, 308
Nucleotide sequences, 288–293

Ornithine transcarbamylase, 287
Ornithine transport, 287

Pentitols
  growth of *Escherichia* on, 46, 48–52, 55–73, 99
  growth of *Klebsiella* on, 3–17, 27–40, 56–64, 75–85, 98–99
  metabolism of, 1–4, 27–40, 50
  structure of, 2, 28
  transport of, 3, 15–17, 28, 35–36, 49, 96, 123, 154, 156
Pentitol operons, 55–85, 96–101
Pentuloses, 2–3, 4, 111
Phenylacetamide, 207–210
$N$-Phenylacetamide, *see* Acetanilide
Phenylgalactoside, 168
Phosphoriboisomerase, 111
Prephenate dehydratase, 286
Prephenate dehydrogenase, 286
Promoter-up mutants, 14
L-1,2-Propanediol
  facilitator of, 138, 141, 148–149, 155–157
  growth of *Escherichia* on, 136–141, 143, 151, 154
  oxidoreductase, 137–151, 157–159
  permease, 145
Propionamide, 190–194, 219
*Proteus mirabilis*, 307
*Pseudomonas acidovorans*, 228
*Pseudomonas aeruginosa*
  amidases, 187–228
  metabolism of acetamide by, 189–193, 203

# INDEX

*Pseudomonas aeruginosa (cont.)*
  metabolism of acetanilide by, 210
  metabolism of butyramide by, 195–198, 205–207
  metabolism of formamide by, 194–196
  metabolism of lactamide by, 198–199
  metabolism of phenylacetamide by, 207–210
  metabolism of valeramide by, 207
*Pseudomonas putida*, 187–189, 226–228
*Pseudomonas saccharophila* growth on D-lyxose, 119

Regulatory mutations, 179–183, 191, 195, 200–203
Repressor evolution, 180–182
Retrograde evolution, 7–8
L-Rhamnose
  growth of *Escherichia* on, 152
  metabolism of, 136–141, 159
  permease, 141
  structure of, 141
L-Rhamnose isomerase, 141, 169
L-Rhamnulokinase, 141, 160
L-Rhamnulose 1-phosphate, 141, 160
*Rhizobium leguminosarum*, 308
*Rhizobium meliloti*, 308
Ribitol
  availability of, 2–3
  growth of *Escherichia* on, 17–19, 46, 50–51, 305
  growth of *Klebsiella* on, 4–19, 27–46, 302–303
  metabolism of, 3–5, 12, 14, 28–52, 116, 136, 139
  structure of, 2, 28, 67, 71
  transport of, 16, 28, 49
Ribitol dehydrogenase, 9–10, 14–15, 28–51, 55–67, 71–85, 103, 117, 121, 129, 139
Ribitol operon (*rbt* operon), 2, 10–11, 34–38, 49, 55–64, 70–87, 96–101, 117, 121, 127–129
Ribitol operon messenger RNA, 72–80
Ribitol operon promoter, 91–92
Ribitol operon repressor, 82, 85, 89, 91
D-Ribose
  availability of, 109
  growth of *Klebsiella* on, 112
  metabolism of, 111–112

L-Ribose
  availability of, 109
  structure of, 110
D-Ribulokinase, 4–5, 10–11, 28, 55–57, 67–72, 79–82, 93, 96, 98, 111–117, 121, 124, 129
L-Ribulokinase, 4, 111–115
D-Ribulose, 2–3, 71, 113–115, 121, 124
L-Ribulose, 2–3
D-Ribulose 1-phosphate, 124, 127–128
D-Ribulose 5-phosphate, 4–5, 28, 111, 121, 124
L-Ribulose 5-phosphate, 4, 13, 129

*Saccharomyces cerevisiae*, 235–240
*Salmonella typhimurium*
  gene duplications in, 287, 293, 295–296, 300–308
  growth on D-arabinose, 128
  growth on L-fucose, 128, 153
leucine biosynthesis by, 260, 263, 267, 271–275, 278, 287
Silent genes, 25–26, 132
*Streptomyces coelicolor*, 308
*Streptomyces rimosusm*, 308
Succinate, 198, 201, 204–205, 209, 212–213
Superproducing, 82, 98, 195

Thioacetamide, 189
Thiogalactoside, 168
Transgalactosylation, 178–179
Transpositions, 296–299, 302–305, 308

Up-promoter mutants, 37, 151, 204
Urea, 194, 210

Valeramide, 192–194, 207–208, 213

Xylitol
  availability of, 2–3
  growth of *Erwinia* on, 18
  growth of *Escherichia* on, 17–18, 46, 48–52, 155–156
  growth of *Klebsiella* on, 3–17, 28–46, 72, 156
  metabolism of, 4–10, 29–39, 139, 156
  mutants which grow on, 6–19, 33–52
  structure of, 2, 28, 67, 71
  transport of, 15–17, 35–36
Xylitol dehydrogenase, 4, 9, 11–13, 29–46

D-Xylose
  availability of, 110
  growth of *Klebsiella* on, 112
  metabolism of, 12, 18, 111, 154
  structure of, 109–110
  transport of, 119
D-Xylose isomerase, 98, 117, 111–112, 121, 156
D-Xylose operon, 118
D-Xylose operon regulation, 112
D-Xylose permease, 155–156
L-Xylose
  availability of, 110
  growth of *Klebsiella* on, 112–115, 121, 125, 129, 131
  isomerization of, 117–118, 121, 125, 129–130, 156
  metabolism of, 113–115

L-Xylose (*cont.*)
  mutants which grow on, 114, 125, 129, 131
  structure of, 109–110
  transport of, 130, 155–156
L-Xylose isomerase, 117–118, 121, 125, 129–130, 156
D-Xylulokinase, 4–7, 12–13, 16–17, 28–29, 55–57, 70–72, 93, 96, 96–98, 111–118, 121, 156
L-Xylulokinase, 4, 13, 113–116, 121, 129–130
D-Xylulose, 2–3, 7, 28, 71, 113–115, 155
L-Xylulose, 2–3, 71, 113–115, 129
D-Xylulose 5-phosphate, 4–5, 13, 17, 28–29, 128
L-Xylulose 5-phosphate, 4, 13

*Yersinia enterocolitica*, 304

**DATE DUE**